KU-226-943

3D Nanoelectronic Computer Architecture and Implementation

Series in Materials Science and Engineering

Series Editors: **B Cantor**, University of York, UK
M J Goringe, School of Mechanical and Materials Engineering, University of Surrey, UK
E Ma, Department of Materials Science and Engineering, Johns Hopkins University, USA

Series in Materials Science and Engineering

3D Nanoelectronic Computer Architecture and Implementation

Edited by

David Crawley, Konstantin Nikolić and
Michael Forshaw

*Department of Physics and Astronomy,
University College London, UK*

Institute of Physics Publishing
Bristol and Philadelphia

© IOP Publishing Ltd 2005

All rights reserved. No part of this publication may be reproduced, stored in a retrieval system or transmitted in any form or by any means, electronic, mechanical, photocopying, recording or otherwise, without the prior permission of the publisher. Multiple copying is permitted in accordance with the terms of licences issued by the Copyright Licensing Agency under the terms of its agreement with Universities UK (UUK).

British Library Cataloguing-in-Publication Data

A catalogue record for this book is available from the British Library.

ISBN 0 7503 1003 0

Library of Congress Cataloging-in-Publication Data are available

Series Editors: **B Cantor, M J Goringe and E Ma**

Commissioning Editor: Tom Spicer
Commissioning Assistant: Leah Fielding
Production Editor: Simon Laurenson
Production Control: Sarah Plenty
Cover Design: Victoria Le Billon
Marketing: Nicola Newey, Louise Higham and Ben Thomas

Published by Institute of Physics Publishing, wholly owned by The Institute of Physics, London

Institute of Physics Publishing, Dirac House, Temple Back, Bristol BS1 6BE, UK

US Office: Institute of Physics Publishing, The Public Ledger Building, Suite 929, 150 South Independence Mall West, Philadelphia, PA 19106, USA

Typeset in LaTeX 2_ε by Text 2 Text Limited, Torquay, Devon
Printed in the UK by MPG Books Ltd, Bodmin, Cornwall

Contents

Preface

Nanoelectronics is a rapidly expanding field with researchers exploring a very large number of innovative techniques for constructing devices intended to implement high-speed logic or high-density memory. Less common, however, is work intended to examine how to utilize such devices in the construction of a complete computer system. As devices become smaller, it becomes possible to construct ever more complex systems but to do so means that new and challenging problems must be addressed.

This book is based is largely on the results from a project called CORTEX (IST-1999-10236) which was funded under the European Commission's Fifth Framework Programme under the Future Emerging Technologies proactive initiative Nanotechnology Information Devices and ran from January 2000 to June 2003. Titled 'Design and Construction of Elements of a Hybrid Molecular/Electronic Retina-Cortex Structure', the project examined how to construct a three-dimensional system for computer vision. Instead of concentrating on any particular device technology, however, the project was primarily focussed on how to connect active device layers together in order to build a three-dimensional system. Because the system was aimed at the nanoscale, we concentrated on techniques which were scalable, some of which utilized molecular self-assembly. But we also needed a practical experimental system to test these technologies so we also included work to fabricate test substrates which included through-chip vias. Considering the multi-disciplinary nature of the project, the end result was remarkably successful in that a three-chip stack was demonstrated using more than one molecular interconnection technology.

For this book, we have extended the scope somewhat beyond that encompassed by the CORTEX project, with chapters written by invited experts on molecular electronics, carbon nanotubes, material and fabrication aspects of nanoelectronics, the status of current nanoelectronic devices and fault tolerance. We would like to thank everyone who contributed to the book, especially for their timely submission of material.

David Crawley, Konstantin Nikolić, Michael Forshaw
London, July 2004

Chapter 1

Introduction

1.1 Why do we need three-dimensional integration?

It is becoming increasingly clear that the essentially 2D layout of devices on computer chips is starting to be a hindrance to the development of high-performance computer systems, whether these are to use more-or-less conventional developments of silicon-based transistors or more exotic, perhaps molecular-scale, nanodevices. Increasing attention is, therefore, being given to three-dimensional (3D) structures, which will probably be needed if computer performance is to continue to increase. Thus, for example, in late 2003 the US Defense Advanced Research Projects Agency (DARPA) issued a call for proposals for research aimed at eventually stacking 100 chips together. However, new processing devices, even smaller than the silicon-based Complementary Metal Oxide Semiconductor (CMOS) transistor technology, will require new connection materials, perhaps using some form of molecular wiring, to connect nanoscale electronic components. This kind of research also has implications for investigations into the use of 3D connections between 'conventional' chips. Three-dimensional structures will be needed to provide the performance to implement computationally intensive tasks. For example, the visual cortex of the human brain is an existing proof of an extraordinarily powerful special-purpose computing structure: it contains $\sim 10^9$ neurons, each with perhaps 10^4 connections (synapses) and runs at a 'clock speed' of about 100 Hz. Its equivalent in terms of image-processing abilities, on a conventional (CMOS) platform, would generate hundreds of kilowatts of heat. The human visual cortex occupies 300 cm^3 and dissipates about 2 W. This type of performance can only be achieved by advances in nanoelectronics and 3D integration. With the long-term goal of achieving such computing performance, a research project funded by the European Commission was started in 1999 and finished in 2003. Much of this book is based on the work carried out in that project, which was given the acronym 'CORTEX'.

Computer designers and manufacturers have been making computers that are conceptually three- or higher-dimensional for many years. Perhaps the most well known is the Connection Machine of 1985, in which 64 000 or fewer processing elements were connected in a 16-dimensional hypercube wiring scheme [1]. However, the individual processing elements in this and other systems were laid out in two dimensions on the surface of silicon chips, and most of the interconnections were in the form of multilevel wiring on-chip or with complex wiring harnesses to join one chip to another [2, 3]. Many stacked-chip systems with edge connections have been designed and built [2, 4–6] but there are few designs, and even fewer fabricated systems, where more than two chips have been stacked in a true 3D block, with short electrical connections joining the face of one chip to the next. The reasons for this are obvious, the main one being that there was no need to produce such systems until performance demands made it necessary to overcome the many technical problems. There will never be an end to such demands for increased performance but the 'visual cortex' represents a useful goal at which to aim.

Progress in CMOS technology will slow dramatically in about ten years' time, partly because of fundamental technical limitations, partly because of the extraordinarily high projected costs. Near-term molecular computing technologies are likely to be limited to two dimensions and to be very error-prone. This book concentrates on one innovative concept—the use of the third dimension to provide high-density, self-aligning molecular wiring. This could eventually lead to the design and construction of hybrid molecular/electronic systems. Our target system here, used as an example of the power and usefulness of 3D structures, would emulate the 3D structure and function of the human retina and visual cortex. This longer-term goal, although speculative, has very good prospects for eventual success. This target system would, for the first time, incorporate high image resolution, computing power and fault tolerance in a small package. Other future applications could include autonomous vehicle guidance, personal agents, process control, prostheses, battlefield surveillance etc. In addition to its potential usefulness for conventional semiconductor circuits, the 3D molecular interconnect technology would also be an enabling technology for providing local connectivity in nanoscale molecular computers.

1.2   Book Summary

Our ultimate aim is to devise technology for a 3D multiple-chip stack which has the property of being scalable down to the nanometre size range. The main results for the first stage in that process are described here.

Our concept is illustrated in figure 1.1 and comprises alternating layers of electronic circuitry and self-assembling molecular networks. The electronic layers perform the necessary computing, the molecular layers provide the required interconnections.

Figure 1.1. A conceptual illustration of complex structures with high-density 3D connections represented by a 3D multiple-chip stack, with distributed interconnects between the layers.

There are many obstacles to be overcome before truly nanoscale electronics can be successful. Not only must new devices be developed but improved ways to connect them and to send signals over long distances must also be developed. One possible way to reduce some of the problems of sending signals over long distances is to have a stack of chips, with signals being sent in the third dimension (vertically) from layer to layer, instead of their being sent horizontally to the edge of a chip and then from the edge of one chip to the next, as is done now with stacked-chip techniques. In addition, it will be necessary to increase the density of vertical connections from one chip to another. Existing techniques such as solder balls or conducting adhesives are limited to spacings between connections of not much smaller than 100 μm. It is, therefore, important to prepare the way for the development of nanoscale circuits by developing new 3D circuit geometries and new interconnection methods with good potential for downscaling in size towards the nanoregion. These two developments were the main tasks of the CORTEX project (EC NID project IST-1999-10236). This book emerged from the results and experiences of the CORTEX project but it contains additional, complementing material, which is relevant for research into 3D interconnect technology.

Formally, the aim of the CORTEX project was to examine the feasibility of connecting two or more closely-spaced semiconductor layers with intercalated molecular wiring layers. Three partners (University of Leeds, University of Strathclyde and CNRS/University of Marseille) developed new self-assembled molecular wiring materials and assessed their mechanical and electrical conduction properties—discotic liquid crystals for the first two partners and several types of polymer-based wiring materials for the third. Another partner, the Technical University of Delft (TU Delft), prepared test chips for the first three partners to make conductivity measurements with but the other goal of TU Delft was to examine ways to make high-density through-chip connections (vias) in silicon chips and wafers. The other partner (University College London (UCL), coordinator) designed the test chips and built test rigs for aligning the small test chips that were used for conductivity measurements. These test chips, which contained a range of interconnected electrode pads with sizes in the range 10–300 μm, were intended to allow the other partners to assess experimentally, at a convenient 'micro' scale, the electrical properties of the self-assembled molecular wires, rather than attempting directly to find the ultimate downscaling limits of any particular chemical structure. UCL also looked at the design of a nanocomputing system to carry out the functions of the human visual cortex. This requires the processing power of a present-day supercomputer to solve (dissipating $\sim$1 MW of heat in the process) but the visual cortex in human brains does it almost effortlessly, dissipating only a few watts of heat in the process.

The project was very successful overall. The University of Leeds developed ways to increase the electrical conductiviy of liquid crystals by more than a million times and they designed improvements in the basic discotic systems that led to a $\sim$10^3 improvement in the charge-carrier mobilities. The University of Strathclyde developed ways to improve the structural rigidity and stability of liquid crystals so that their conducting properties would be enhanced: the results demonstrated that the basic concept for molecular scaffolding of discotic liquid crystals (DLCs) was valid and could be used as a way to modify a range of DLC properties. CNRS/University of Marseille developed three very promising polymer-based conducting materials with conductivities of 50 S m^{-1} or more, including a new directional electropolymerization technique (DEP). TU Delft examined and developed two ways to produce high-density through-chip connections (Cu/Si), with lengths anywhere between tens to hundreds of micrometres and diameters anywhere between 2 and 100 μm or more. Finally, UCL, apart from designing the test chips and test rigs and helping the other partners to carry out their measurements, examined the structures that might be needed to implement the connections in a nanoscale implementation of the visual cortex. They also examined possible ways to avoid some of the heat dissipation and fault-tolerance problems that may occur with 3D chip stacks containing more than 10^{11} nanoscale devices. During the course of the project, a three-chip molecular wiring test stack was designed, fabricated and tested with different types of molecular wires.

It must be emphasized that it was not the aim of the project (or this book) to solve all of the problems associated with nanoelectronic devices. In particular, there are some fundamental unanswered questions about heat dissipation. Present-day workstation and PC chips, with about 10^8 devices, dissipate $\sim$100 W. Although improvements in CMOS performance are expected, it is extremely unlikely that a CMOS system with perhaps 10^{11} or 10^{12} devices will ever dissipate less than 1–10 kW when running at maximum clock frequencies. Although many potential nanodevices are being investigated around the world, it is not clear which, if any, of them will have power dissipation levels that are comparable with, for example, that of synapses in animal brains. Heat dissipation problems with 'flat' two-dimensional (2D) chips are already a major problem and it is known that heat dissipation in 3D chip stacks is even more difficult to handle.

In order to make any progress with the CORTEX system design, it was therefore necessary to assume that nanodevices will become available, whose heat dissipation would approach the minimum value allowed on theoretical grounds and, thereby, eventually allow the CORTEX system to be built. Fortunately, as part of the theoretical work that was carried out on the CORTEX project, some new constraints on nanoscale devices and architectures were analyzed: these offer some potentially new ways to develop nanodevices and nanosystems with operating speeds that approach the ultimate limits for classical computing systems.

In chapter 2 of this book, we consider how device constraints will place limitations on how devices could be put together to form circuits and nanocomputing systems that will be more powerful than final-generation CMOS systems. The consideration of 3D structures will be a necessary part of the future task of designing operational nanoelectronic systems. Architectural aspects of 3D designs are discussed. Finally, we examine the connection requirements—the 'wiring'—for a possible ultra-high performance nanocomputer that would carry out the same sort of data-processing operations that are carried out by the human visual cortex. Chapter 3 considers the key question of thermal dissipation in 3D structures. In addition, packaging issues are also addressed.

The search continues for a nanodevice technology that could provide better performance than CMOS, measured in terms of more operations per second for a given maximum heat dissipation per square centimetre. This search has mainly been concentrated on devices that could eventually be made much smaller than CMOS, perhaps even down to the molecular scale. A brief survey of the current status of the field of the nanoelectronic devices is given in chapter 4 of this book. Here we try to cover not only the physical aspects of the proposed devices but also to summarize how far each of these new concepts has progressed on the road from device to circuit to a fully functional chip or system. The next chapter (chapter 5) is intended to address only molecular electronics concepts, as molecular-scale devices represent the smallest possible data-processing elements that could be assembled into 3D structures.

Chapter 6 presents information on the material and fabrication aspects of nanoelectronics, including some recently developed device fabrication technologies such as nano-imprint lithography and liquid-immersion lithography.

The main purpose of chapters 8 and 9 is to investigate the use of molecular wiring as a possible way to pass signals from layer to layer in a future nanocomputer. These chapters contain many of the main results from the CORTEX project. In addition, we have invited a group from the University of

Cambridge, experts on carbon nanotube wires, to write a chapter (7) summarizing the progress achieved by using this technique for 3D interconnectivity.

It has been clear for some time that the use of very small devices will bring certain problems. Such devices will not only have to operate in or near the quantum regime but much effort will have to be devoted to their fabrication and reliable assembly onto chips with very large numbers of devices, much more than the 10^9 devices per cm^2 that will be achievable with CMOS. Furthermore, because of manufacturing defects and transient errors in use, a chip with perhaps 10^{12} devices per cm^2 will have to incorporate a much larger level of fault tolerance than is needed for existing CMOS devices. The questions must then be asked: how much extra performance will be gained by going to new nanoscale devices? And what architectural designs will be needed to achieve the extra performance? In chapter 12, we examine the main ideas proposed so far in the area of nanoelectronic fault tolerance that will affect the attainment of the ultimate performance in nanoelectronic devices.

1.3   Performance of digital and biological systems

Before we leave this introductory chapter, we provide a short presentation of the ultimate limits to the performance of conventional, 'classical', digital computers. This presentation does not explicitly take any account of the computing structures or architectures that might be needed to reach the boundaries of the performance envelope but conclusions may be drawn that suggest that some sort of parallel-processing 3D structure may be the only way to reach the performance limits and it provides clues about what the processing elements will have to achieve. This hypothetical architecture may be quite close to what was investigated during the CORTEX project.

Before we attempt to answer the questions about speed and architectures, it is helpful to consider figure 1.2, which illustrates, in an extremely simplified form, the performance and the size of present-day computing systems, both synthetic (hardware) and natural ('wetware'). Current PCs and workstations have chips with 50–130 million transistors, mounted in an area of about 15 mm by 15 mm. With clock speeds of a few GHz, they can carry out 10^9 digital floating point operations per second, for a power dissipation of about 50–100 W on the main chip. By putting together large numbers (up to 10^4) of such units, supercomputers with multi-teraflop performance (currently up to 35×10^{12} floating-point ops s^{-1}) can be built but at the expense of a heat dissipation measured in megawatts (see, for example, [8]).

The performance of biological computing systems stands in stark contrast to that of hardware systems. For example, it is possible to develop digital signal processing (DSP) chips which, when combined with a PC controller and extra memory, are just about able to mimic the performance of a mouse's visual system (e.g. [9]). At the other extreme, the human brain, with approximately 10^9

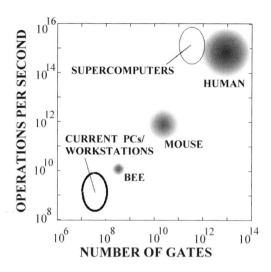

Figure 1.2. Processing speed *versus* number of devices for some hardware and wetware systems. The numerical values are only approximate.

neurons, each with approximately 10^4 synapses and running at a 'clock speed' of about 0.1 kHz, has as much processing power in a volume of 1.5 l as the biggest supercomputer and only dissipates about 15 W of heat. Although it must be emphasized that these numbers are very approximate, there is no doubt that biological systems are extremely efficient processing units, whose smallest elements—synapses—approach the nanoscale (e.g. [10]). They are not, however, very effective digital processing engines.

Having briefly compared the performance of existing digital processing systems, based on CMOS transistor technology, with that of biological systems, we now consider what performance improvements might be obtained by going to the nanoscale.

It is obvious that improvements in performance cannot be maintained indefinitely, whatever the device technology. Consider figure 1.3, which compares the power dissipation and 'clock speed' of a hypothetical chip, containing 10^{12} devices, all switching with a 10% duty factor. The 'operating points' for current and future CMOS devices are shown, together with those of neurons and synapses (we must emphasize the gross simplifications involved in this diagram [10, 11]).

The bold diagonal line marked '50 kT' represents an approximate lower bound to the minimum energy that must be dissipated by a digital device if the chip is to have less than one transient error per year (k is Boltzmann's constant and T is the temperature, here assumed to be 300 K). Thermal perturbations will produce fluctuations in the electron flow through any electronic device, and will, thus, sometimes produce false signals. The exact value of the multiplier

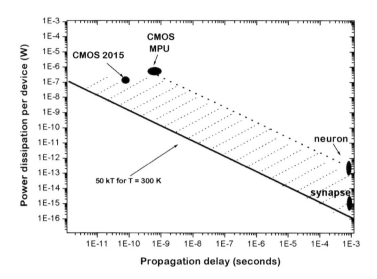

Figure 1.3. Power dissipation per device *versus* propagation delay for current and future CMOS and for wetware devices. The two diagonal lines are lines of constant switching energy per clock cycle. The '50 kT' thermal limit line is the approximate lower bound for reliable operation, no matter what the device technology. Note that a neuron has (very approximately) the same energy dissipation as a present-day CMOS device in a main processor unit (MPU).

depends on the specific system parameters but it lies in the range 50–150 for practical systems, whether present day or in the future. Devices in the shaded area would have a better performance than present-day CMOS. How can it be then that neurons ('wetware'), operating with a 'clock frequency' of about 100 Hz, can be assembled into structures with apparently higher performance than the fastest supercomputers, as indicated in figure 1.2? The answer is in three parts. First, existing computers execute digital logic, while wetware executes analogue or probabilistic logic. Second, synapses are energetically more efficient than CMOS transistors and there are $\sim 10^{13}$ synapses in human brains. Third, brains use enormous amounts of parallelism (in three dimensions) but conventional workstation CPUs do not, although programmable gate logic arrays (PGLAs), application-specific integrated circuits (ASICs), digital signal processing chips (DSPs) and cellular neural network (CNN) chips can achieve varying levels of parallelism, usually without the use of the third dimension.

1.3.1 System performance constraints

Only those devices whose energy dissipation per clock cycle lies within the shaded area in figure 1.3 will offer any possible advantages over current CMOS technology. An ultimate performance improvement factor of more than 1000 over current CMOS is potentially achievable, although which device technology will reach this limit is still not clear. It is, however, unlikely to be CMOS. Furthermore, the device limits are not the only factor that determine the future possible performance but system limits also have to be considered [7]. Meindl *et al* [7] consider five critical system limits: architecture, switching energy, heat removal, clock frequency and chip size. Here we present our analysis of some of these critical factors in figure 1.4. This figure is almost identical to figure 1.3 but now the vertical axis represents the power dissipation per cm^2 for a system with 10^{12} devices ('transistors') on a chip. We choose 10^{12} because this is approximately equivalent to the packing density per cm^2 for a layer of molecular-sized devices but equivalent graphs could be produced for 10^{11}, 10^{13} or any other possible device count.

System considerations such as maximum chip heat dissipation and fluctuations in electron number in a single pulse, can have significant effects on allowable device performance. The effect of these constraints is shown in figure 1.4 for a system with 10^{12} devices cm^{-2} and assuming a 1 V dc power supply rail.

Figure 1.4 shows that limiting the chip power dissipation to 100 W cm^{-2} for a hypothetical chip with 10^{12} devices immediately rules out any device speeds greater than ~ 1 GHz. A chip with 10^{12} present-day CMOS devices (if it could be built) would dissipate ~ 100 kW at 1 GHz. To reduce the power dissipation to ~ 100 W, a hypothetical system with 10^{12} CMOS devices would have to run at less than 1 MHz, with a resultant performance loss.

Because of the random nature of the electron emission process, there will be a fluctuation in the number of electrons from pulse to pulse [12]. It is possible to calculate the probability of the actual number of electrons in a signal pulse falling below $N_e/2$ (or some other suitable threshold)[1]. When this happens the signal will be registered as a '0' instead of a '1'—an error occurs. It turns out that, for a wide range of system parameters, a value of $N_e = 500$ electrons or more will guarantee reliable operation of almost any present-day or future system. However, if N_e drops to 250 (for example), then errors will occur many times a second, completely wrecking the system's performance.

The shaded area in figure 1.4 is bounded below by the electron-number fluctuation limit and it is bounded above by the performance line of existing

---

[1] Let N_e be the expected (average) number of electrons per pulse. If electron correlation effects are ignored (see, e.g., [12]) then the distribution is binomial or, for values of N_e greater than about 50, the distribution will be approximately Gaussian with a standard deviation of $\sqrt{N_e}$. It turns out that the probability of obtaining an error is not sensitive to the number of devices in a system, or to the system speed. However, it is very sensitive to the value of N_e, the number of electrons in the pulse.

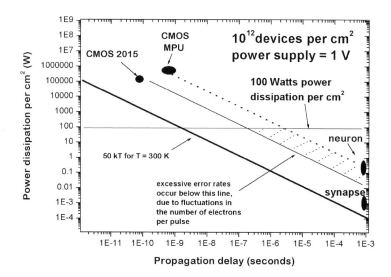

Figure 1.4. The power-delay graph for a hypothetical chip with 10^{12} digital electronic devices, using a 1 V dc power supply. The middle diagonal line in this diagram represents the approximate lower bound to the energy that a digital electronic signal can have, if errors are not to occur due to fluctuations in the numbers of electrons in a pulse. This line corresponds approximately to 500 electrons per pulse. The upper, dotted, diagonal line approximately represents the range of power-delay values that could be achieved with current CMOS devices.

CMOS devices. In other words, the shaded area is the only region where improvements over existing CMOS devices would be achievable. The previous lower bound of 50 kT switching energy cannot be achieved without catastrophically frequent signal errors. How can the parameters be modified to allow the ideal lower 50 kT boundary to be reached?

In order to approach the ultimate system performance, the 'power supply rail' voltage has to be reduced drastically. For example, if it is reduced from 1 V to 25 mV, it would theoretically be possible to produce a chip, containing 10^{12} devices, all switching at 10^8 Hz, while dissipating no more than 100 W. There are, however, several factors that would prevent this ideal from being achieved. Depositing 10^{12} devices in an area of a few cm^2 would inevitably be accompanied by manufacturing defects, which would require some form of system redundancy: this is discussed in a later chapter. There would be resistance capacitance (RC) time-constant effects, associated with device capacitances, which would slow the usable clock speed down. A more severe limitation would perhaps arise from current-switching phenomena. The current through an individual device would be 4 nA but the total power supply current requirement for the whole chip would be 4000 A switching 10^8 times a second. Even if the problems of power supply design are discounted, intra-circuit coupling from electromagnetic radiation would be extraordinarily severe. It would, therefore, be necessary to accept some reductions in the total power dissipation, the clock frequency or a combination of both. The operating point in figure 1.4 would have to move down diagonally, parallel to the $50\,kT$ boundary, until a slower but more practical system could be devised.

A full discussion of exactly what system parameters to choose in order to have a high-performance computing system that approaches the limits imposed by physical and technological constraints is outside the terms of reference of this book. The discussion of the last few paragraphs makes no assumptions about the tasks that the hypothetical chip can solve (and it ignores some very important questions, such as the ratio of logic processing elements to memory elements). However, it is known that there are many classes of problem for which there is much to be gained by parallelism, whereby different parts of a chip (or system) operate relatively slowly but simultaneously, thereby increasing the speed with which a problem can be solved. One such problem is that of mimicking the operation of the visual cortex, which is what this book is about.

References

[1] Hillis W D 1985 *The Connection Machine* (Cambridge, MA: MIT)
[2] Al-sarawi S, Abbott D and Franzon P 1998 A Review of 3D Packaging Technology *IEEE Trans. Comp. Pack. Manuf. Technol.* B **21** 2–14
[3] Nguyen T N and Sarro P M 2001 Report on chip-stack technologies *TU-Delft DIMES Report*
[4] Lea R M, Jalowiecki I P, Boughton D K, Yamaguchi J S, Pepe A A, Ozguz V H and Carson J C 1999 A 3-D stacked chip packaging solution for miniaturized massively parallel processing *IEEE Trans. Adv. Pack.* **22** 424–32
[5] George G and Krusius J P 1995 Performance, wireability, and cooling tradeoffs for planar and 3-D packaging architectures *IEEE Trans. Comp. Pack. Manuf. Technol.* B **18** 339–45
[6] Zhang R, Roy K, Koh C-K and Janes D B 2001 Stochastic interconnect modeling, power trends and performance characterization of 3-D circuits *IEEE Trans. Electron. Devices* **48** 638–52
[7] Meindl J D, Chen Q and Davis J A 2001 Limits on silicon nanoelectronics for terascale integration *Science* **293** 2044–9
[8] http://www.top500.org/top5/2002
[9] Burt P J 2001 A pyramid-based front-end processor for dynamic vision applications *Proc. IEEE* **90** 1188–200
[10] Koch C 1999 *Biophysics of Computation: Information Processing in Single Neurons* (New York: Oxford University Press)

[11] Waser R (ed) 2003 *Nanoelectronics and Information Technology* (New York: Wiley) pp 323–58

[12] Gattabigio M, Iannaccone G and Macucci M 2002 Enhancement and suppression of shot noise in capacitatively coupled metallic double dots *Phys. Rev.* B **65** 115337/1–8

Chapter 2

Three-dimensional structures

D G Crawley and M Forshaw
University College London

2.1   Introduction

The main purpose of the CORTEX project was to investigate the use of molecular wiring as a possible way to pass signals from layer to layer in a future nanocomputer but it was also necessary to try to assess what form this future three-dimensional (3D) system might take. This chapter considers some of the many factors that would have to be considered before such a system could be developed. The chapter is divided into four main parts. Section 2.2 outlines some of the processing operations that are carried out in the visual cortex and considers how they might be carried out using a set of image-processing algorithms in a digital logic processing system. Section 2.3 examines some aspects of 3D architectural considerations, in particular the question of how to reduce power dissipation by switching PEs on or off as required, and how the processing elements in the layers in a 3D stack might be laid out. Section 2.4 examines the bandwidth requirements for transmitting sufficient amounts of data from one layer to another in a 3D stack, in order to carry out sufficient operations to emulate the visual cortex. A provisional system specification for the CORTEX system is then given and it is shown that interchip molecular wiring connections with very modest bandwidths would be sufficient to transfer all of the necessary information from one layer to another. The later processing stages of the CORTEX system would almost certainly use several processing units, more or less equivalent to present-day workstation main processing units but with access to large amounts of memory, so a brief discussion of memory latency requirements is provided. The chapter finishes with a short description of the experimental setup that was used by the other CORTEX partners to make electrical measurements on molecular wires.

2.2 Parallel processing—simulation of the visual cortex

Over the last 30 years, many types of CMOS chips have been developed, with differing levels of parallelism. At one extreme is the conventional microprocessor, with some parallelism (multiple arithmetic units, pipelined data processing etc). At the other extreme are dedicated Fourier transform chips or 'Cellular Nonlinear Network' (CNN) chips, where many small identical units work in parallel on a single chip (current CNN designs, intended mainly for image processing, have about 16 000 identical processor units).

What is to stop conventional CMOS from being miniaturized to the degree at which a complete 'visual cortex' system can be implemented on a single chip? It has been shown earlier (see figure 1.2, chapter 1) that processing speeds at the 10^{15} bit-ops/second level and device counts at the 10^{12}–10^{13} level are needed to equal the processing power of the human brain and the human visual cortex occupies a significant fraction of the brain. Speeds and device counts of this level are only available in supercomputers, whose power demands are at the megawatt level. Improvements in CMOS device technology are unlikely to achieve the 1000-fold reduction in device power that would be needed to reach the limit imposed by thermally-induced errors (see figure 1.3, chapter 1).

What operations are implemented by the visual cortex? The visual cortex is a set of regions of the brain, many of which are contiguous, but with connections to other parts of the brain. Each of the regions implements a complex set of functions: for example, the largest region, V1, combines signals from the corresponding regions for the retinas of the left and right eyes, for subsequent use in binocular depth estimation but it also responds to features of different size, contrast and spatial orientation. Other regions respond to colour or to directed motion, and these 'low-level' processing operations are followed by higher-level processing. Its structure, which is still not completely understood, is partly hierarchical, with many layers that communicate with one another in feed-forward, feedback and lateral modes. Excellent summaries of its properties are provided in books by two of the foremost investigators of the brain, Hubel [1] and Zeki [2].

In the present context, the detailed structure of the animal visual cortex is of relatively little concern. What is relevant is that the retina of each human eye has about 7×10^6 colour-sensitive cones, used for high spatial acuity vision at high light (daylight) levels, and about 120×10^6 rods, used for low light level (night) vision (e.g. [3]). Thus, the equivalent digital input data rate is, very approximately, $2 \times 130 \times 10^6$ bits of information every 20–40 ms—about 10^{10} bit s^{-1}. This may be compared with the $\sim 10^9$ bit s^{-1} in a high-resolution colour TV camera. However, a significant amount of low-level processing is carried out in the retina and subsequently in the lateral geniculate nucleus (LGN), so that perhaps 2×10^6 signals from each eye are fed to the cortex for subsequent processing. The angular resolution of the 'pixels' of the human eye varies from the high resolution of the fovea (~ 1 mrad over a few degrees) to less than one degree at the edge of the field of vision. There is no high-speed synchronizing clock in the human brain but the 4×10^6 bits (approximately) of information emerging from the LGN are initially processed by cortical neurons with a time constant of perhaps 20 ms. The data rate at the input to the visual cortex is, therefore, approximately 2×10^8 bit s^{-1}. Thereafter, some 2×10^8 neurons, with a total of perhaps 2×10^{12} synapses, are used to carry out all of the visual processing tasks that are needed for the animal to survive.

How do these figures compare with digital image-processing hardware and software? There is a big gap. For example, dedicated CNN image-processing chips with mixed analogue and digital processing, can carry out 11 low-level image-processing operations on 128×128 pixel images at a rate of about 2000 images/s, using four million transistors and dissipating about 4 W [4]. Note that here the term 'low level' means a very low level instruction such as 'shift an image one pixel to the left' or 'add two images together and store the answer', so that the effective data-processing rate is, very approximately, $>10^9$ bit/s (bit-op). To give another example, a dedicated front-end processor for dynamic vision applications used pyramid-based image operations to maximize the processing efficiency and achieved a total processing rate of 80 GOP (8×10^{10} byte or bit-op s^{-1}) using a DSP chip with 10 million transistors, an unspecified amount of external memory and a workstation controller [5]. The total power dissipation (DSP chip plus memory plus workstation) was not specified but was probably >100 W.

How many digital operations would be needed to implement the functions of the visual cortex? For the purposes of illustration, we assume that the image size is 768×640 three-colour pixels, running at 25 frame/s, i.e. a data rate of 3.8×10^7 byte s^{-1}. Again, for the purposes of illustration, we assume that the following image-processing operations have to be carried out:

- contrast enhancement and illumination compensation,
- edge enhancement at four space scales,
- motion detection (of localized features),
- bulk motion compensation,
- feature detection (100 features),
- pattern matching (1000 patterns) and
- stereo matching.

Note that these operations are not exactly the same as those that are believed to be implemented in the animal cortex, although they are broadly similar. Although neural network paradigms have been used extensively as models for computer-based image processing, the development of machine perception techniques has mainly followed a different path (e.g. [6, 7]). This has largely been due to the availability of the single-processor workstation and the relative lack of computing systems with thousands of small processing elements, all operating in parallel and thereby partly mimicking some of the operations of the human brain. Similarly, the development of image-processing algorithms has not been straightforward:

until quite recently, constraints on computer power have often forced researchers to use quite crude algorithms, simply to obtain any sort of results in a finite processing time. For example, the vision-processing chip described in [5], when carrying out image correlation (matching), truncates the image pixels to black/white (0/1) in order to speed up the calculations.

There are many different algorithms that can be used to carry out each of the various operations listed previously. One major problem in comparing algorithms is that the so-called computational complexity of an algorithm—the formula that describes how the number of numerical operations that are needed to process an image of N by N pixels (for example) depends on N—is not necessarily a completely reliable measure of how fast an algorithm will run. For example, an operation might take of order N^2 operations with one algorithm and of order $N \log_2(N)$ operations with another. This would imply that the second algorithm would be better than the first if $N > 2$. However, for the important task of finding edges at different space scales, it has been shown that the constant of proportionality (implied by the term 'of the order of') for the two algorithms may be such that the N^2 algorithm runs much faster than the $N \log_2(N)$ algorithm over a wide range of space scales (see e.g. [8]). In addition, the use of parallelism—having many calculations carried out simultaneously— can provide massive speedups in time at the expense of additional hardware. In the present context—the processing of images in real time to extract information that can be used for guidance, for tracking or for object identification—the animal cortex is undoubtedly close to optimum in its use of processing power. To simplify the discussion, we shall assume that only two types of digital processing engines are available—either von Neumann single processors ('workstations') or a collection of small parallel processing elements (PEs) arranged in a 2D grid, each of them carrying out the same sequence of operations in synchrony but on different pieces of data (an example of the SIMD (Single Instruction, Multiple Data) processor type.

To simplify the discussion further, we will only analyse the operation of illumination compensation and contrast enhancement. Figure 2.1 illustrates the phenomenon of colour compensation in human vision. To the human eye, the grey square is visibly darker than the white square below it. However, the brightness of the centre of the grey square is, in fact, identical to the centre of the white square: a complex series of operations are carried out in the human brain to compensate for variations in uniformity and colour of the scene illumination. A good description of the process is given in [1].

There are several models of the illumination compensation mechanism: one popular though somewhat controversial model is the Retinex algorithm (e.g. [9]). This can be implemented by carrying out a local contrast adjustment for each colour (R,G,B) over the image but at a range of different space scales, as illustrated in figure 2.2. The filtered images are then weighted and combined to produce a colour-adjusted (and contrast-adjusted) image. On a 2 GHz workstation a (non-optimized) version of the algorithm takes about 4 min. An optimized

Figure 2.1. Illumination compensation: the upper grey square appears to be much darker than the white square below it but, in fact, the number of photons reflected per second from the surface of the paper is the same at the middle of both squares.

Figure 2.2. Original monochrome image (top left) and three images, filtered at different space scales to adjust for local intensity variations. For a colour image, weighted combinations of the images in three colours would be combined to produce an illumination-compensated image.

version would take about 10 s, and a dedicated DSP chip (with a controlling workstation) would take about 0.5 s, i.e. not quite real time. However, by using the apparent extravagance of a dedicated array of 768 by 640 PEs, arranged in a rectangular, four-connected grid and running at 1 MHz, it would be possible to produce an illumination-compensated three-colour image every 10 or 20 ms, i.e. in real time.

Similar analyses can be carried out for the other processing operations described earlier. Edge enhancement would require, very approximately, the same processing power as for illumination compensation. Global motion compensation can be carried out relatively easily—it is done in some home video cameras—but the correction of distorted images would require almost the same processing effort as that for illumination compensation. Motion detection of localized features ('Is an object being thrown at me?') is easy to implement but object tracking is considerably more complicated, depending on the complexity of the object to be tracked, and overlaps with feature detection and pattern matching. If the object is very simple—a circle or a square, say—then a workstation MPU can carry this out in real time but for more complicated objects—for example, people walking around—then the processing requirements are at least comparable to the illumination compensation task and probably exceeds it (e.g. [7]).

The 'features' referred to in 'feature detection' could be things like corners with different included angles or objects of a certain size and aspect ratio, moving in eight different directions. This stage of processing would be very similar in purpose to some of the operations carried out by region V1 of the cortex. Alternatively, the 'features' could be things like 'noses', 'eyes', 'mouths' and other elements that are needed for face recognition. This brings us on to the 'pattern recognition' phase. Humans can recognize thousands of different classes of object, ranging from faces to vehicles and from sunsets to medieval texts. To build and program any sort of computer to match this level of processing power is far beyond present-day research capabilities. Here we suppose that only one or two classes of objects have to be recognized—for example a limited number of faces, a limited number of types of vehicle or a relatively limited range of texts written in a relatively uniform handwriting style. To simplify the discussion even further, it suffices to say that each of these tasks requires identification of low-level features, then combining them in various ways and comparing them against prototype patterns stored in some form of database.

Finally, there is stereo matching. This is easy in principle but extraordinarily hard in practice to implement well. It requires iterative nonlinear spatial distortion over local regions, repeated matching operations, a variety of interpolation and extrapolation methods to bridge over regions where data are missing or of low contrast and, ideally, frame-to-frame matching to build-up a 3D space model (for example, from a moving vehicle).

How many conventional workstations would be needed to carry out all of these operations? The answer is seven for very much non-real-time operation and about 50 for operation at about once a second (i.e. a total maximum processing power of about 50 Gflops). Using dedicated DSP chips for each process, then one controller and 15–20 chips would be needed for real-time processing (i.e. a total processing power of about about 500–1000 Gops), together with perhaps

100 Mbyte of memory. If a 3D stack of 2D SIMD processor arrays were available, then approximately ten layers would be needed. Most of the layers would carry out the image-based operations: they would each contain $\sim$500 000 PEs, each with about 2000 logic devices and perhaps 2000 byte of local memory, giving a total of $\sim 10^{10}$ devices per layer. The remaining layers would have a similar number of devices but in the form of a much smaller number of more powerful processors, with large amounts of associative memory to provide the pattern recognition functions. The layers would be arranged in the form of a 3D stack, with through-layer interconnections in the upper image-based layers from each PE in one layer to its counterpart in the next. The layer-to-layer signals would be buffered, probably by CMOS drivers/receivers.

How much heat would these systems dissipate? The answer is several kilowatts, if they were to be fully implemented in CMOS, whether now or in the future. However, new nanodevices will only be of use in high-performance applications if they can outperform CMOS and/or have a lower heat dissipation rate. It is well known that heat dissipation is a severe problem in existing stacked-chip assemblies—values nearer 5 W cm^{-2}, rather than 100 W cm^{-2}, would be necessary to avoid thermal runaway in the centre of a chip stack, unless complicated heat transfer mechanisms such as heat pipes are invoked. With ten layers and 10^{10} devices/layer, a total of about 10^{11} devices, it can be shown that it would in principle be possible to operate devices at a propagation delay of about 10^{-8} s (at the thermal limit) or at 10^{-7} s per device at the electron number fluctuation limit for a 25 mV power supply. Allowing a factor of 10 for the conversion from propagation delay to a nominally square pulse gives a maximum digital signal rate of 10 and 1 MHz respectively.

Could a set of small PEs, each containing $\sim$2000 gates running at 1 (or even 10) MHz, carry out sufficient program steps to implement the necessary image-processing functions in real time? The answer is yes: a 10 000 PE array, with each bit-serial PE having 300 gates running at a 2 MHz clock frequency, has been programmed to implement a function similar to that shown in the lower right-hand part of figure 2.2 at about 10 frames/s [9]. A 2000-gate PE should easily handle most of the low-level operations needed for illumination compensation, motion detection, feature detection and so on.

2.3   3D architectural considerations

In this section, we consider some structural and architectural aspects of a 3D computational system as envisaged in the CORTEX project.

One possible arrangement for a 3D vision system is shown in figure 2.3. The image sensor layer is not considered in detail here but note that it is probable that the raw image data from the sensor would need to be distributed to all the SIMD and MIMD layers.

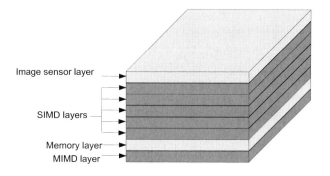

Image sensor layer

SIMD layers

Memory layer
MIMD layer

Figure 2.3. A possible CORTEX system structure.

The first part describes a possible means of reducing power consumption in 3D systems by using local activity control in SIMD arrays—some quantitative results are described and discussed. In the second section, we examine how SIMD layers might be implemented in a 3D stack and this leads naturally to a brief consideration of the implementation of fault tolerance. Finally, the third section describes how, in a Multiple Instruction stream, Multiple Data stream (MIMD) system, the increased memory bandwidth and reduced latency made possible by 3D interconnections might be used to implement a high-performance cacheless memory system. An MIMD subsystem, containing a small number of independently acting, relatively powerful processors, might be used for implementing the higher-level non-image-based functions of the visual cortex.

2.3.1 Local activity control as a power reduction technique for SIMD arrays embedded in 3D systems

Members of the CORTEX project successfully demonstrated the fabrication of molecular 'wires' which might be used to form the vertical signal interconnections between integrated circuit layers in a 3D stack. The ultimate goal is to construct a 3D nanoelectronic computing system for vision applications which might be structured as illustrated in figure 2.3. The figure shows a system in which the uppermost layer is an image sensor and low-level image-processing functions are performed on the data generated by the image sensor by the SIMD layers. Intermediate- and high-level operations are performed by the MIMD layer at the bottom of the stack, which is shown as having an associated memory layer immediately above. Putting the MIMD layer at the bottom of the stack would enable it to be in thermal contact with a heatsink. The remainder of the discussion in this section is confined to consideration of the SIMD layers.

Heat dissipation in 3D systems is a serious problem [10–12]. Whereas a single 2D chip may have the underside of its substrate bonded directly to a heatsink in order to provide sufficient cooling, this is not easily (if at all) possible

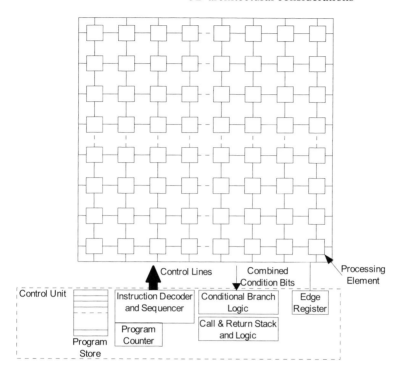

Figure 2.4. SIMD processor array.

for a layer in the centre of a 3D stack. Although a number of techniques, such as CVD diamond films [11], copper vias [12] and liquid cooling [10] have been suggested as possible approaches to conducting heat away from the layers in a 3D stack, none appears to provide a complete solution to the problem. It would seem that a number of techniques, both in power reduction and improved heat dissipation, will need to be combined in order to construct practical systems. Rather than focusing on cooling techniques, we concentrate in this section on a possible architectural technique to reduce power consumption in parallel processing arrays. Heat dissipation is considered in more detail in chapter 3.

Local activity control is a well-known method of achieving a degree of local autonomy in SIMD processor arrays [13]. In such processor arrays (figure 2.4), every PE performs the same instruction as it is broadcast across the array from a control unit. Often, each PE contains a single-bit register whose contents determine whether or not the PE is active—i.e. whether or not it performs the current instruction. In practice, this has often been implemented by using the contents of the single-bit activity register to control whether a result is written to the PE's local registers.

Table 2.1. Comparison of instruction sequences with and without activity control.

With activity control		Without activity control	
Load activity registers	activity $\leftarrow r_D$	Perform operation:	$r_c \leftarrow r_a \langle op \rangle r_b$
Perform operation:	$r_c \leftarrow r_a \langle op \rangle r_b$	AND with mask:	$r_c \leftarrow r_d \wedge r_c$
Reset activity registers:	activity $\leftarrow 1$	Invert mask:	$r_d \leftarrow \neg r_d$
		AND data with inverse mask:	$r_e \leftarrow r_d \wedge r_a$
		OR data with transformed data:	$r_c \leftarrow r_e \vee r_c$

2.3.1.1 Local activity control

The activity register mechanism effectively constitutes a somewhat complex masking operation. Table 2.1 compares a sequence of instructions using the activity control with another sequence which does not use the activity control but generates the same result. Register d (r_d) contains the mask and register a (r_a) contains data which should only be modified at locations where the corresponding mask value is set. Register c (r_c) contains the result in both cases.

From table 2.1, it may clearly be seen that activity control may be used to reduce the number of instructions executed and, hence, improve performance. This performance improvement has been estimated to be around 10% [13].

The activity control, however, may also be exploited as a mechanism for reducing the power consumed by the SIMD array during operation. Rather than simply controlling whether or not results are written to the local registers, the activity register may be used to control the power to the PE logic. Since the register contents (and the contents of local memory, if distinct from the local registers) must be preserved, the power to this section of the PE must be maintained. The logic used for computation, however, may be powered down when the activity is disabled. As a technique for reducing power consumption, the use of activity control to switch the power to PE logic has the following advantages:

(i) It has already been shown to enhance performance in SIMD arrays, so some form of activity control would be likely to be included as a matter of course.
(ii) It is applicable to many technologies, including advanced CMOS, where leakage currents (as opposed to dynamic power dissipation) may form a large part of the total power consumption.

However, the technique is not without some possible disadvantages. Depending on the technology used for the implementation of the activity control and its current requirements, the PE logic may need a large device to control the power. This could consume significant area and would itself dissipate some heat. Careful consideration would need to be given to the power savings which could be made using this technique, since they are dependent on both the algorithm and data used as well as on how the algorithm is coded so as to maximize the use of the activity control.

2.3.2   Quantitative investigation of power reduction

In order to investigate the effectiveness of this technique, some simulations were performed in order to measure how much power might be saved when low-level image-processing tasks are run on such a machine. The simulator used was developed at UCL and is particularly flexible in that the number of PEs in the array, their interconnections, instruction set, number of registers and word-length of the PEs may easily be changed. The algorithms selected for this task were template matching, the Sobel operator and median filtering.

The SIMD array processor [14] which was simulated using 16-bit PEs linked to their nearest neighbours to form a four-connected array as shown in figure 2.4. Each PE consists of a 16-bit arithmetic logic unit (ALU) capable of addition, subtraction and logical operations. Each PE also has 64 words of memory in a three-port register file so that two operands may be read whilst a third is written. One ALU operand is always supplied from the register file whilst the other may be sourced from either the second read port on the register file, immediate data supplied by the control unit or from a neighbouring PE to the north, south, east or west (selected using another multiplexer). The result from the ALU may either be written to the register file or to a register whose output is connected to the four neighbouring PEs. The PE also contains a status register comprising carry (C), zero (Z), overflow (V) and activity (A) bits. The C, Z and V bits are set according to the result of the ALU operation whilst the activity bit determines whether the PE is active. The A bit may be loaded from one of the C, Z or V bits. The status register may also be loaded with immediate data from the control unit. The PE structure is illustrated in figure 2.5.

Figure 2.4 shows the entire array including the control unit. The control unit comprises a program store, call/return stack, program counter, conditional branch logic and an edge value register.

2.3.2.1   *Simulations*

Three simulations were performed in order to obtain some indication of the power savings which might be attained by using the activity bit to control power to the PE logic. The first consisted of a template-matching program, the second consisted of a median filter operating on a 3×3 neighbourhood and the third was edge detection.

For the template-matching programs, a square array of 64×64 PEs was used whilst for the median filter and edge detection an array of 128×128 PEs was

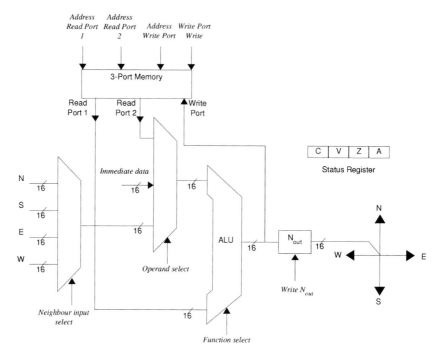

Figure 2.5. Sixteen-bit Processing Element.

used. This was because the template-matching program executed a very much larger number of instructions than the median filter or edge detection so a smaller array was used to reduce the execution time.

The simulator incorporates the ability to count how many instructions of each type are executed when a program is run. Other information may also be recorded and so, for this work, the simulator was made to count the number of PEs which were active for each instruction. Instructions which only execute in the control unit (for example branch, call and return) are assumed to have no PEs active because these instructions can be executed much more quickly than instructions which use the PEs: this is because there is no need to broadcast control signals across the array.

For each algorithm, two different programs were written, one of which made use of the activity control whilst the other did not. Each program was run and the total number of PEs which were active during the execution of the program was recorded as described earlier.

It was possible to make two comparisons: one between the program which made use of the activity control and that which did not; and one between the program which made use of the activity control assuming that inactive PEs were powered down and the same program assuming that inactive PEs still consumed power. The second comparison is arguably the more relevant, since (as stated earlier) activity control improves performance and is, thus, likely to be incorporated for that reason.

One interesting point to note is that an instruction was included which, if any bit in an word was set to '1', set all bits in the resultant word to '1'. This instruction was found to be extremely useful in conjunction with the 16-bit PEs, especially when generating masks.

2.3.2.2 Template matching

The template-matching program took as input a pair of images, a template and a scene image. Both the template and the target took the form of a cross. The template image was shifted over the array and at each location the absolute difference between the two images was calculated and the volume calculated from this result was stored in a register in the PE corresponding to the position of the shifted template. After the program has completed execution, the location of a value of zero in the register containing the volume results indicated the position where the template match occurred. Two versions of the program were written: one using local activity control and the other using no local activity control. During the simulations, the total number of PE cycles (a 'PE cycle' is the number of PEs active during an instruction cycle) was recorded for each instruction, as shown in table 2.2. Instruction mnemonics are shown in the leftmost column. Notice that the PE cycle count is zero for instructions which are not mapped over the array and execute only in the controller (branches, calls and so on).

The column labelled 'No Activity' indicates the number of PE cycles used by the version of the program which did not use local activity control. Of the two columns beneath the caption 'Activity', the leftmost column indicates the number of PE cycles used based on the assumption that inactive PEs are 'powered down' whilst the rightmost column indicated the number of PE cycles used, assuming that even inactive PEs still consume power. The results indicate that if it were possible to use the local activity control to 'power down' inactive PEs, then power consumption would be 81% of that needed for the case when no local activity control is used and 87% of that needed where local activity control is used but inactive PEs are not 'powered down'. This represents power savings of 19% and 13% respectively (strictly speaking, this is the saving of energy, since only a single template-matching operation has been considered—in practice, however, the program would run continuously on a stream of images and the value would indeed represent the saving in power).

2.3.2.3 Median filter

The median-filter algorithm which was implemented was that described by Danielsson [15] and subsequently by Otto [16] and operated on a 3 × 3 neighbourhood. The total number of instructions executed for this program was

Table 2.2. PE cycle counts for the template-matching programs.

Instruction	No Activity	Activity	
br	0	0	0
brz	0	0	0
brnz	0	0	0
brc	0	0	0
brnc	0	0	0
jmp	0	0	0
call	0	0	0
ret	0	0	0
repeat	0	0	0
ldi	1638500	1638500	1638500
ld	2073900	81945900	82765100
ldnin	109061772	109253300	107328000
ldnout	107328000	107328000	107328000
ldedgei	0	0	0
add	53248000	675840	53657600
sub	409600	409600	409600
xor	0	0	0
and	54886400	409600	409600
or	54476800	53657600	53657600
not	27033700	409700	409700
cpl	409600	409600	409600
shl	0	0	0
wset	27853000	0	0
wclr	0	0	0
ldstat	0	0	0
ldact	0	0	0
Totals:	438419272	356137640	408013300

two orders of magnitude less than for the template-matching program, so a larger array of 128×128 PEs was used in the simulation.

The algorithm used in this example operates on a grey-scale image and replaces each pixel with the median calculated over the 3×3 neighbourhood. An example is given in figure 2.6, which shows an input image containing salt-and-pepper noise and the resultant image after median filtering. As in the template-matching program described in the previous section, two versions of the program were written: one utilizing local activity control and the other not.

Using local activity control to 'power down' inactive PEs gives 86% of the power consumed when using no local activity control and 91% of the power

Figure 2.6. Input (left) to and result (right) from the median filter.

Figure 2.7. Input (left) to and result (right) from the Sobel gradient operator.

consumed when using local activity control but not powering down inactive PEs. This corresponds to power savings of 14% and 9% respectively.

2.3.2.4 *Edge detection*

An edge detector using the Sobel gradient operator [17] was implemented. Again two versions of the program were written, one using local activity control and the other not using local activity control. Figure 2.7 illustrates the effect of the Sobel operator in providing a means of edge detection. Like the median filter, the program does not require many instructions to be executed, so a 128 × 128 array of PEs was used in the simulation.

It was found that using local activity control to 'power down' inactive PEs gave 79% of the power consumed when using no local activity control and 98% of the power consumed when using local activity control but not powering down inactive PEs. This corresponds to power savings of 21% and 2% respectively.

2.3.2.5 *Conclusions*

The use of local activity control for power reduction in SIMD arrays may prove to be a useful technique. Figure 2.8 shows the percentage of power saved for each of the three functions by using local activity control. The technique does not provide a complete solution to the problem of power generation and heat dissipation in 3D systems but it could contribute to the successful implementation

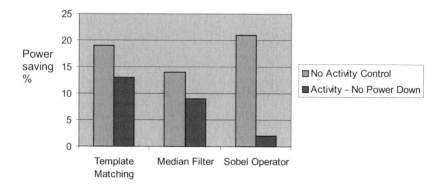

Figure 2.8. Power saving obtained by using local activity control of processors in an SIMD array *and* also switching them off. The comparisons are made with an array where none of the processors is disabled in any way ('No activity control') and with an array where the activity control is enabled but the PEs are not actually switched off completely ('Activity—no power down'). The comparisons are made for the three low-level image-processing operations described earlier.

of a vision computing system. It should be noted that, for the examples given here, no attempt was made at optimization—algorithms were implemented in the most straightforward way. It seems likely that a power saving of around 10% might be expected without any particular extra effort. It is possible that better power savings might be achieved by using some optimization techniques but it is not easy to predict what might be achieved as the results are highly dependent on the algorithm, program coding and, possibly, the operand data.

The technique is not without disadvantages: some means of controlling the power to the logic circuitry for each PE must be provided—this will come at some cost in terms of area and possibly speed, since it seems unlikely that power could be switched arbitrarily quickly. The power switching devices would themselves also dissipate some heat.

Further investigation would need to be aimed at a particular technology for implementation and the details would have to be carefully studied.

2.3.3   SIMD implementation in 3D systems

SIMD arrays could be used to perform low-level image-processing tasks. Such tasks could consist of, for example, edge detection, median filtering, thresholding, Fourier transformation, histogramming, object labelling and so forth.

The SIMD arrays would require 1 PE per sensor pixel. Ideally, it would be possible to implement one complete SIMD array on a single physical layer of the stack, as this would be the most straightforward arrangement.

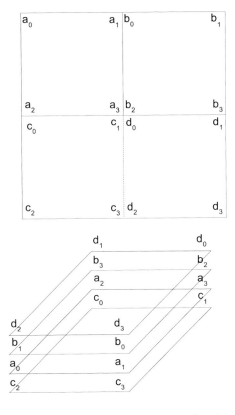

Figure 2.9. Implementing a large array on four layers.

However, depending on the number of pixels needed and the complexity of the PEs, it may become necessary to split a single SIMD processor array over several physical layers. Figure 2.9 shows one way in which this might be achieved. The upper part of figure 2.9 represents a 2D array of PEs which is divided into four quadrants whose corners are defined by the subscripted letters a, b, c and d. This 2D array may be transformed into the four-layer structure shown in the lower part of figure 2.9 if one imagines a 'cut' being made along the line (shown broken) described by c_1c_3 (equivalently d_0d_2) and the array then being 'folded' along the lines a_1a_3, b_2b_3 and c_0c_1. This transformation preserves the adjacency of all the edges of the four subarrays except that of the 'cut' separating the edges c_1c_3 and d_0d_2; however, even these edges lie on the same face of the stack—connections need merely to pass through the two middle layers to join the two edges. This method has the advantage that each physical layer could be identical.

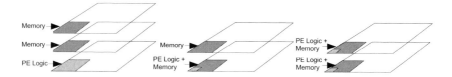

Figure 2.10. Possible distributions of PE logic and memory between layers.

2.3.3.1 Other techniques

Techniques such as that of Zhang and others [18] in which CMOS devices are used with all n-channel devices in one layer and all p-channel devices in another are not considered here. This is because (a) in the CORTEX project the resistance of molecular wires forming the interconnections between layers is likely to be too great to allow the CMOS circuits to operate at high speed and (b) the technique is unlikely to be applicable to technologies other than CMOS—the CORTEX project was concerned with downscaling to nanometre dimensions, where alternative technologies having very different properties to CMOS might be used.

One radically different approach might be to use a 3D array of PEs rather than a series of 2D SIMD arrays. In this arrangement, the PEs form a 3D locally connected array. As in a 2D array, all PEs perform the same operation (unless local autonomy control such as an activity mask disables certain PEs). This could have applications in processing sequences of images: for example, it might be possible to assign one image from the sequence to each 'plane' of the 3D structure and extract motion vectors from the z-direction in a somewhat similar manner to edge detection on a single image. This technique would need a great deal of effort to develop new simulators, design techniques and algorithms.

2.3.4 Fault tolerance

Another technique which might be appropriate is illustrated in figure 2.10. Here, in the leftmost part of figure 2.10, a single PE in an array is highlighted to show how it could have its logic (shown light grey) on one layer and the associated memory (shown dark grey) on one or more vertically adjacent layers. Earlier work [19] has indicated, however, that the PE logic is likely to occupy considerably less area than the memory. A suitable structure is shown in the middle section of figure 2.10, in which the PE logic is integrated on the same layer as some of the memory whilst the remainder of the memory associated with that PE is vertically adjacent on the next layer. More memory could be added using additional layers. If the PE logic consumes only a small fraction of the area occupied by the memory, it could be possible to make the layers identical and utilize the duplicated PE logic to implement a fault tolerant SIMD array at low cost, as shown in the rightmost part of figure 2.10. Note that, in the CORTEX project, we are really only concerned with soft, or transient, errors. This is because the techniques

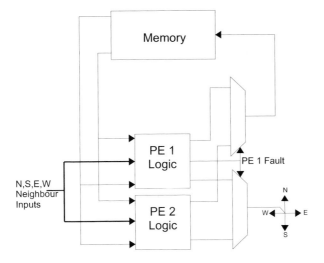

Figure 2.11. Fault-tolerant Processing Element structure.

for vertical interconnection in the CORTEX project are intended to allow all the chips comprising the layers in the 3D stack to be individually tested before assembly into the stack. Assembly into the stack and the formation of vertical interconnections between layers in the stack requires no further high-temperature processing. It should, therefore, be possible to ensure a high yield of working stacks.

The essential idea of the fault tolerant structure using two identical layers is shown in figure 2.11. The two PE logic blocks are labelled PE1 and PE2 and their outputs are connected to a pair of multiplexers. The upper multiplexer selects whichever of the two PEs is connected to the memory write port, whilst the lower multiplexer selects which of the two PEs is connected to the output to the neighbouring PEs. In the absence of a fault, PE1 has its outputs connected to the memory and to the neighbouring PEs. PE1 also has an output which controls the multiplexers. This output indicates whether PE1 has detected an error in its outputs and is used to enable the outputs from PE2 to be connected to the memory and neighbouring PEs instead. Note that there is an assumption that, in the event of PE1 detecting a fault, that PE2 is fault-free. In practice, since both PE1 and PE2 are identical, PE2 would also generate a signal in the event of it detecting a fault on its outputs but this signal could only be used to indicate a system failure. There is no means of generating a correct result in the same processor cycle if both PE1 and PE2 detect faults on their outputs at the same time.

Some of the necessary modifications to the PE are shown in figure 2.12—for simplicity, the modifications needed in order to access the memory on the two layers as a contiguous block are not shown. In essence, the ALUs of the two

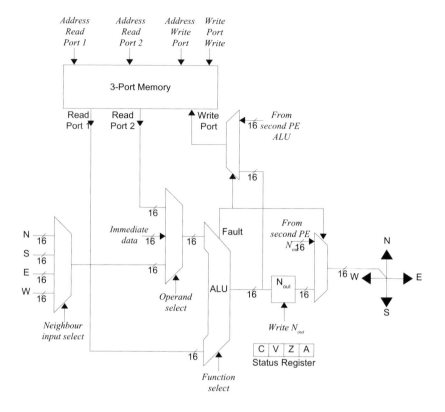

Figure 2.12. Processing Element modified for fault-tolerant operation.

PEs need to be self-checking (in fact, only one of them needs to be self-checking but the point of this discussion is that the layers are identical) and the multiplexers need to be incorporated into each PE because both layers must be identical. When a PE detects that the result from its ALU is incorrect, the result from the other PE is used. There is a wide variety of self-checking techniques which might be used [20–22] but all of them, including simple parity schemes, add complexity and significant area to the ALU.

Thus, whilst the idea of using identical layers in the 3D stack to implement the desired amount of memory and achieve low-cost fault tolerance is superficially attractive, in practice the fault tolerance is not low cost. Considerable additional area is required to implement the self-checking ALUs as well as the multiplexers. More importantly, however, running two self-checking PEs will dissipate more than twice the power of a single, non-self-checking PE. This is likely to be the real obstacle preventing the use of this technique.

A possible solution would be to still use identical layers in the stack and only implement one-half of the PE logic in each layer. Connections between the two layers would enable a complete PE to be constructed. If a simple self-checking scheme could be implemented without a prohibitive increase in complexity, then a time-redundancy scheme might be used to achieve fault tolerance. For a real-time system, this might not be acceptable, however. An alternative would be not to use any explicit fault tolerance at all and simply rely on the inherent robustness of the array processor itself [19].

2.4   3D-CORTEX system specification

2.4.1   The interlayer data transfer rate

The amount of information that would have to be transferred from one layer to the next can be estimated quite readily. With N PEs on each of the early processing layers ($N \sim 500\,000$), and a frame rate of 25 frames/s, filtered images such as those shown in figure 2.2 would have to be passed at a rate of approximately four RGB images per frame—an interlayer data bandwidth of $500\,000 \times 25 \times 4 \times 3 \times 8 = 1.2 \times 10^9$ bit s^{-1}. The data transfer bandwidth between the later layers is more difficult to estimate, but it is likely to be smaller, because only information about detected object features has to be passed to the pattern recognition layers. We assume conservatively that it is one-tenth the rate for the earlier layers, i.e. about 10^8 bit s^{-1}.

Inter-chip data transfer rates of 10^8 or 10^9 bit s^{-1} are modest, even by present-day standards. At first sight, it would appear to be quite simple to have edge connections to transfer the data from one layer to another. However, this would be very undesirable, because the data have to be collected from $500\,000$ uniformly distributed PEs in each layer, then passed from one layer to another, then re-distributed on the next layer. This would involve large high-speed 2D data buses and line drivers on each layer, with an accompanying heat dissipation of perhaps 1 W per layer. However, if direct connections were available from each PE in one layer to the corresponding PE in the next layer, then the data rate per connection would be very low—perhaps only a few kbits s^{-1}. The total data transfer rate would remain the same but the accompanying heat dissipation would drop dramatically, to perhaps tens of milliwatts per layer. For the final layers, if they were to contain only one or a few large-scale processing engines, then edge connections might be a realistic option.

Figure 2.13 illustrates the difference in the inter-layer data distribution geometries between edge-connected and through-layer connected layers.

With each layer having dimensions of 16 mm by 13 mm (for example), then each PE would occupy 20 μm $\times$ 20 μm. To ensure that the interlayer connections do not take up more than a small fraction of the PE area, then they must have dimensions in the range 1 μm $\times$ 1 μm to 4 μm $\times$ 4 μm. In addition to a CMOS driver/receiver combination for each PE, provision must be made for global program control lines and global clock distribution and these would also be implemented most readily in CMOS. Fortunately, because the system would

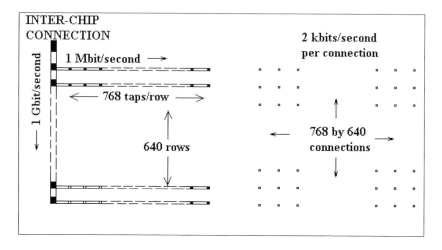

Figure 2.13. Data buses and data rates for single off-layer data connection (left) and for multiple through-layer data connections (right).

run at ∼1 MHz, the total power dissipation of these extra CMOS devices would also be quite small.

It is now possible to provide a general system specification for the CORTEX system—see table 2.3.

Note that, although the devices are not specified, the assumed dimensions of the chips (16 mm by 13 mm) 'force' the device sizes to be quite small: 20 nm × 20 nm. This is not a fundamental constraint. It would be quite possible, for example to use much larger layer sizes—for example, 160 mm by 130 mm. Of course, if the layers were still made of silicon, then the structure would be very fragile but there is no fundamental reason why other substrate materials could not be used: the only requirements are that they can support the active devices, that they can be fitted with high-density through-layer connections and that they be strong. It is relevant to note that device dimensions of 200 nm by 200 nm are at the upper end of the 'nano' scale. The use of large-area layers would also allow the connection pads for the molecular wires to be increased in size to perhaps 10 μm by 10 μm, or even 20 μm by 20 μm.

It is also very relevant to note that the larger the device, the better the chances of its manufacturing reliability being improved. There are major uncertainties about the ability of fault-tolerant techniques having any benefits for nanoscale systems, even if the manufacturing failure rates of the devices were to be the same as for present-day CMOS circuits. There is much to be gained by making the devices as large as possible, provided (i) that they dissipate no more heat in total than 1 W (or a very few watts) and (ii) that their manufacturing reliability is high.

Table 2.3. CORTEX system specification.

CORTEX:
A 3D chip stack using nanoelectronic devices and high-density interlayer connections to implement the basic functions of the human visual system

---

SYSTEM SPECIFICATION
Number of layers: 12
First layer: Sensor array, with e.g. 768 by 640 RGB pixels, perhaps incorporating A/D converters running at 25 frames/second and generating data at 75 bytes/second/pixel,
Seven or eight low level processing layers: each containing ~500 000 Single Instruction, Multiple Data (SIMD) processing elements (PEs), with four nearest neighbour in-layer connections running at ~1 Mbits/second and two 'vertical' between-layer connections running at ~2 kbits/second. These layers would implement functions such as illumination compensation, edge enhancement, motion correction, detection of moving features, stereo matching etc.,
One or two 'high level' processing layers, each incorporating one or more powerful Multiple Instruction, Multiple Data (MIMD) processors and memory, running at 1MHz, to implement pattern matching and pattern recognition, and
One or two memory layers, depending on the complexity of the pattern recognition task(s).

Dimensions of each layer: **16 mm by 13 mm**
Number of PEs per layer: **~500 000**
Number of devices per PE: **>10 000**
Operating frequency :   **1 MHz**
Intra-layer data transfer **rate per PE** ('horizontally', in-layer): **1 Mbit/second**
Inter-layer data transfer **rate per PE** ('vertically', between neighbouring layers): **2 kbits/sec.**

Size of connection pads for vertical data transfer: **1 μm $\times$ 1 μm to 4 μm $\times$ 4 μm**
Horizontal separation between pads: **20 μm**
Digital data bandwidth per pad: **2 kbits/second**

Maximum data **processing rate** (complete system): **$\sim 5 \times 10^{12}$ byte ops/second**
Maximum total system power dissipation (excluding sensor array): **1 – 2 watts**
System volume: ~1 cubic centimetre

Probable nanoelectronic device: not specified, device size 20 nm x 20 nm

---

Probable vertical interconnection technique: molecular inter-layer connections

---

It has so far been assumed (without proof) that the device heat dissipation can be reduced to near the fundamental limits described earlier. Even if this is the case, heat dissipation in a 3D stack must always be of concern. Section 2.3.1 examined some methods for reducing the power dissipation in an array of SIMD processors by optimizing the construction of the processors, so that inactive

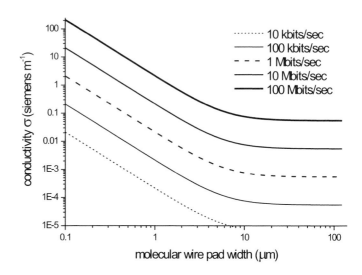

Figure 2.14. Plots to show the molecular wire conductivity σ that is needed to achieve a given data transmission bandwidth between a CMOS-driver–receiver combination, connected by a through-chip via and a molecular wire, as a function of the size of the connecting pad (assumed to be square). Molecular wire length = 5 μm.

processors can be switched off to reduce the heating. Some alternative fault-tolerant techniques for PE logic/memory are also examined in the next section, while chapter 12 discusses fault-tolerance techniques in more general terms.

The hypothetical 'large'-chip stack, with dimensions of perhaps 160 mm by 130 mm by 1 or 2 mm, has a rather inconvenient shape. In section 2.3.3, some possibilities for 'folding' the low-level processing layers into four are examined. This would reduce the footprint area by a factor of four, at the expense of a four-fold increase in the number of layers.

Figure 2.14 shows what conductivity levels are needed to transmit digital signals at various rates between electrode pads of different sizes using drivers and receivers in 0.18 μm CMOS. This diagram is optimistic, in that it does not take into account any effects such as molecular wire capacitance; fringeing field effects in liquid crystal wires or possible fluctuations in the wire distribution with polymer wires. The measured values of the conductivity for polymer-based wires lie in the range $\sigma = 1$–100 S m^{-1}, see chapter 8, which would allow very large bandwidths (>100 Mbit s^{-1}). In contrast, the conductivity values for discotic liquid crystals are, at present, significantly lower ($\sigma \sim 10^{-5}$–10^{-3} S m^{-1}, chapter 9). At present, this would limit the data bandwidth for liquid crystal wires to perhaps 1 Mbit s^{-1} for pad sizes of 10 μm or larger. However, it is clear that the bandwidths that would be needed to transmit signals from one layer to another would, in any case, be very modest—a few tens of kbit s^{-1} at most—and that the pad sizes do not have to be extremely small—perhaps 1 μm at the smallest and, more probably, 4 μm or even larger. It is, therefore, clear that both discotic liquid crystals and polymer methods would be good potential candidates for use in a multiple-chip stack for use in the CORTEX application, although both approaches need further investigation and development.

2.4.2 MIMD memory latency and bandwidth

Computer vision tasks almost invariably require manipulation of a high-level representation of properties of the original image and it is unlikely that the output from the system would consist simply of images having passed through a pipeline of low-level image-processing operations. Examples might be face recognition (where the output is the name of the person whose face has been recognized), industrial inspection (where a manufactured part is either accepted or rejected as being defective—a Boolean output could be used to indicate this) or target tracking (where the output could be the coordinates of the object being tracked).

The manipulation of objects other than images suggests the use of a microprocessor or an MIMD machine. Like the SIMD arrays, the memory and processor could be placed on separate physical layers. Unlike the SIMD arrays, however, the processor (or processors) would be likely to consume a larger amount of area in comparison with the memory. Also, unlike a conventional microprocessor system where multiple levels of cache are traditionally used to attempt to hide the latency of the main memory, a 3D system should (in theory, at least) not need a cache memory. Instead, the large number of connections available from processor (or processors) to memory should enable a wide bandwidth and latency reduction. Latency reduction is particularly important if dynamic RAM (DRAM) is used since the latency is greater than that of static RAM (SRAM).

Whilst the wide bandwidth to main memory is clearly available via the use of a large number of connections in the vertical dimension, latency reduction techniques are possibly not so obvious. One option might be to split the memory into a number of independently addressed banks, each with its own data input/output (I/O), as illustrated in figure 2.15. Here, the memory is split into a number of smaller blocks, each of which has its own address decoder, storage elements, sense amplifiers and connections to the microprocessor. A separate address decoder is used to select the appropriate bank. The scheme could be made more complex so that, for example, contiguous addresses are not in the same bank. However, work by Yamauchi and others [23] indicates that there is an increase in area as the number of banks in the memory system increases—changing from 4 to 32 banks causes the area to increase by a factor of 1.8 for a 256 Mbit DRAM. This is not surprising since the amount of I/O circuitry needed increases linearly with the number of banks. Additional column address decoders

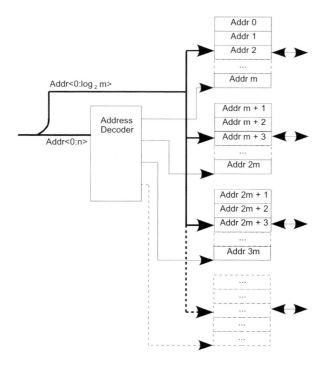

Figure 2.15. Banked memory system.

are also required. If the reduction in latency was such that the cache was no longer necessary, however, such area increases could be an economic proposition.

A further advantage of the multibank approach could be obtained if the memory system enabled simultaneous access to more than one bank at a time. If we consider the data storage space of a modern load-store microprocessor architecture we could envisage a system similar to that shown in figure 2.16. The figure shows a number of memory banks connected via a crossbar switch to the machine registers. The crossbar connects memory bank data lines to machine registers. Several registers may be loaded from (or have their contents stored into) main memory in a single cycle.

The load and store addresses are generated in the usual way from the machine registers. Common addressing modes are register-plus-displacement and register-plus-index: the first consists of adding the contents of a register to a constant (defined in the instruction) whilst the second consists of adding the contents of two registers. The 'Address Calculation' units shown in the figure are used to perform these functions—although in this system several addresses may be generated simultaneously.

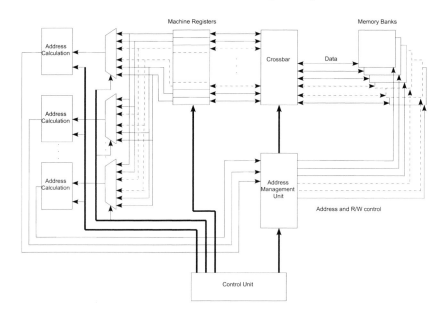

Figure 2.16. Use of the memory structure in a microprocessor.

One problem with parallel memory accesses is that locations in the same bank cannot be accessed simultaneously. Accesses to the same bank must be serialized—this is one purpose of the 'Address Management Unit' (AMU) shown in figure 2.16. The situation is further complicated by the need to consider data dependencies; if, in program order, the contents of one register are stored in at a location in memory and subsequently the contents of that memory location are read into a second register, then the two memory accesses must occur in the correct order. This is an example of a read-after-write (RAW) data dependency [24]. The AMU must also deal with this type of dependency and thus needs information from the control unit in order to sequence the operations correctly. The AMU also controls the crossbar so that the data are transferred between memory and the correct register.

Given the area costs associated with splitting the memory into multiple banks, it would be important to determine the optimum number of banks to use. At first sight, it would seem pointless to use more banks than there are registers but the more banks there are, the lower the probability that simultaneous attempts to access the same bank will occur.

(a)

(b)

Figure 2.17. Pictures of the CORTEX test rig used to align the two layers of silicon. The second picture shows a close-up of the chip itself, showing how electrical connections are taken off the chip.

2.5   Experimental setup

For completeness, we include here a short description of the experimental setup that was used for measuring the properties of the molecular wires. Figure 2.17(*a*) shows a picture of the test rig; on the right-hand side can be seen the arm for manipulating the chip, whilst on the left is a microscope used for visualizing the alignment of the chips. Figure 2.17(*b*) shows a close-up of the rig and shows the configuration of electrical connections onto the chip. The CORTEX test chips were deliberately designed to be as simple as possible, with the sole aim of providing the partners with a series of chips with electrode pads (either aluminium or gold). In all cases, the bottom chips contained metallic electrode pads, where

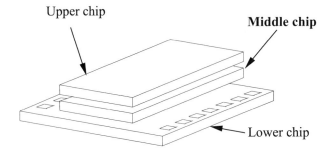

Upper chip

Middle chip

Lower chip

Figure 2.18. Three-chip stack: the lower chip is 5 mm by 5 mm.

the molecular wire connections would be located. The electrodes on the first chips were very large (100 and 300 μm^2) but they were reduced in size in the later chips, with the smallest being 20 μm. In the 'two-chip' stacks, the upper chips simply contained electrode pads, of the same size as those on the lower chips, but joined in pairs, so that a continuous conducting path could be made from the outside world, onto the lower chip, through a molecular wire connection to the top chip, then back from the top chip to the lower chip and so back to the outside world. In the 'three-chip' stack (figure 2.18), the middle chip was intended simply to provide a sandwich between the upper and lower chips. Now there would be two molecular wire layers, joined by 'straight-through' electrical leads (vias) between the two surfaces of the middle chip.

After overcoming many technical difficulties, TU Delft was able to produce some middle chips: these were supplied to the other partners and successful measurements were made on three-chip stacks with two layers of molecular wires (see chapter 8).

References

[1] Hubel D H 1987 *Eye, Brain and Vision* (New York: Scientific American Library)
[2] Zeki S 1993 *A Vision of the Brain* (Oxford: Blackwell Science)
[3] Bruce V, Green P R and Georgeson M A 1996 *Visual Perception* (Hove: Psychology Press)
[4] 2002 ACE16K: A 128 × 128 Focal plane analog processor with digital I/O *Proc. 7th IEEE Int. Workshop on Cellular Neural Networks and Their Applications* pp 132–9
[5] Burt P J 2001 A pyramid-based front-end processor for dynamic vision applications *Proc. IEEE* **90** 1188–200
[6] Zavidovique B Y 2002 First steps of robotic perception: The turning point of the 1990s *Proc. IEEE* **90** 1094–112
[7] Heisele B, Verri A and Poggio T 2002 Learning and vision machines *Proc. IEEE* **90** 1164–77

[8] Forshaw M 1988 Speeding up the Marr–Hildreth edge operator *Comput. Graphics Vision Image Process.* **41** 172–85

[9] Cowan J D and Bressloff P C 2002 Visual cortex and the Retinex algorithm *Proc. SPIE* **4662** pp 278–85

[10] Vogel M R 1995 Liquid cooling performance for a 3-D multichip module and miniature heat sink *IEEE Trans. Comp. Pack. Manuf. Technol.* A **18** 68–73

[11] Ozguz V, Albert D, Camien A, Marchand P and Gadag S 2000 High power chip stacks with interleaved heat spreading layers *Proc. 50th IEEE Electron. Components Technol. Conf.* pp 1467–9

[12] Yamaji Y, Ando T, Morofuji T, Tomisaka M, Sunohara M, Sato T and Takahashi K 2001 Thermal characterization of bare-die stacked modules with Cu through-vias *Proc. 51st IEEE Electron. Components Technol. Conf.* pp 730–7

[13] Fountain T J and Freeman H (ed) 1988 Introducing local autonomy to processor arrays *Machine Vision—Algorithms, Architectures and Systems* (New York: Academic)

[14] Duff M J B and Fountain T J (ed) 1986 *Celluar Logic Image Processing* (New York: Academic)

[15] Danielsson P E 1981 Getting the median faster *Comput. Graphics Image Process.* **17** 71–8

[16] Otto G P 1984 ALgorithms for Image Processing on the CLIP4 cellular array processor *PhD Thesis* University College, London

[17] Gonzalez R C and Wintz P 1987 *Digital Image Processing* 2nd edn (Reading, MA: Addison-Wesley)

[18] Zhang R, Kaushik R, Koh C-K and Janes D B 2001 Stochastic interconnect modelling, power trends, and performance characterization of 3-D circuits *IEEE Trans. Electron. Dev.* **48** 638–52

[19] Crawley D G and Forshaw M R B 2001 Design and evaluation of components for an 'intelligent eye' chip using advanced microelectronic technology *Final Report, DARPA Advanced Microelectronics Program* grant no N0014-99-1-0918

[20] Lala P K 2001 *Self-Checking and Fault-Tolerant Digital Design* (San Mateo, CA: Morgan Kaufmann)

[21] Mitra S and McCluskey E J 2000 Which concurrent error detection scheme to choose? *Proc. 2000 Int. Test Conf. (Atlantic City, NJ, October 3–5)* pp 985–94

[22] Lo J-C, Thanawastien S, Rao T R N and Nicolaidis M 1992 An SFS Berger check prediction ALU and its application to self-checking processor designs *IEEE Trans. CAD* **11** 525–40

[23] Yamauchi T, Hammond L and Olukotun K 1997 The hierarchical multi-bank DRAM: A high performance architecture for memory integrated with processors *Proc. 17th Conf. on Adv. Res. in VLSI* pp 303–19

[24] Patterson D A and Hennessy J L 1990 *Computer Architecture: A Quantitative Approach* (San Mateo, CA: Morgan Kaufmann)

Chapter 3

Overview of three-dimensional systems and thermal considerations

D G Crawley
University College London

3.1 Introduction

Research into three-dimensional (3D) electronics has continued for more than three decades and a very large number of techniques have been studied. But, until recently, there has been relatively little commercial exploitation for volume applications. Most of the motivation has previously come from a desire to reduce the physical size of systems, for example for aerospace applications, hearing aids or products such as Flash memory and mobile phones. It is also interesting to note that more work is being directed towards the development of CAD tools for the design of integrated circuits intended to be part of a 3D system.

However, as discussed in chapter 1, it is only comparatively recently that the problems associated with wiring delays [2] have become sufficiently serious for more attention to be directed towards 3D systems. Effort has been focused towards the implementation of systems containing very large numbers of switching devices, where 3D techniques may be used to reduce the maximum lengths of interconnections and, hence, minimize the delays associated with them.

A serious consideration for any 3D system is power dissipation. In a 2D integrated circuit, it is possible to cool the substrate by placing a heatsink in direct contact with it and provide a path for heat dissipation with low thermal resistance. In a 3D system, this is not possible and a number of techniques for cooling have been described in the literature, some of which are discussed in the second part of this chapter.

3.2   Three-dimensional techniques

Three-dimensional systems may be divided into four separate classes, each representing a distinctly different approach to their construction:

(i)   3D multi-chip modules (MCM),
(ii) stacked chips using connections at the edges of chips,
(iii) 3D integrated circuit fabrication and
(iv) stacked chips using connections across the area of chips.

   For completeness, a description of each class is given here but because the final two classes are likely to be of direct relevance to nanoelectronics, more attention is given to them. However, some knowledge of 3D MCMs is of interest because some of the cooling techniques may be of relevance to future systems. Folded substrates may also provide an economical assembly technique in the future especially for systems using organic semiconductors or for molecular electronics.   Reference [1] provides a comprehensive review of the first two classes.

3.2.1   Three-dimensional multi-chip modules

This technique uses chips which are mounted on a separate substrate or substrates. The 3D structure may then be formed either by folding (in the case of a single substrate) or by stacking (in the case of multiple substrates). This approach has some advantages: conventional, commodity chips may be used and the separate substrates used to mount them may be of high thermal conductivity and possibly equipped for forced cooling by a gas or liquid. In the case of a single flexible substrate which may be folded, the technique provides an effective method of producing a system with a small footprint at minimal cost.
   Conversely, this approach has a number of disadvantages. In particular, because of the use of the additional substrates, (i) wiring lengths are not reduced by as large a factor as systems in which chips are stacked directly and (ii) wire sizes are larger so the density of connections is lower than for directly stacked chips and the capacitance greater. These two factors lead to lower performance than might be achievable in a system using directly stacked chips.
   An example of one possible structure of a 3D MCM is shown in figure 3.1. This illustrates a key feature in that each integrated circuit is mounted on a separate substrate. These substrates contain the horizontal metal interconnections (or wiring) which connect signals to the vias which pass through the substrates and, thence, to the vertical interconnections through the vertical spacers. In this example, the electrical connections between the integrated circuits and substrates are made via solder balls (such integrated circuits are often referred to as 'flip-chips'). The substrate material has a high thermal conductivity and may be in thermal contact with each integrated circuit and, hence, permits heat dissipation via a heat sink mounted in contact with the edges of the substrates. If the spacers

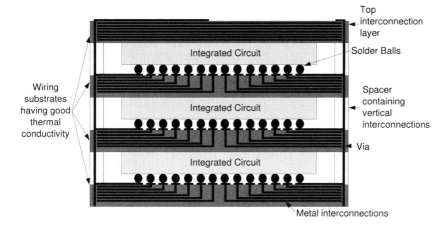

Figure 3.1. Example cross section through a 3D multi-chip module (MCM).

do not completely fill the entire circumference of the stack, then it would also be possible to use forced cooling by pumping a fluid (gas or liquid) into the spaces between layers in the stack.

3.2.2 Stacked chips using connections at the edges of chips

These are 3D systems in which connections between chips in the stack are made by connections at the periphery of the stack. The connections can be made in a great variety of different ways [1] but figure 3.2 shows an example somewhat similar to that described in [3]. The integrated circuits in the stack are modified so that their input/output (I/O) pads are connected to the edge of the die. The vertical connections may then be made on the sidewalls of the stack, as shown in figure 3.2. The integrated circuits in the stack are thinned and then assembled into a stack using, for example, epoxy resin. The ends of the horizontal leads are exposed and then metal is deposited on the vertical sidewalls and patterned to form the vertical interconnections. A cap at the top of the stack allows external connections to be made via bond pads.

Although this type of approach allows integrated circuits which have not necessarily been specially designed for 3D applications to be used in the stack, it has a number of disadvantages. The number of vertical interconnections is limited by the length of the perimeter of the integrated circuits in the stack and, in practice, has often been less than this, as only one or two sidewalls have been used. Also, since signals must be routed to the periphery of the stack for vertical interconnection, the track lengths are very much greater than for systems in which vertical connections are made over the area of the integrated circuits. Compared with 3D MCMs, cooling the integrated circuits in the stack is much more difficult as they are completely encapsulated.

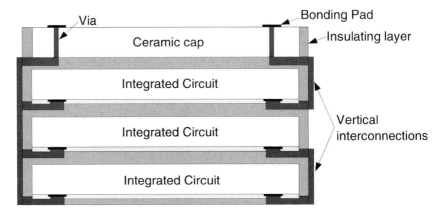

Figure 3.2. Example chip stack with vertical connections made at the edge of the stack.

3.2.3   Three-dimensional integrated circuit fabrication

An alternative approach is to build-up a number of active devices and interconnect layers on a single substrate. An example of this has been developed by Matrix Semiconductor [4, 5]. Here, thin film transistors (TFTs) are fabricated using polysilicon layers grown on layers of oxide and chemical mechanical polishing (CMP) is used to planarize the layers. The advantage of this approach is that it uses only a single silicon wafer and it is, therefore, very much cheaper than the other techniques described here and is compatible with a standard semiconductor fabrication process. The disadvantages are that the TFTs are slower than conventional MOS transistors and the problem of heat dissipation remains. The possiblity of faulty devices may be handled by using fault tolerant techniques.

3.2.4   Stacked chips using connections across the area of chips

In this approach, vertical interconnections between layers in the stack are made over the whole area of each chip in the stack. In order to connect more than two chips together, it is also necessary to make vertical connections or 'vias' through each chip in the stack.

   Although the most obvious method is to provide a vertical 'wire' to connect chips in the stack, this is not the only possibility. Signals may be communicated between layers by inductive or capacitive links. However, power must be supplied to any active circuitry and this means that current-carrying conductors of some description will be necessary to connect to most, if not all, the layers.

   Figure 3.3 shows the cross section through a stack of three vertically interconnected chips, similar to the structure used in the CORTEX project. The upper two chips have through-chip vias to connect the upper surfaces of the chips

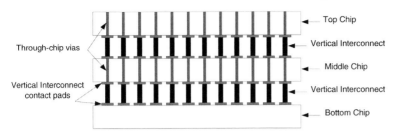

Figure 3.3. Example of a stacked chip structure showing vertical interconnections.

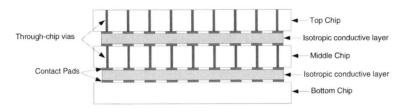

Figure 3.4. Example of a stacked chip structure using an isotropic conductor to provide the vertical interconnections.

(where the active devices and local wiring are located) to their lower surfaces in order to connect to the contact pads. Through-chip vias are described in detail in chapter 11. The vertical interconnections may take many forms, including 'molecular wires' formed using the polymer and discotic liquid crystal techniques described in chapters 9 and 8.

Another interesting vertical interconnection method might be to use an isotropic conductive layer [6–8] to connect the stacked chips, as shown in figure 3.4. This represents an interesting case in that other interconnection techniques having the general structure of figure 3.3 are very likely to have some crosstalk between the vertical interconnections. Therefore, the results from this technique provide a useful 'benchmark' for comparison with other techniques in which the conduction is anisotropic.

Metallic connections may also be used and a number of different techniques have been developed. An example [9] is shown in figure 3.5. Here, it is not possible to distinguish a vertical interconnection 'layer'. Because the vertical connections are formed by metal deposition into deep trenches, they can be considered as extended vias which also connect the stacked chips.

Another important distinction between this technique and those described earlier, is that fabrication of the stack must be done at the wafer level as opposed to assembling a stack from dice which have already been cut from their wafers because it is necessary to perform process steps such as etching and metallization. This introduces a point which may be of importance: individual chips may be tested before assembly into a stack so that the yield of working stacks is improved

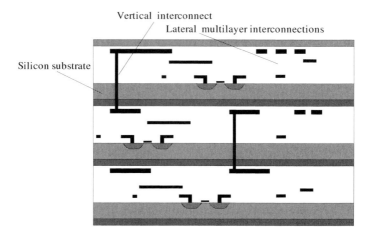

Figure 3.5. Schematic cross-section view of a vertically integrated circuit, after [9].

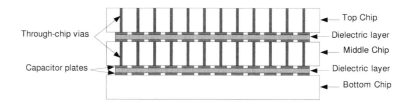

Figure 3.6. Example of a stacked chip structure using capacitive vertical signal paths.

because only working chips are used. When wafers are bonded together and later sawn into stacks, a stack may contain one or more faulty layers which, unless fault tolerant techniques are used, may cause the entire stack to fail. However, this may be of little significance where the yield of working chips is high and the number of layers in the stack is small.

Instead of providing a conductive path through which to transmit signals between layers in a stack, capacitive coupling has been investigated [10]. This technique might be used to implement a stack such as that shown in figure 3.6. This technique has the advantage of not requiring vertical conducting pathways for the signals at all so that alignment of the layers in the stack might be simplified, provided that crosstalk due to stray fields is not problematic. But it would still be necessary to distribute power to all layers in the stack, so some other connections would presumably be required. Also, because of their capacitive nature, the signal paths can only transmit ac signals so it is necessary to design and implement an appropriate signalling scheme. This could introduce problems of excessive power dissipation.

Finally, an inductive signalling system has also been proposed [11] in which vertical signal transmission is implemented by means of metal spiral inductors.

Essentially, the system consists of an array of air-cored transformers with one 'winding' on the lower chip and the other on the upper chip. Again only ac signal transmission is possible, so some other means of distributing power is required and an appropriate signalling scheme is needed.

3.2.5   Three-dimensional computing structures

A number of 3D computing systems have been realized, using both MCMs [23, 24] and stacked wafers or dice. Brief descriptions of non-MCM work are given here.

Little *et al* [25] describe the Hughes 3D Computer—an SIMD system consisting of a 32×32 array of processing elements (PEs) constructed using a five-layer stack. Each layer in the stack was a wafer. Interconnections between layers were provided using 'microbridges'—small U-shaped contacting metal springs. Interconnections between the upper and lower surfaces of the stack were formed using 'thermomigration vias' where aluminium was allowed to diffuse through the wafer to form a p-type conducting via through the n-type wafer. Parts of each single-bit PE were on separate layers with most of the processing occurring in 'shifter' and 'accumulator' layers. Each of these layers had a 16-bit serial memory with the 'shifter' layer being used to communicate between neighbouring PEs (in a four-connected mesh) and the 'accumulator' layer having a single bit adder and 'other logic'. In order to obtain a good yield, each functional cell on each layer consisted of two identical circuits which operate in parallel but further details of the fault tolerance are not given. The clock frequency was 10 MHz and the controller (program store, program counter etc) was external to the stack.

A 3D implementation of the Associative String Processor (ASP) is described by Lea *et al* [3]. This implementation uses the Irvine Sensors technology in which each layer consists of a die which is 're-routed' so that connections to each layer are brought out to one or, as in this case, two edges. The wafers are thinned by grinding so each layer is 250 μm thick. Layers are laminated together using adhesive with a 1 μm thickness of adhesive between each layer. Vertical connections between layers are formed by bringing the edge connections on each layer out through polyimide passivation to patterned metallization on the stack face. A 'cap' layer is used as the top layer of the stack to which wire bonds may be connected. There were five chips in the stack in addition to the cap layer.

Each chip contained 256 associative processing elements (APEs) which incorporated a 64-bit data register, a 6-bit activity register and a bit-serial arithmetic logic unit (ALU). The register storage at each APE is implemented as a fully-associative content addressable memory. Communications between APEs is by a communication network which also allows faulty APEs to be bypassed so that the system is fault tolerant. Power dissipation was 5 W with a predicted temperature rise of 16 °C. Only 32 of the possible 256 APEs were enabled on each layer, giving a total of 160 APEs out of a possible 1280 in the stack. Of six fabricated stacks, one was fully functional and three were partially functional.

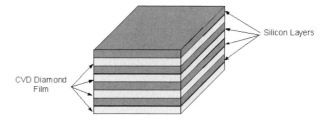

Figure 3.7. Stack with CVD diamond heat spreading layers. The CVD diamond technique could be difficult to adopt where area connections between layers are needed (as in the CORTEX project) since (i) vias would need to pass through the diamond and (ii) thick layers are not compatible with downsizing to nanoscale dimensions.

3.3   Thermal aspects of 3D systems

Because of the low ratio of surface area to volume in a cube, heat dissipation is a serious concern for 3D systems. As mentioned earlier, it is possible to provide additional heat dissipation in 3D MCM systems, since there is generally some thermally conductive substrate which can be joined at its periphery to a heatsinking structure or other cooling system. It is also possible to use liquid cooling in such a system if the modules in the stack are peripherally connected [15, 16].

3.3.1   Technological approaches

Systems using directly stacked chips are more difficult to cool because the substrates are generally thin and have low thermal conductivity. This makes it difficult to thermally connect a heatsink so that the centre of the cube is efficiently cooled. An interesting attempt to circumvent this problem by a group from Irvine Sensors and UCSD is described in [17]. This approach uses chemical vapour deposition (CVD) to form a diamond layer on the back of each silicon layer in order to conduct heat away from the centre of each layer—this is referred to as a 'heat spreading layer' as illustrated in figure 3.7. Diamond has a thermal conductivity of 2000 W m^{-1} K^{-1} whilst silicon has a thermal conductivity of 150 W m^{-1} K^{-1}. The authors constructed a stack using test chips which included heaters and diodes for temperature sensing. The system was air cooled. Compared to other work, the layers used in this stack were quite thick—200 μm for the silicon and 300 μm for the diamond. Using this technique, the authors claim that a power dissipation of between 80 and 100 W could be achieved with a cubic volume of about 16 cm^3 (at a peak temperature of around 73 °C). This would imply a stack of about 50 layers of active devices. It should be noted that, in this work, the connections between layers were made on one face of the cube.

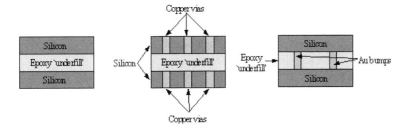

Figure 3.8. Structures studied by Yamaji *et al* [19]

An alternative possibility might be to use active device layers of isotopically enriched or purified silicon. The isotopically enriched silicon crystal has a greater fraction (99.86%) of the isotope ^{28}Si compared with naturally occurring silicon (92.22%). Because of the reduced isotope-scattering of phonons, it has a thermal conductivity which is reported to be 20% greater than the naturally occurring mixture of isotopes of silicon at room temperature [18] but less at higher temperatures. This would probably not have as good a performance as the CVD diamond technique but it would be more suited to area connections.

The effect of using through-wafer copper vias on the thermal performance was studied by Yamaji *et al* [19]. A laser flash technique was used to measure the thermal conductivity of various pairs of silicon dice joined by a layer of epoxy 'underfill' as shown in figure 3.8. A single 625 μm-thick silicon die patterned with aluminium was also measured. The underfill thickness varied between 5 and 150 μm. The structures with copper vias (figure 3.8, middle) had the vias distributed over the area of the die. Vias were (a) 50 μm^2 on a 100 μm pitch, (b) 20 μm^2 on a 240 μm pitch and (c) 10 μm^2 on a 60 μm pitch. The gold bump structures (figure 3.8, right) were at the periphery only. Silicon die thickness for the stacked structures was 50 μm.

In the laser flash technique, the sample is placed in a vacuum chamber and the output from a pulsed laser is applied to one face of the sample. An infrared thermometer is then used to measure the transient temperature on the opposite face of the sample. The change of temperature over time is then used to calculate the thermal conductivity.

The results gave a figure of 148 W m^{-1} K^{-1} for the silicon die alone, which is in good agreement with the accepted figure of 150 W m^{-1} K^{-1} for silicon. However, they also measured a value of 0.33 W m^{-1} K^{-1} for the underfill epoxy—almost three orders of magnitude lower. The results for the stack with copper vias were, thus, the same as for the stacked dice with no copper vias because of the experimental accuracy—the poor thermal conductivity of the underfill epoxy dominated the results. The result for the stack with gold bump connections was 0.7 W m^{-1} K^{-1} but because the laser flash method is designed to only measure 1D conductivity and the widely and irregularly spaced gold bumps make the layer inhomogeneous, the authors question the accuracy of this result and present some simulation results in order to obtain a more accurate value.

It would be expected that the thermal conductivity of a silicon die with copper vias would be somewhere between the value for silicon ($150 \text{ W m}^{-1} \text{ K}^{-1}$) and less than that of copper ($400 \text{ W m}^{-1} \text{ K}^{-1}$) depending on the density of the vias. This work highlights the fact that total thermal resistance is determined by the poorest conductor in the system. One may note that, at the very smallest nanoscales, thermal conductivity is itself quantized but this should not prove significant at the space scales envisaged here.

3.3.2   Architectural approaches

In addition to providing technological solutions to the cooling problem in 3D stacks, it is possible to structure the system so as to improve the heat distribution. In the limiting case, it might be possible to reduce the power of the system to a point where heat dissipation is no longer of serious concern (or can at least be achieved with relative ease). However, whilst this approach has been used for memories and even some computing structures, account must be taken of thermal considerations in a high-performance computing system, especially as the total number of devices in a nanoscale system is likely to be very large whilst the total volume is small—hence, the power density is also likely to be large. Mudge [20] provides a useful discussion of power consumption issues in CMOS microprocessor-based systems and Zhang *et al* [21] estimate 3D stack power dissipation by considering interconnect capacitance.

One possibility, mentioned briefly in [14], is to arrange the system so that the layers with the highest power consumption are placed at the top or bottom of the stack, where they can contact a heatsink. Layers with lower power consumption (such as memory) may be placed nearer the centre of the stack. It should be noted, though, that the top of the stack may, of necessity, be occupied by an image sensor in an intelligent camera application. This concept could be applied to the CORTEX project. Using the assumption that power consumption is related to clock frequency (as is the case for CMOS technologies), the structure shown in figure 3.9 might prove appropriate. Here, low-level image-processing operations (such as edge detection, median filtering, thresholding etc) on images produced by the image sensor are performed by layers of SIMD arrays (possibly arranged as a pipeline). Because these arrays operate on a whole image at once, it might be possible to operate them at a lower clock frequency than the MIMD processors used for high-level image-processing tasks. The MIMD processors would be placed in the layer at the bottom of the stack in contact with a heatsink with the necessary memory in a layer immediately above. Connections through and between layers would be provided by the interconnection technologies described elsewhere in this book.

Finally, the work by Inoguchi *et al* [22] makes the assumption that different processing elements (PEs) on the same layer of a 3D stack might operate at

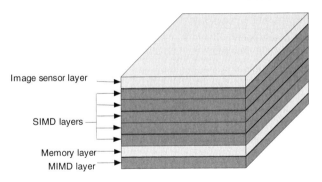

Image sensor layer

SIMD layers

Memory layer
MIMD layer

Figure 3.9. A possible parallel computing system structure for vision applications.

different temperatures and suggests schemes whereby the hotter PEs are located towards the edge of each layer where they might be more efficiently cooled. Their work stems from the concept of fault tolerance where a mesh of PEs is reconfigured so as to replace faulty PEs with otherwise redundant spares. The unused spares are idle and assumed to generate less heat so they are located toward the centre of each layer.

3.4 Conclusions

This chapter has illustrated a number of different techniques for implementing 3D computing structures. It is clear that only a few of the techniques mentioned in section 3.2 will be appropriate as device dimensions decrease to a few nanometres. Then it will be necessary to be able to form very large numbers of vertical interconnections at an extremely high density in order to connect the vast numbers of switching devices in the system. At the same time, power dissipation from the system is likely to become a serious problem and it appears that, almost certainly, a number of different approaches will be required in order to overcome this. New cooling techniques, low power switching devices, low power interconnect and low power computer architectures will all be necessary for successful implementation.

References

[1] Al-sarawi S, Abbott D and Franzon P 1998 *IEEE Trans. Comp. Pack. Manuf. Technol.* B **21** 2
[2] Bohr M T 1995 *Proc. Int. Electron. Devices Meeting Tech. Dig.* pp 241–4
[3] Lea R M, Jalowiecki I P, Boughton D K, Yamaguchi J S, Pepe A A, Ozguz V H and Carson J C 1999 *IEEE Trans. Adv. Pack.* **22** 424
[4] Lee T H 2002 *Scientific American* **286** 52
[5] Knall N J and Johnson M *US Patent* 6,653,712

[6] Pan W, De Tandt C, Devisch F, Vounckx R and Kuijk M 1999 *Proc. Elec. Perf. Electron. Pack.* pp 29–32

[7] Pan W, Devisch F, De Tandt C, Vounckx R and Kuijk M 2000 *Proc. Interconn. Technol. Conf.* pp 228–30

[8] Devisch F, Maillard X, Pan W, De Tandt C, Vounckx R and Kuijk M 2002 *IEEE Trans. Adv. Pack.* **25** 92

[9] Kühn S A, Kleiner M B, Ramm P and Weber W 1996 *IEEE Trans. Comp. Pack. Manuf. Tech.* B **19** 4

[10] Kanda K, Danardono D W, Ishida K, Kawaguchi H, Kuroda T and Sakurai T 2003 *Proc. IEEE Int. Solid State Circ. Conf.* p 186

[11] Mizoguchi D, Yusof Y B, Miura N, Sakurai T and Kuroda T 2004 *Proc. IEEE Int. Solid State Circ. Conf.* p 142

[12] Mick S, Wilson J and Franzon P 2002 *Proc. IEEE Cust. Integr. Circ. Conf.* p 133

[13] Nguyen T N and Sarro P M 2001 *Report on Chip-Stack Technologies* TU-Delft DIMES Report December

[14] Goldstein H 2001 *IEEE Spectrum* **38** (8) 46

[15] Vogel M R 1995 *IEEE Trans. Comp. Pack. Manuf. Technol.* A **18** 68

[16] George G and Krusius J P 1995 *IEEE Trans. Comp. Pack. Manuf. Technol.* B **18** 339

[17] Ozguz V, Albert D, Camien A, Marchand P and Gadag S 2000 *Proc. 50th IEEE Electron. Comp. Technol. Conf.* 1467

[18] Ruf T, Henn R W, Asen-Palmer M, Gmelin E, Cardona M, Pohl H J, Devyatych G G and Sennikov P G 2000 *Solid State Commun.* **115** 243

[19] Yamaji Y, Ando T, Morofuji T, Tomisaka M, Sunohara M, Sato T and Takahashi K *Proc. 51st IEEE Electron. Comp. Technol. Conf.* p 730

[20] Mudge T 2001 *IEEE Computer* **34** 52

[21] Zhang R, Roy K, Koh C-K and Janes D B 2001 *IEEE Trans. Electron. Devices* **48** 638

[22] Inoguchi Y, Matsuzawa T and Horiguchi S 2000 *Proc. 4th. Int. Conf./Exhib. On High-Perf. Comput. in the Asia–Pacific Region* **2** 1087

[23] Segelken J M, Wu L J, Lau M Y, Tai K L, Shively R R and Grau T G 1992 *IEEE Trans. Compon. Hybr. Manuf. Tecnol.* **15** 438

[24] Terrill R *Proc. 1995 Int. Conf. Multichip Modules SPIE* **2575** 7

[25] Little M J, Etchells R D, Grinberg J, Laub S P, Nash J G and Yung M W 1989 *Proc. 1st Int. Conf. Wafer Scale Integration* p 55

Chapter 4

Nanoelectronic devices

K Nikolić and M Forshaw
University College London

Many new physical phenomena have been suggested as a basis for new data-processing devices. However, in order to achieve any practical value, each new idea has to be successful not only at the device stage but it is also necessary to develop and test a series of increasingly complicated structures based on the new device. The first step is to design and fabricate prototypes of a single device on the basis of a proposed basic physical effect and its theoretical description. The next step requires two or more of these devices to be connected into a simple circuit. If it is possible for individual devices to form circuits, then we try to make circuits which perform some useful functions such as logic gates or memory cells. Eventually, we need to integrate increasingly larger assemblies of logic gates or blocks of memory cells which form electronic chips. Our intention here is to discuss the advances in some of the more developed concepts for new devices. Some of the devices discussed here may not, at present, have some characteristic dimension in the below 100 nm range but they should all, in principle, be scalable and the general intention for all of them is to reduce their size into this region.

4.1 Introduction

The silicon-based metal-oxide–semiconductor field effect transistor (MOSFET) technology dominates the electronics industry. The long reign of the MOSFET has been possible due to the ability of the device to be incrementally scaled down as appropriate production technology was developed. Now the minimum feature size (the gate length) of MOSFET devices is reaching the sub-100 nm dimensions yielding a whole transistor area of about 1 μm^2 but they are becoming increasingly hard to make. If the current rate of scaling continues, some time before 2015 MOS transistor will reach its physical scaling limits [1] but the technology will probably continue to provide an ever-increasing computing power, through various material, circuit and architecture improvements (such as 3D structures).

At the same time as the MOSFET technology was developing, there were many different proposals for alternative or complementary devices to MOSFET. Although the MOSFET technology is, by far, the most mature and heavily invested in area, there are still many areas where other device technologies could offer an improvement on the conventional technology, especially for some specific applications. Since MOSFET has, as does every other technology, its limits in respect to device size, speed, reliability, power dissipation, integrability, cost of production, etc, it is possible to focus on one or more of these limits and try to produce better devices and systems in that respect. The Semiconductor Industry Association (SIA) Roadmap and its successor, the International Technology Roadmap for Semiconductors [2], has long provided a guide to the development of conventional electronics devices and it has recently included a chapter on emerging devices. The European Commission has published its own Roadmap for Nanoelectronic Devices [3]. However, only a limited number of alternative devices to the MOSFET have so far been proposed, designed or fabricated [4, 5]. Generally, after some new physical phenomenon (most often quantum mechanical, such as resonant tunnelling or single-electron tunnelling) was identified, people have tried to utilize it and make a new electronic device. In applications, some of the devices are mainly intended to carry out Boolean logic operations and some are intended to be used as memory devices for the storage of binary data. Some are potentially capable of use in both logic and memory devices. The devices that embody various physical effects vary greatly in their state of development from idea to device, then circuit, then system and, finally, to full-scale chip implementation.

Although the vast majority of nanoelectronic devices are based on some quantum mechanical effect, they are intended to implement the classical computing principles of the universal Turing machine or some variation of this concept [6]. However, there are other information-processing paradigms; for example, 'neural networks' or 'quantum computing' [7]. Neural network systems are based on analogies with the extraordinarily successful processing that occurs in biological neural systems. It is possible that some nanoelectronic devices will find some role in neural network concepts. Quantum computing is based on certain basic concepts from quantum physics which allow for a fundamentally new way of coding and processing information. Quantum computers are systems which use the manipulation of quantum-mechanical states (entanglements) [7, 8] and their interference as a means of *computation*, rather than as the operating principle of a digital electronic device. However, currently only a few classes of algorithms exist [8, 9] (e.g. integer factorization, data search), where quantum computing appears to offer significant speed advantages over classical computing. In addition to computing, this concept offers some unique functions, such as quantum cryptography and quantum teleportation [9, 10].

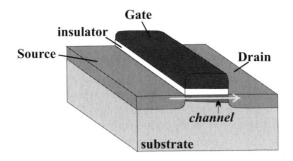

Figure 4.1. The metal-oxide–semiconductor field effect transistor (MOSFET). The flow of current from a source electrode to a drain electrode is controlled by the voltage on a gate electrode.

4.2 Current status of CMOS

The idea of using a perpendicular electric field to control the charge concentration and, hence, the semiconductor material's resistance and current through it—the field-effect transistor (FET) concept—was patented by J E Lilienfeld in 1926. One of the main implementations of this concept was the metal-insulator–semiconductor (MIS) structure. However all initial attempts to create a transistor effect in this way were unsuccessful, because the semiconductor–insulator interface was not sufficiently clean and free from surface states which prevented the electric field from penetrating into the semiconductor. The first working structure was made by Kahng and Attala in 1960 in the Bell Laboratories, when silicon oxide was used for the insulating layer—that was the first MOSFET, (see figure 4.1). In the meantime, in 1947, the first transistor had been demonstrated by Bardeen and Brattain but it operated differently, i.e. it was a bipolar junction transistor. In 1958, Kilbi (Texas Instruments) and Noyce (Fairchild) fabricated the first integrated circuits [11]. After the first successful demonstration of MOSFET, the complementary MOS (CMOS) structure, which consists of two MOSFET transistors of opposite types, was created in 1963. CMOS technology represents the basis of modern electronic chips.

The great advantage of semiconductor structures such as the MOSFET was the potential for scaling down individual devices. The scaling of devices combined with the scaling of interconnects has allowed more and more components to be placed on a single chip. At the same time, smaller devices, in general, operate faster and consume less power. These trends have so far produced an exponential increase in the performance and functionality of computers, as predicted by Gordon Moore in 1965 [12]. The current definition of Moore's law states that the number of devices per square centimetre doubles approximately every 18 months and this empirical rule has been successfully implemented by the semiconductor industry for decades. The first integrated circuit had only one transistor, whereas the latest models of Pentium chips have nearly 100 million transistors. (However, a similar rule applies to the size of software packages, e.g. the first version of MS Word had 27 000 lines but more recent versions have several million lines of code.) It is expected that Moore's law will hold at least until 2010 [2], when MOSFETs which have been scaled down below a 30 nm gate length may see production [13]. Gate lengths of 15 nm have already been reported for some experimental transistors [14]. Due to its technological importance, the phenomenon of scaling of MOS transistors has been exhaustively examined [15]. Naturally, MOSFET technology, as any other technology, has its scaling limits and a few comments on this subject follow.

All devices used for conventional computing are subject to fundamental physical limits [16, 17]:

(i)   thermodynamic—two (binary) states have to be separated by an energy barrier of the order of kT, otherwise the device function is affected by thermal fluctuations; and

(ii) quantum—the Heisenberg; uncertainty principle links the speed of change between two states and the energy difference between the states,

In addition, there is also a limit in signal propagation in a system set up by the speed of light (in vacuum).

Further limits for every technology are imposed by the physical properties of the materials that are used for making devices: maximum density of stored energy, carrier speed, thermal conductance, etc (see [16]). Then we have limits created by the nature of specific devices. For example, the scaling of MOS transistors is limited by the following factors.

- The thickness of the gate insulator d_{ox}, see figure 4.1, for example, if the insulator is SiO_2, for $d_{ox} \lesssim 1$ nm, the leakage current (which exponentially increases as d_{ox} decreases) creates too much heat [18], furthermore, the thickness value must prevent dielectric breakdown.

- Channel length: If the channel length is too short, charge will flow through the device regardless of the gate voltage, i.e. whether device is ON or OFF.

- Doping: As the device size is reduced, the number of dopant atoms also decreases and the fluctuations in the dopant atom number increases, leading to increased variations between individual devices.

- Electric fields in the device: The values for the minimum supply voltage ($V_{DD} \approx 0.9$ V) and threshold voltage ($V_{T0} \approx 0.24$ V) approximately satisfy the requirements imposed by the subthreshold current leakage and the desire for high performance operations (e.g. high gate overdrive) [76] but note that the specified values are more practical limits than physical ones.

Finally, some technological and practical issues are very important:

- Lithography. The main tool in fabricating CMOS devices is optical lithography. The current technology uses 193 nm eximer lasers to produce a minimum feature size of ~130 nm. However, further reduction in the MOSFET size will require other techniques, such as extreme ultraviolet, x-ray or electron beam lithography,

• Cost. With the decrease in device size and increase in chip integration and complexity, there has been a steady increase in the cost of chip fabrication facilities (fabs), implying that there must be some limit when further CMOS miniaturization becomes economically impractical.

There are attempts to push some of the limits by proposing different MOSFET geometries, such as vertical replacement gate (VRG) MOSFETs [19] or materials (high ϵ_r gate insulator, etc) but they cannot significantly alter the limits of this technology [4]. It is predicted [16] that the technological size limits of the (double gate) MOSFET will be approximately: 1 nm oxide thickness, 3 nm channel thickness and 10 nm channel length. The system limits, induced by the use of MOSFET devices and created by the architecture, interconnect delays and heat removal, will put the clock frequency to about 10 GHz and $V_{DD} = 0.5$ V for a silicon chip with 1 billion MOSFETs (of 50 nm channel length) and with 50 W cm^{-2} of heat removal [16].

4.3 New FET-like devices

It is a natural extension of the MOSFET idea to examine possible implementations of FETs using molecular materials, which might allow a further reduction in the size of the devices, beyond the powers of most advanced lithographic methods. Molecules, or groups of connected molecules, probably represent the smallest practical nanoelectronic elements that can ever be used in large, complicated digital circuits. Molecular electronics represents an important part of the drive towards nanoelectronics [20, 21] and it was considered in particular in chapter 5. A significant part of molecular electronics examines FET-type devices with molecules and here we shall briefly mention the main ideas.

When individual molecules are considered for transistor action, we should first address the question: 'Can individual molecules conduct current and if yes, how much?'. One of the first experiments on single molecules was performed by Reed *et al* [22] who showed that a single benzene ring with thiol end groups in contact with metal leads could transmit up to 10^{12} electrons per second at 5 V. Furthermore, the conductance of a molecule could be controlled by an external electric field [20], thereby creating a molecular switch or transistor. Molecular conductance is the subject of intense theoretical investigation, which extends into the modelling of molecular device behaviour [23, 24].

4.3.1 Carbon nanotubes

Carbon nanotubes (CNTs) are today the most promising (macro)molecular structures with possible nanoelectronics application [25]. The use of CNTs for

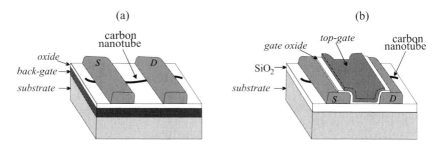

Figure 4.2. A carbon nanotube field effect transistor (CNTFET): (*a*) bottom gate, usually with thick gate oxide, hence lower transconductance, with the CNT exposed to air; (*b*) top-gate electrode, a very thin gate insulator layer allows higher transconductances and the CNT is completely covered by the insulator.

vertical interconnects is discussed in chapter 9 and here we focus only on the device aspects. Carbon nanotube field effect transistors (CNTFET, see figure 4.2), with the CNT as a channel connecting two electrodes, have been demonstrated with bottom gate [25, 26], top gate [27] and vertical gate electrodes [28]. Single-wall CNTFETs (SWCNT-FETs) are found to be p-type if the nanotube is exposed to air [29] or oxygen but they could be n-type if, for example, they are annealed in vacuum [30]. SWCNTFETs exhibit very good electrical characteristics, comparable to silicon MOSFETs: typical currents of $I_{ON} \sim 3\ \mu A$ and $I_{OFF} \sim$ 0.2 nA, a transconductance of $\sim 3\ \mu A/V$ (top gate). The top-gate configuration can offer even better dc performance than nanosize Si p-type MOSFET. However, the mechanism of the current modulation in CNTs is not yet fully understood. Initially, it was assumed that the gate electric field alters the conductivity along the nanotube (in analogy to a conventional MOSFET) but some recent results point to the Schottky barriers, formed at the contact between metal electrodes and the CNT as the source of the transistor action (via modulation of the contact resistance) [31].

The CNTFET technology has already progressed to the next level by producing simple circuits: NOT, NOR, NAND and AND logic circuits [26, 29] have been demonstrated and arrays of FETs made of single-walled nanotubes have been constructed [29, 32]. Different types of memory cells have been proposed: flip-flop type [26], CNTFET (with CNT bundles) utilizing the hysteretic effect in the drain current in respect to the gate voltage [33], CNT floating-gate memory, etc.

CNTs are very interesting for potential applications in nanoelectronic devices because of their size, relative simplicity for fabrication and for good control of the electronic properties (by controlling the geometry and nanotube surface treatment). However, there is still no good device reproducibility and yield and methods for higher levels of integration have not yet been developed.

4.3.2 Organic molecules

Many different organic molecules, especially macromolecules, have been examined for possible use in nanoelectronic devices [20, 34, 35]. Different approaches to contact individual molecules or a small number of molecules have been developed (for some more recent attempts, see [36, 37]). In general, the measured conductance of a molecule is quite low (up to $\sim 10^{-7}$ S, i.e. ~ 10 MΩ).

It was inevitable to see experiments on DNA molecules, connected to two electrodes and with or without a third (gate) electrode. Despite initial controversies about the electronic nature of individual DNA molecules (whether they behave as insulators, semiconductors, metals or even superconductors), after systematic investigation by Storm *et al* [38], it is widely accepted that they are insulators (with a high resistance of at least 10 TΩ) regardless of the base-pair sequence, type of contacts with the electrodes or some other parameters. A similar conclusion was established for small bundles of DNA, longer than 40 nm.

Finally, we should mention transistors based on organic semiconductors and polymers. Polymer electronics is a relatively well-established technology and an emerging commercial competitor. This technology offers low-cost, flexible and relatively simple to manufacture devices. However, these devices cannot as yet be qualified as nanoelectronic because the size of the current polymer transistors is in the range of several micrometres. There is scope for scaling down and devices with submicrometre minimum feature size have been made (vertical-channel polymer FET with channel length of 0.7 μm [39]). Polymer/organic transistors offer a lower performance than conventional MOSFETs, mainly due to a much lower carrier mobility (which is in the region from 10^{-4} up to 10 cm^2 V^{-1} s^{-1} [40]).

4.3.3 Nanowires

Various nanowires [41, 42] and nanotubes [43] based on inorganic materials have been synthesized, with the idea of using them in nanoelectronic devices. Semiconductor nanowires (e.g. GaP, GaN, InP and Si) with 10–30 nm diametres are produced by combining fluidic alignment and surface-patterning techniques [41]. This manufacturing technique allows a relatively simple circuit assembly technique but avoids the need for complex and costly fabrication facilities. With p- and n-doped nanowires used as building blocks and by making structures consisting of crossed nanowires, basic semiconductor nanodevices have been assembled: diodes, bipolar transistors and FETs, as well as simple circuits such as complementary inverter-like structures [44], OR, AND and NOR gates [41]. A bit full 1-adder circuit has been demonstrated [41]. The Si nanowire FET performance may offer some substantial advantages over conventional planar Si MOSFETs [45] or polymer FETs [46]. Vertical nanowire FET fabrication, with CuSCN wires grown in cylindrical openings ($\sim$100 nm in diameter) in thin polymer films, has been reported [47]. Building on the success of the CNTs,

UNIVERSITY OF HERTFORDSHIRE LRC

various inorganic nanotubes have been created (e.g. from boron nitride, gallium nitride, MgO, ZnS, ZnCdS, [43]).

4.3.4 Molecular electromechanical devices

Instead of a (gate) electric field, mechanical deformation can be used to modulate the molecular conductance. Transistor-like effects have been demonstrated in the case of the C_{60} molecule [48] and CNT [32], where a scanning tunnelling microscope (STM) tip has been used to compress molecules. If the deformation tool (e.g. the tip) is part of a piezoelectric element, then the mechanical action is controlled by an applied current which allows hybrid electromechanical circuits to be created. In principle, this type of device can operate at room temperature. The properties of small-scale and large-scale memory–logic circuits of such molecular electromechanical transistors have been analysed in theoretical simulations [49]. The main challenges with this technology are: high device leakage currents, problems with positioning molecules in nanojunctions and problems with the control of the position of the deformation tool.

An IBM Zurich group has recently proposed a new type of storage device called a 'Millipede' [50]. A huge array of atomic force microscope (AFM)-style cantilevers perform write, read and erase operations on a thin polymer film, by pressing the polymer with the heated tip and creating small indentations on the film ('write' operation) and subsequently sensing the presence/absence of the indentations ('read' operation). A CNT can be used as an indenter stylus.

4.4 Resonant tunnelling devices

The resonant tunnelling diode (RTD) is the most well developed 'quantum-mechanical' device. The basic operational principle of the RTD arises from the discrete character of the available energy levels in confined structures, due to the wave nature of electrons. The energy level separations have to be $\Delta E \gg$ kT, for the operational temperature, in order to utilize the effect that the transmission probabilities through the confined part (quantum well) of the RTD depend on the energy of incoming electrons, see figure 4.3. The potential barriers which create the RTD's quantum well are normally very thin (1–2 nm) so that the overall transmission probability is not too low. Usually, an RTD is a semiconductor heterostructure and, so far, they have been realized in many III–V compound systems [51]. Silicon-based RTDs are even more desirable, since silicon devices dominate the microelectronics industry today. Significant progress has been reported on Si/SiGe heterostructure RTDs [52]. Silicon structures with Si-oxide barriers should be a natural choice for silicon RTDs but methods for the growth of crystalline Si overlayers on top of oxide barriers are still being developed [53]. Magnetic RTDs are based on the II–VI (Zn, Mn, Be)Se material systems, containing dilute magnetic material in the quantum well [54].

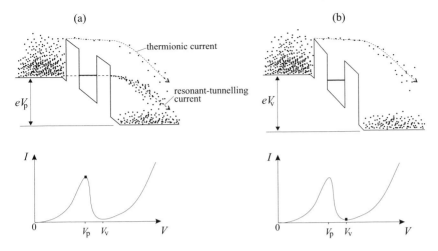

Figure 4.3. The (electron) band structure and I/V curves for a typical RTD system for (*a*) a peak current and (*b*) a valley current.

RTDs are two-terminal devices but they can be transformed into three-terminal transistor-type devices (RTT) if a third electrode which controls the quantum well part of the RTD structure is added [55]. However, in circuits, RTDs are usually combined with ordinary transistors, for example heterostructure RTDs with heterostructure FETs (HFET).

RTD structures could consist of only one atom or molecule between two nanowires (electrodes) e.g. see, [56]. An RTT device has, in addition, an STM tip as a gate electrode [57].

4.4.1 Theory and circuit simulation

RTDs show negative differential resistance (NDR) in the I/V characteristics, see figure 4.3. This feature has been successfully exploited for many applications, for example in very high frequency oscillators and amplifiers as well as in digital systems with Boolean or threshold logic [58], low-power memory cells [59, 60], multivalued and self-latching logics and even in neural networks [61]. The advantage of RTDs, apart from their very short electron transit times and, therefore, very high switching speeds, is that they can operate at relatively low voltages (0.3–1 V). They are also more immune to surface noise effects, since there are no gate electrodes.

Electron transport through RTDs depends on the barrier geometry (i.e. the band structure), as well as on the charging and electron scattering processes. A comprehensive device simulator has been developed for quantitative device modelling in layered heterostructures, the so-called NEMO (NanoElectronic MOdelling tool) [62]. NEMO-1D provides the I/V characteristics for the given

(a) RTD circuit model

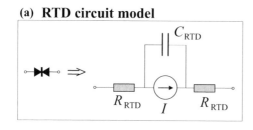

(b) TSRAM Cell

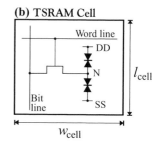

Figure 4.4. (*a*) RTD circuit model [64] and (*b*) the RTD memory cell unit, after ref. [60].

structure and can be used to investigate and design individual RTDs. For circuit simulations one needs a circuit model for the RTD and, generally, a physics-based model [63, 64] has been used in creating such a model, see figure 4.4(*a*). This circuit model is then used in SPICE simulations of arbitrary circuits containing RTD devices.

4.4.2 Memory

Apart from their nanoelectronic characteristics, RTD memories are also potentially interesting as high-speed, dense, low-power, on-chip memories, for fast compound semiconductor microprocessors. DRAMs based only on HFETs exist but the problem is that they have a very high subthreshold leakage current in comparison with MOSFETs. Among several suggested designs for a memory cell based on RTDs perhaps the simplest and most successful was the tunnelling-based SRAM (TSRAM) cell that uses only one transistor and two RTDs [59, 60], figure 4.4(*b*). The RTDs operate in a latch configuration to create two stable voltage levels at the storage node. A bit is stored as a charge on the node (N) between two RTDs, i.e. at the source node of the pass HFET. The RTDs must have only nanoampere valley currents in order to reduce the static power dissipation, so the cells have to be sensed dynamically (e.g. by the DRAM). However, no data refreshing is necessary, which means significant power savings in comparison with conventional DRAM. Small (4 bit × 4 bit) tunnelling-based SRAM memory arrays have been fabricated and a 1-kbit chip has been designed (by Raytheon Systems) [60] and reported in the open literature. The performance of large-scale memory circuitry using RTD-HFET devices has also been investigated using HSPICE simulation [65].

4.4.3 Logic

In recent years, in addition to the standard Boolean logic, threshold logic gates, based on the high-speed logic family of monostable–bistable (MOBILE) transition RTD elements, have been proposed. Parallel adder units, containing

~20 RTD-HFET elements have been built, tested and reported [66] in the open literature. There were some problems with the fabrication of a chip containing a 1-bit full adder. Probably, at this level, more sophisticated and expensive equipment is needed, i.e. facilities of production-line standard.

In summary, RTD technology offers the potential advantage of (a) a higher circuit speed; (b) a reduced number of circuit components and, hence, reduced wiring and circuit complexity; (c) low operational voltages (0.3–1 V) and lower power dissipation; (d) better scaling (miniaturization) properties in comparison to MOSFETs, since a reduction in the RTD area is mainly limited only by surface leakage currents. However, there are no simple designs as yet for memory cells and logic gates using only RTDs and this is the case for many other applications of RTDs in digital circuits. RTDs are usually combined with conventional transistors and, therefore, the advantage of their good scaling properties have small effect on the scaling of the entire chip as well as better power consumption if combined with HFETs. If RTD chips are based on III–V compound semiconductors, although very fast, they are not compatible with conventional silicon production lines. As such, they may find their place in special applications such as analogue-to-digital signal conversion or in optoelectronic communication systems. Provided that material and fabrication problems with Si-based RTDs and RTTs can be overcome, then RTD production could be CMOS compatible, thus making RTD technology more interesting for chip manufacturers.

4.5   Single-electron tunnelling (SET) devices

Single-electron tunnelling (SET) devices consist of one or several small conducting islands, separated from each other and from external electrodes by tunnel junctions (or capacitors). If the total island capacitance is very small ($C_\Sigma \sim$ 1–100 aF) and the tunnel resistances R_T are larger than $R_q = h/e^2 = 25.8$ kΩ, then effects related to individual electron tunnelling to/from the island are observable. Small capacitance means that the charging energy of the island $e^2/2C_\Sigma$ is greater than kT (for a given temperature) and high barriers around the island (i.e. $R_T \gg R_q$) mean that the electron is well localized on the island. The electrostatic potential of the island may change considerably when an electron tunnels onto (or from) the island, affecting further tunnelling to/from the island. SET devices have been intensively studied for possible applications in metrology, instrumentation and for computing [67]. A model of a transistor based on SET is shown in figure 4.5.

In the 1960s, a series of experiments on thin granular metal films [68] showed the suppression of current at low voltages (an effect now known as the Coulomb blockade), which represented the first experimental detection of SET events combined with charge discreteness effects. Single-electron charging effects in electron tunnelling, namely Coulomb blockade and Coulomb

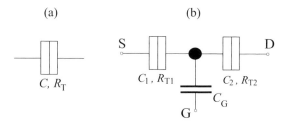

Figure 4.5. (*a*) Symbol for a tunnelling junction. (*b*) SET transistor. The filled circle represents the isolated island, connected to the source and drain side by two tunnelling junctions and to the gate by a capacitor.

oscillations, were first detected experimentally in small-area metal tunnel junctions in 1987 [69].

4.5.1 Theory

Every SET device contains an island, whose electrical connection to the rest of the world goes through tunnel junctions or capacitors. Each junction can be characterized by its tunnel resistance R_T and capacitance C, figure 4.5(*a*). The theory of SET was established in the 1970s and 1980s [70, 71], later named the 'orthodox' theory. The relative simplicity of the theory is based on the assumptions that the island size is not considered (i.e. the electron energy spectrum is continuous), the tunnelling time is negligible in comparison to other time scales in the system and simultaneous tunnelling events ('co-tunnelling') are excluded. The tunnelling probability rate, according to the 'orthodox' theory [71], through the junction of resistance $R_T/R_q \gg 1$, is

$$\Gamma = \frac{R_q}{R_T} \frac{\Delta F/h}{\exp(\Delta F/kT) - 1} \tag{4.1}$$

where ΔF is the difference in the free energy after and before the electron has tunnelled through the junction. Therefore, a tunnelling event has high probability only if the free energy decreases.

4.5.2 Simulation

The detailed time evolution of a circuit which contains tunnel junctions is a stochastic process, because the tunnelling events are random processes. Circuits with SET devices allow for different charge configurations depending on the signal sources and each circuit state has its probability determined by the energy landscape of the system [72]. Due to the Coulomb blockade phenomenon, electrons are effectively trapped in the system's local energy minima and some quantum increase at the voltage sources (or sometimes capacitances) is required for an electron to be able to tunnel through the junction—the transition rate is given by equation (4.1). Co-tunnelling events enable electrons to tunnel effectively through two or more junctions, making metastable states unstable and, therefore, introducing errors into device behaviour. The evolution of a discrete set of states can be expressed by a Master Equation (ME) [73]. In the case of SET circuits, this number of states might be quite large. Alternatively, one can use a Monte Carlo (MC) approach, where, for all possible tunnel events, the tunnel times (τ) are calculated as $\tau = -\ln(r)/\Gamma$, where $r \in (0, 1)$ is a random number. The event with the shortest time is chosen. A new charge configuration is obtained as well as a new potential distribution and then, by repeating this procedure many times, the macroscopic behaviour of the circuit is determined.

The main SET circuit simulators are:

- SIMON (SIMulations of Nanostructures) (TU Wien, 1997 [72]). This is an MC-type simulator with a graphical circuit editor. Based on the orthodox theory, it also features simulation of co-tunnelling events, stability plots, energy-dependent density of states and a single-step interactive analysis mode. The latest version (SIMON 2.0) allows for the following circuit elements: tunnel junctions, capacitors, (normal) resistors, voltage and current sources.
- SENECA (Single Electron NanoElectronic Circuit Analyser) (SUNY, Stony Brook, 1995 [73]). This is a computer algorithm for analysing dynamics and statistics of SET devices and circuits containing tunnelling junctions, capacitances and signal sources. The probabilities of the possible states of systems are calculated using the Master Equation simulator. The program can calculate error rates due to co-tunnelling.
- MOSES (Monte Carlo Single Electron Simulator) (SUNY, Stony Brook 1995 [74]). This is an MC simulator of SET circuit dynamics with a text interface, based on the orthodox theory. The circuit may be an arbitrary configuration of capacitors, resistors, tunnel junctions and voltage sources.

These simulation packages can be improved for more accurate device simulation (e.g. for smaller island structures, energy level spacings and transmission probabilities are required, inclusion of other effects such as electron-electron interaction, etc) but this would only increase the simulation time, which already grows very quickly with the size of a circuit.

Other more efficient simulations than MC calculations would be possible if a SPICE-type circuit model of SET devices could be introduced. However, macromodelling of SETs is, in general, problematic because the tunnelling events are random and the evolution of the system depends on the free energy of the whole system. Only under certain conditions can the I/V characteristics of an SET transistor be independent of the other SETs in the circuit, e.g. if the SET output capacitance is large enough [75]. In this case, an empirical circuit model has been proposed [75] but it is still difficult to be sure that this model describes the strongly coupled nature of SET circuitry properly. Recently, a new concept in

SET circuit design, based on artificial evolution [76], has been proposed, but this concept is probably as computationally expensive as MC calculations. Finally, the new SPICE-level SET model developed at TU Delft [77] will enable SET devices to be included in SPICE simulators with realistic circuit parameters. The circuit parameters of SETs are extracted from experimental measurements [77].

4.5.3   Devices and circuits

Many techniques exist for creating single-electron devices and some of them are described in figure 4.6.

The simplest circuits which use SET are: SET transistors, SET traps, single-electron turnstiles and pumps, SET oscillators and negative differential resistance elements, see figure 4.7. Each of these circuits have been demonstrated experimentally.

An SET transistor consists of an island connected to the source and drain electrodes through tunnel junctions and capacitively coupled to a gate electrode. Individual electrons tunnel to and from the island but the tunnelling is suppressed for low biases because the energy of the system is greater if the electron is on the island—an effect known as the Coulomb blockade, see the first I/V curve in figure 4.7(*a*). The gate electrode enables the background charge of the island (which forms due to impurities and stray capacitances) to be controlled and, modulates the current through the transistor—Coulomb oscillations, figure 4.7(*a*). An interesting SET transistor has been made by using two crossed multiwalled CNTs [85]. Practical SET-transistor circuits have been demonstrated, e.g. an electrometer with subelectron charge sensitivity [69].

An SET trap (or box), figure 4.7(*b*), is an isolated ('trapping') island connected to other SET junctions and capacitively coupled to a neighbouring electrode(s). By applying an appropriate bias voltage, one or more electrons can be attracted to or pushed out of the trapping island [86]. A basic turnstile circuit [87] consists of an array of tunnel junctions with a gate electrode capacitatively coupled to the central island, figure 4.7(*c*). By applying a periodic gate voltage of frequency f, with the maximum voltage above the Coulomb blockade, an electron is pulled into the central island and then pushed out by decreasing the gate voltage. In this way, one electron is transferred through the circuit in each period. Single-electron pumps [88], have several gate electrodes, in contrast to the turnstile, figure 4.7(*d*), and the voltage applied to each electrode forms a potential wave which carries an electron from source to drain. In theory, these devices are able to generate very precise electric currents by transferring single electrons one by one through the circuit ($I = ef$ or, generally, $I = nef$). Therefore, with precise 'clocking', they could be used in metrology as a quantum standard for electrical current. However, thermal activation errors limit the operating frequency (f) and, hence, only very small currents (a few pA) have been achieved with SET turnstile and pump circuits so far. This is still not enough for a current standard but metrological accuracy has been theoretically predicted [89]. Another

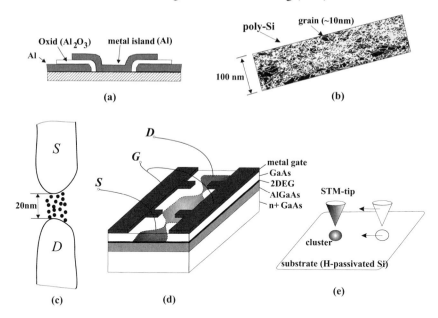

Figure 4.6. Techniques for creating SET devices. (*a*) Shadow mask evaporation is used for fabricating high-quality metallic, Al/Al$_2$O$_3$/Al, tunnel junctions, [78]. Problem: unwanted features are also generated, the technique is restricted to a few materials. (*b*) Ultra-thin films (e.g. 1–5 nm thick poly-Si film [79]) or nanowires [80] of semiconductors or (*c*) granular thin films of metals [81]. Usually arrays of isolated, nanometre size quantum dots are formed, due to the varying thinness of the film or small particles. Problem: the exact position and size of the islands are not controlled. (*d*) Two barriers and an electron island are formed in the 2D electron gas in a heterostructure by applying a negative voltage on the patterned split gate [82]. Quantum dots produced in this way are still relatively large, therefore only low-temperature operation is possible. (*e*) Molecular-scale fabrication produces very small structures in a controllable way by using an STM/AFM tip for manipulation of metallic clusters (such as Au, Ag, Sb) [83], nano-oxidation of Ti films to TiO$_2$, nano-oxidation of Nb [84], positioning two multiwalled carbon nanotubes to cross [85], etc. The technique yields very small islands ($\sim$5–50 nm) and, therefore, improved operational temperatures (including room temperature) are possible but, at present, fabrication is very slow and, therefore, it is not yet suitable for mass production. Apart from these techniques, electron-beam lithography is widely used for creating small structures but the resolutions achieved so far ($\sim$10 nm) restrict the operating temperatures to $T < 1$ K.

interesting metrological application is the possibility of precisely charging a cryogenic capacitor using SET current sources.

The circuit shown in figure 4.7(*e*), consisting of two capacitively coupled islands and tunnelling junctions, was demonstrated to exhibit negative differential

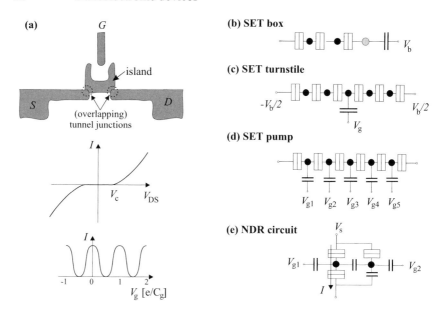

Figure 4.7. (*a*) The layout of an SET transistor and I/V characteristics when gate voltage V_g is fixed (Coulomb blockade) and when the bias voltage V_{DS} is fixed (Coulomb oscillations); (*b*) SET trap (the trapping island is grey); (*c*) SET turnstile; (*d*) SET pump; and (*e*) NDR SET circuit, after [90].

resistance (NDR) [90] but only for very low temperatures. A typical application for NDR elements is in small signal amplification and in oscillators. It could also be used as a memory cell.

4.5.4 Memory

Chips for information storage are perhaps the most promising application of single-electron devices. The main reasons are that memory chips normally have a very regular structure and relatively simple algorithms for error correction. There are nearly a dozen proposed memory cell designs based on the Coulomb blockade, with different operational temperatures and immunity to the random background charges (for a review, see [91]). Here we describe only a couple of ideas presented in the literature so far. Although they are called 'single-electron' devices, SET memory devices, in fact, usually rely on storing or sensing more than one electron for reliable operation (but they are all based on the Coulomb blockade effect).

The typical idea for an SET memory is to construct a floating-gate-type device, see figure 4.8, such as the NOVORAM (non-volatile RAM [67]), which is based on the storage of electrons in a floating gate ('nano-flash' device). Yano *et al* [79] have constructed a single-electron memory which consists of a nanowire of

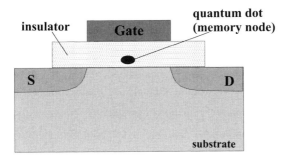

Figure 4.8. Single Electron Floating Gate Memory. Sufficiently high gate voltage can cause charge to tunnel to/from the memory node ('write'/'erase' process). Presence/absence of charge on the memory node will affect, or even suppress, the current through the transistor ('reed' process).

ultrathin poly-Si, see figure 4.6(*b*), between two electrodes and a gate wire under it. The thickness of the film varies (1–5 nm) and forms randomly distributed barriers and small localized areas where charge can be trapped. The conductivity of the nanowire is modulated by the trapped charge and, hence, charge can be detected. A 128-Mbit memory chip based on Yano-type memory cells in combination with CMOS peripheral circuits has been demonstrated (Hitachi, 1998 [79]). The cell size was $0.15~\mu\text{m}^2$, read time 1.2 μs, room-temperature operation but less than a half of the cells were operational. Another memory type was proposed by a Cambridge group [92], who have fabricated a 3×3 bit memory array of integrated SET/MOSFET cells. The memory node was the gate of a split-gate MOSFET which is connected to the word line by a very narrow silicon wire, which behaves as a multiple tunnel junction SET.

4.5.5 Logic

One of the basic blocks of conventional logic circuits is the inverter. An SET inverter circuit has been demonstrated [93], based on an GaAs SET transistor and a variable load resistance (operation temperature 1.9 K), as well as an SET inverter with a voltage gain [94], based on metallic, $\text{Al}/\text{Al}_2\text{O}_3/\text{Al}$, tunnel junctions (but only at temperatures below 0.14 K). By making some alterations to the circuit, NAND and NOR logic gates might be produced. An attempt to realize a NAND logic gate experimentally, by using SET transistors based on silicon nanowires (at temperature 1.6 K) has been reported [95] but the results point to severe problems (low ON/OFF voltage ratio, small output current, need for individual adjustment of each transistor, etc). One major problem with SET logic devices is how to control the unpredictable effects of charge hopping in the substrate.

A number of SET digital circuit designs have been suggested [96] based on conventional circuit design methods. One example of this is the single-electron adder that has been developed at Siemens [97]. This theoretical proposal uses a cellular automaton functionality and Coulomb blockade devices for adding binary numbers. SET devices are probably not well suited for driving long interconnects or a larger number of devices, due to their high tunnelling resistance and, therefore, high RC time constant. Hence, the cellular automata concept seems to be more appropriate for information-processing functions implemented in SET devices.

4.6   Other switching or memory device concepts

4.6.1   Magnetoelectronics

Electron transport in ferromagnetic materials depends on the electron spin, due to the interaction between the spin and the resultant magnetic moment of the material. Electron scattering depends on the electron energy and the additional energy term due to spin depends on the spin orientation [98, 99]. Fabrication of multilayer magnetic nanostructures allows this effect to be utilized. The resistance for the current flow both parallel and perpendicular to the layers is smaller if the layer magnetizations are parallel than if they are antiparallel. This difference can result in giant magnetoresistances (GMRs): structures that use this effect are called spin valves [100, 101]. Similarly, the spin-dependent tunnelling between magnetic layers separated by an insulator also can lead to significant magnetoresistance and this effect is used in another device—the Magnetic Tunnelling Junction (MTJ) [102,103]. When a spin valve is sandwiched between two semiconductor layers, a new hybrid ferromagnet–semiconductor device is obtained: the spin-valve transistor (shown in figure 4.9). This device may be very important for the future development of magnetoelectronics. In all these devices, by applying an external magnetic field, the magnetization of the layers can be changed from parallel to antiparallel and the conductance of the spin valve and, therefore, the current through the device is changed (see diagram in figure 4.9).

GMRs and MTJs have been successfully used for magnetic sensors in the read/write heads in the hard disc drives, as well as for the magnetic random-access memory (MRAM) [100]. Magnetic memory devices are now used for high-density, low-power, high-speed, reliable (radiation hardened), non-volatile information storage (when the magnetic field is switched off, the magnetization orientation may be retained for a long period). In addition, MRAM offers non-destructive readout (unlike similar ferroelectric memories), very high retention times (which eliminates the need for data refreshing) and durability. Magnetoelectronic memory devices using MOSFET transistors in each memory cell are rapidly approaching full commercial use [98, 104].

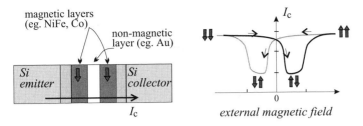

Figure 4.9. Spin-valve transistor (top view) where magnetic fields are used to control spin-polarized electron transport and a representative current characteristic as a function of the applied magnetic field. The arrows show the magnetization direction in the magnetic layers, after [101].

4.6.2 Quantum interference transistors (QITs)

Electron transport in nanoscale quantum devices can be simplified and described in terms of one-electron ballistic or quasi-ballistic transport. Electrons are scattered by impurities or device edges but if the phase change is not a random variable, then scatterings are elastic and the electron wave-function retains coherence. The scattering region is connected to the reservoirs which serve as sources/drains of electrons with an equilibrium energy distribution. If a voltage is applied to a reservoir, the electrochemical potential of the reservoir is changed, causing the net flux of electrons through the device, i.e. the electrical current between the reservoirs. The simplest device would consist of two reservoirs and a perfectly even lead between them. If electrons are confined to a quasi 1D geometry, the electron waveguide regime occurs, when well-defined electronic sub-bands or discrete modes are formed, equivalent to those in optical waveguides. Each propagating mode contributes to the conductance of the waveguide (or quantum wire) up to $2e^2/h = 77.5\ \mu$S (hence, the minimum resistance for the *monomode* regime is 12.9 kΩ) [105, 106]. Conductance quantization has been confirmed experimentally [107].

If leads are not perfect or some more complicated structure is connected to the reservoirs, then elastic scatterings of electron (probability) waves occur, which give rise to observable quantum interference effects. T-shaped waveguide-like structure [108, 109] (see figure 4.10) is a very simple alternation of a perfect lead structure and offers a possible way to exploit quantum interference effects in order to achieve device function, such as switching. The electron transmission probability and the current are controlled simply by the length of the stub and the stub length is changed by the gate voltage V_G [110, 111]. Such nanostructures have been experimentally demonstrated [110], by using a 2D electron gas (in a modulation-doped heterostructure) and a set of electrodes to create the lateral confinements.

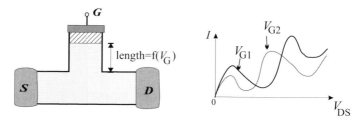

Figure 4.10. Quantum interference transistor (QIT) use the quantum-interference effect in coherent electron transport to modulate electron transmission and current.

Theoretical investigation into quantum interference structures predicted very high switching frequencies in the THz region [108]. Furthermore, the quantum-interference nature of the transistor action requires very low power, which could allow extremely large-scale integration and low power-delay products. Possible applications of these structures for analogue-to-digital conversion, electron directional couplers, far-infrared photodetectors, etc have been experimentally investigated [112, 113]. However, there are still many problems before electron waveguide devices can be developed fully. First, impurities and boundary roughness significantly affect electron transport: conductance quantization is easily destroyed by strong scattering and the Anderson wavefunction localization at impurities poses a serious restriction on the high carrier mobility predicted for these waveguides [114]. The threshold voltage and other transistor parameters become very sensitive to the device geometry, since interference is a periodic rather than an on–off effect. Hence, device and circuit reproducibility are very difficult to achieve. Room temperature operation is only possible for lateral confinements of ∼ 10 nm, and quantum interference transistor (QIT) devices still show a lack of voltage gain. Study of coherent transport and quantum interference is important for molecular electronics. Quantum interference might complicate the operation of molecular circuits, because the operation of each individual device will be very sensitive to its surroundings.

4.6.3 Molecular switches

Two-terminal devices might seem more natural for molecular-scale systems than three-terminal ones, because of the technological difficulties in manipulating small structures. Furthermore, the chemical assembly of molecular devices usually results in a periodic structure. Hence, a two-terminal switch, electronically reconfigurable, where a relatively high voltage is used to close or open the switch, but a relatively low voltage to read was proposed, see, e.g., [115]. Groups at Yale and Rice Universities published results on a different class of molecules that acted as a reversible switch [116]. The chip architecture is based on the Field Programmable Gate Array scheme [117]. This chemically self-assembled structure is an example of an bottom-up approach in designing nanoelectronic circuits. The main idea is to have as simple as possible hardware (which is very important for molecular-scale systems) and to push all complex details of the circuit into the software.

The molecular-switch concept has been demonstrated experimentally by Heath *et al* [117] but on the micrometre scale. Further downscaling has been achieved in the HP labs (Quantum Science Research) very recently [118] in producing a 64-bit memory array of less than 1 μm^2 in area. However, the proposed downscaling to a single-molecule-type switches could be affected by quantum interference effects in molecular-scale circuits [49]. More details about these concepts are given in chapter 5.

4.7 Quantum cellular automata (QCA)

Quantum cellular automata (QCA) offer an alternative computing architecture to CMOS technology [119]. QCA systems consist of arrays of cells. Each cell affects its neighbouring cells through an (electric or magnetic) field and they normally have no other connections. Typical cellular automata (CA) systems are electronic QCAs (EQCA) and magnetic QCAs (MQCA). Furthermore, Josephson Junction Persistent Current Bits (JJPCB) and even Rapid Single Flux Quantum (RSFQ) circuits have been described as cellular automata. For encoding logic states, these systems use: EQCAs, the spatial distribution of electric charges within a cell [120–122]; MQCAs, the direction of magnetic moments [123]; JJPCBs, the direction of the current in a superconducting loop [124, 125] and RSFQs, single-flux-quantum voltage pulses [126, 127], see figure 4.11. Mutual cell interactions can be used to propagate and compute binary [125, 128] (or, in some cases, quantum) information. Every CA cell is supposed to evolve into a stable state that is determined by the properties of the neighbouring cells (and perhaps by external fields). Arrays of CA cells can be seen as computing circuits, where the state of an array is mapped to a computation. The potential advantages of this concept over conventional transistor-based logic are: high speed, insensitivity to electrical disturbances and cosmic rays, ease of fabrication, very good scaling potential (possibly down to molecular sizes), room temperature operation and low power. Unfortunately, these advantages do not all exist in the same device.

4.7.1 Electronic QCA

Typical electronic QCA cell design consists of four quantum dots (Lent and Tougaw [121]). The cells are electro-neutral (but have two free electrons) and can be in a polarized or non-polarized state, depending on the tunnelling potential between quantum wells. There are two polarized states due to Coulomb repulsion between electrons, see figure 4.11(*a*), which can be used for representing binary information.

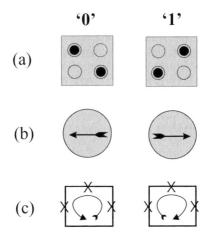

Figure 4.11. One possible encoding scheme of the binary states ('0' and '1') in (*a*) the electronic QCA cell, four coupled quantum dots at corners of a square, with two electrons in the system (black dots) with two stable configuration states; (*b*) the magnetic QCA cell, magnetic moment orientation (represented by arrows); and (*c*) the Josephson junction persistent current loops, the arrows represent the direction of the superconducting current and × represents the Josephson junction.

4.7.1.1   Theory

A qualitative description of the four-dot QCA cell is possible by the use of a Hubbard-type Hamiltonian (which contains a few empirical parameters which describe the potential energy due to charges outside the cell, the tunnelling energy and Coulomb interactions within the cell) and then the quantum states of the cell are found by solving a time-independent Schrödinger equation, see [121]. A quantitative expression for the cell polarization is defined by using the ground-state eigenfunction. The polarization increases as the tunnelling energy increases in respect to the Coulomb energy within the cell. The cell–cell interaction can be quantified by calculating the polarization induced on one cell by the polarization of the other. By using appropriate distances between the cells, the driven cell response (polarization) can be controlled to be highly nonlinear and bistable. The next step is to calculate the response of an array of cells and this can be done by simply extending this method, by calculating the ground state of the whole array of cells for the given polarization of the driver cell. Since the direct product space expands exponentially with the number of cells, approximate methods have been developed for longer arrays. Similarly ground states can be calculated for any combination of cells and input potentials on the edges of the system, which should map into the logical solution of the given computation problem.

The Hubbard-type Hamiltonian modelling of QCAs is relatively simple but gives a good qualitative picture. Attempts to use more detailed models of QCAs, with self-consistent calculations of the charge distribution, run into convergence problems when the electrostatic interaction between the cells becomes comparable with the quantum confinement energy [129]. By developing the so-called configuration-interaction method, Macucci *et al* [130] were able to examine the fabrication tolerance for QCAs which allows correct operation. They have found an extreme sensitivity of the intercell interaction to any disorder in the quantum dot diameter or intercell separation. In principle, by using molecular devices, the extreme sensitivity problems might be avoided.

Although the final result of the ground-state computation does not depend on the detailed evolution of the system, QCA dynamics are important for assessing the speed of such systems. An analytical model for the operational frequency of EQCA systems has been proposed [131, 132].

4.7.1.2 Simulation

Despite considerable work on the physics of QCAs, the development of reliable circuit models is, at the moment, incomplete. However, there are several simulation packages which enable EQCA circuit analysis at a basic, quantum-mechanical level or functional logic level:

- AQUINAS (A QUantum Interconnected Network Array Simulator) (University of Notre Dame, 1997 [133]). A low-level QCA simulation program (with proper graphical user interface (GUI)), where smaller assemblies of QCAs can be simulated at the quantum mechanical level. The simulation itself is performed by solving the Schrödinger equation for the array in a series of discrete time steps. The size of array that can be simulated in AQUINAS is limited. There are two different levels of accuracy: the 'two-state cell' approximation uses a reduced basis state vector, whereas the 'quantum dot cell' approximation uses the full state vector. The simulations themselves take anything from a few minutes to 2 hours, depending on the complexity of the arrays and chosen approximation.
- SQUARES (Standard QUantum cellular automata Array Elements) (University College London, 1998 [134]). This was an initial attempt to standardize QCA designs by introducing standard QCA circuit elements in the form of modular tiles with space for 5×5 QCA cells, figure 4.12. A high-level simulation tool allows simulations of large assemblies of QCAs but only at a logical level. The propagation of information in the individual components has been simulated using a spreadsheet (e.g. MS Excel). The formalism has major advantages from an engineering perspective in that it both standardizes the interconnection rules between circuit elements and allows them to be treated as black boxes. The major disadvantages over low-level designs come in the extra cost both in terms of access times and spatial redundancy (unused space inside modules).
- Q-BART (Quantum Based Architecture Rules Tool) (University of Notre Dame, 2000 [135]). This is a novel architectural simulator used for designing

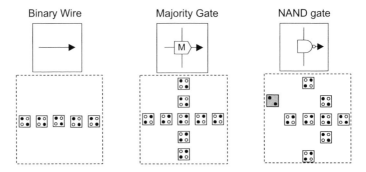

Figure 4.12. The layout of some of the SQUARES elements, after [134].

QCA systems and testing the function performed by the system and data flow. Clocking zones are not yet included.

- Two QCA circuit simulators have been developed at the Università di Pisa. The Static Simulator [136] is based on a quasi-classical approximation. Electrons are treated as classical particles with the possibility of tunnelling between neighbouring dots and the energy of each configuration is evaluated on the basis of classical electrostatics. For small circuits, the full configuration space is explored to determine the ground state, while, for larger circuits, the simulated annealing technique is used. Thermodynamic statistics of circuit behaviour as well as error probabilities can also be derived. The Dynamic QCA Circuit Simulator relies upon a Monte Carlo procedure that enables the user to reproduce the time-dependent evolution of a complex single-electron circuit (cells are represented in terms of capacitors and tunnelling junctions) connected to voltage sources with a piecewise linear time dependence. Different levels of approximations can be used, up to the inclusion of the full capacitance matrix of the system. Co-tunnelling is taken into consideration, too, with an approximate expression for two-electron events. This simulator can treat both clocked and unclocked QCA circuits and allows also the investigation of shot noise in QCA cells, as well as of the cross spectrum between the currents flowing in different branches of the circuit.

4.7.1.3   *Devices and circuits*

Work on QCAs is still essentially theoretical and very few experimental devices have been fabricated [122], although the fabrication of quantum dots (which are the basic elements of solid-state electronic QCAs) is well developed. An experimental demonstration of a four-dot QCA cell has been reported [137]. The device was composed of four metal dots (a pair of dots connected via a tunnel junction and capacitatively coupled to another dot pair) and operated at temperatures below 50 mK.

There are currently only a limited number of functional QCA circuit designs in the literature. The fundamental circuit elements have been proposed: binary wire, fan-out, inverter and majority gate (e.g. see [131]). Gates often have more than one working design. A logic gate based on a four-dot QCA has been experimentally realized [139], where AND and OR logic functions were demonstrated by operating a single majority gate. The majority gate in this experiment consisted of a single (four-dot) QCA cell, gate electrodes (which simulate the polarization of input QCA cells) coupled to each of dots and two single-dot electrometers for signal detection. A clocked QCA two-stage shift register was recently experimentally demonstrated [138].

4.7.1.4 Memory, logic

A QCA memory design has been proposed by Berzon and Fountain [134]. The proposed 1-bit addressable SRAM cell design is shown in figure 4.13. The correct operation of the circuit has been verified by a simulation in Excel.

A full-adder design has been proposed by Lent and Tougaw [120]. In [134], an adiabatically clocked ripple carry adder was described. The adiabatic clocking is necessary for circuit directionality and for thermodynamic effects which might see the excited states rather than the ground state as the preferred final state (for non-zero temperatures). The 'clock' is provided in the form of the tunnelling potential with a ramp wave-form. In this way the circuit is split into phase regions and need only be analysed within a region, instead of searching for the ground state of the whole system.

The proposed circuits are designed by direct analogy with conventional CMOS designs and that might be the reason for the relative complexity. The hope is eventually to move away from CMOS architectures and to develop a coherent approach to QCA circuit design.

The QCA concept is usually characterized as a computation without current. However, it is still not clear whether it is possible to avoid using current and conventional (or some other type of) electronics in providing clocking for a QCA system, which is essential for synchronized computing and for introducing directionality into QCA dynamics. CA designs might be useful as a circuit architecture for molecular chips, since the QCA concept is viable at the molecular level [140, 141].

4.7.2 Magnetic QCA

Magnetic QCAs rely on a property of very small ferromagnetic structures (nanomagnets), namely that the electronic spins act coherently as a 'giant spin' [142]. The word 'quantum' (in QCA) here refers to the quantum mechanical nature of the short-range exchange interaction, which tends to align the spins

Memory Cell

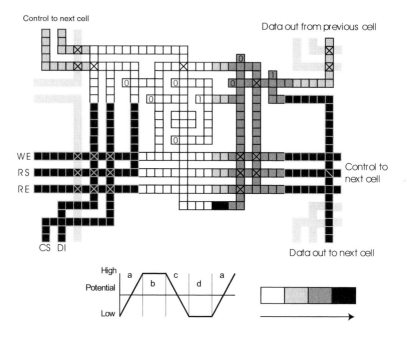

Figure 4.13. The 1-bit addressable SRAM cell design in QCAs. The four-phase clock is shown at the bottom of the figure. The grey levels represent the four phases of the adiabatic clock system. WE=Write Enable, RE=Read Enable, RS=Row Select, CS=Column Select, DI=Data In. Cells labelled X represent a wire crossover.

of neighbouring atoms and wins over the competing, longer range, classical magnetostatic interaction. The direction of cell magnetic moment can represent a logical '1' or '0' state, figure 4.11(*b*). It is possible to arrange that only two directions are energetically favorable, either through shape anisotropy, or magnetostatic interactions between the dots.

Cowburn and Welland [123] demonstrated one of the first MQCAs in the form of a 'wire' of 69 magnetic QCA 'dots', figure 4.14(*a*). The elements were fabricated from a 10 nm thick Supermalloy (a type of $Ni_{80}Fe_{14}Mo_5X_1$, where X is some other metal), each 110 nm in diameter. Binary signal propagation has been demonstrated at room temperature.

An alternative MQCA system was proposed by Parish and Forshaw [143], called the bistable MQCA (BMQCA), where the basic cell is in the shape of a planar elipse. The shape anisotropy provides two stable ground states; hence, each particle can store a binary information. The bistability of the BMQCA allows implementation not only of planar wires (both ferromagnetically

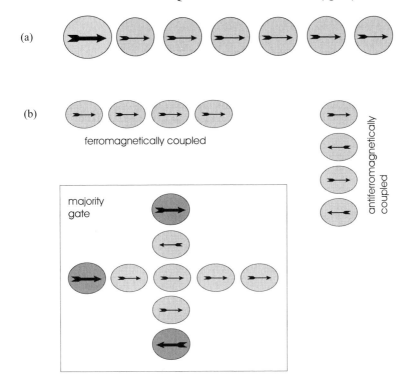

Figure 4.14. (*a*) An MQCA 'wire' proposed by Cowburn and Welland [123]: elongated input dot and a set of circular dots. The propagation of a binary signal down the wire is possible with the assistance of an external magnetic field (not shown). (*b*) Bistable MQCA wires and majority gate, after [143].

and antiferromagnetically coupled) but also vertical wires and majority gates, figure 4.14(*b*).

The theory of micromagnetics is well established [144], but too complex and computationally demanding to be an efficient tool in analysing arrays of micromagnets. Many numerical micromagnetic simulators have been developed but their reliability for any arbitrary case has been questioned [145]. One of the more widely used simulators is the OOMMF (Object Oriented Micromagnetic Framework, mainly developed at NIST) which is in the public domain [146].

A SPICE macromodel has been developed [147] for simulating interacting nanomagnet arrays and this can also be used in the case when these arrays are embedded in microelectronic circuits. The SPICE simulation tool for nanomagnetic arrays is important for the simulation and design of structures with a higher complexity.

In theory, the ultimate size of this type of device could be molecular, because it is theoretically possible to use the spin of an electron as a means for encoding binary information [148]. However, such small devices would be extremely sensitive to a wide range of disturbing effects.

4.7.3   Rapid single-flux quantum devices

Rapid single-flux quantum (RSFQ) devices [126] rely on the use of Josephson junctions in superconducting circuits, and picosecond-long voltage pulses that can be produced when a magnetic flux quantum is transferred from one junction to another. These are not nanoscale devices, although there is no physical reason why they could not be constructed down to 300 nm in size [149] or even less and RSFQ devices also provide an excellent test-bed for the implementation of various CA-based architectures. Small circuits running at 20 GHz are already commercially available. Bit rates of 750 Gbit s^{-1} with flip-flops have been achieved, circuits with 10 000 devices have been built [127, 150] and random access memories [151], adders and multipliers [152] have also been demonstrated. The current generation of devices is mainly based on low-temperature (4–10 K) superconductors but high-temperature ($>$50 K) superconductor technology is also developing. The main advantages of the RSFQ technology are: very high speed, reduced thermal noise (in comparison with room-temperature circuits) and low power (some RSFQ circuits dissipate over 10^5 times less power than the equivalent semiconductor version). This low dissipation is quite important, since the very dense packing of circuits required, e.g. for a petaflop computer, makes it very difficult to cool high-power circuits. The biggest technological challenge is to construct complex circuits by optimizing the fabrication methods for manufacturing a large number of Josephson junctions with a small parameter spread. The lack of power gain has been a serious problem for very large integration but this may now have been overcome [152]. Another big drawback is the need for cooling, so that unless major progress is achieved in making circuits with high-temperature superconductors, these devices will be limited to those applications where the disadvantages of cooling can be tolerated.

4.7.4   Josephson junction persistent current bit devices

Josephson junction persistent current bit (JJPCB) devices also use a superconducting loop, with three Josephson junctions around its perimeter. This can support two opposite circulating currents, whose relative magnitude can be controlled by external magnetic fields, figure 4.11(*c*). A micrometre-sized working device has been built [124, 125]. It was originally proposed (and demonstrated) as being capable of storing a quantum bit (qubit) but it has also been proposed for use as a 'classical' binary logic device ('classical qubit' or 'cubit') [128]. Nanoscale superconducting quantum bits based on Josephson junctions could be the building blocks for future quantum computers, because they combine the coherence of the superconducting state with the control possibilities of conventional electronic circuits.

Although simple arrays of CAs are theoretically capable of universal computation [153], in practice such arrays might not make efficient use of space and they would perhaps also be quite slow. In order to improve their performance, it is necessary to control the directionality of the information propagation (between neighbouring cells) and the reliability of the computation. To do this, additional control elements, clocking signals and fields and/or changes of geometry are needed. Even with such modifications, current EQCA circuit architectures occupy a lot of space (see figure 4.13), because they are confined to two dimensions and largely represent a direct translation of CMOS-based circuits. The implementation of CAs with electrical, magnetic or superconducting systems has not yet made all the progress that was originally expected but research is continuing, as this architecture is a strong contender for use in molecular-scale systems.

4.8   Discussion and conclusion

The crucial tests for every device come at the system level. System functionality, speed, power dissipation, size, ease of circuit design and manufacture are now the most important parameters, whatever the technology. For example, detailed studies show that the incorporation of devices into large circuits always produces a drastic reduction in the effective system speed, by a factor of 100 or 1000, compared to the speed of individual devices [132]. Hence, there is a need for systematic investigation into every newly proposed technology at the system level.

A condensed summary of the advancement of different concepts of (nano)electronic devices is given in the status table shown in figure 4.15 [154]. The status table provides a relatively easy comparison of the development of some of these new technologies with that of the industry standard, MOSFET. At the top of the chart are those devices which can be purchased commercially or to special order (such as MRAM), although some of these are only available in the form of small circuits. There follows a group of devices that are still experimental but are relatively easy to make in very small numbers, although their performance is not completely reliable. In the lower half of the chart are some devices that are proving very hard to fabricate in more than one-off quantities. At the bottom of the chart are mono-molecular transistors, which are still extremely hard to design and fabricate but will probably be the smallest useful electronic devices that will ever be made and nanoscale quantum computing devices. The boxes indicate the development status for each device and circuit size, from white (no information, or undeveloped) to black (available commercially or to special order). One very important development factor, which is frequently overlooked, is the availability of adequate simulation tools: these are shown with dotted boxes.

There are some potentially serious challenges to the manufacture of useful circuits containing large numbers of devices that approach the nanoscale. A major concern is related to device reliability, both in manufacture and in

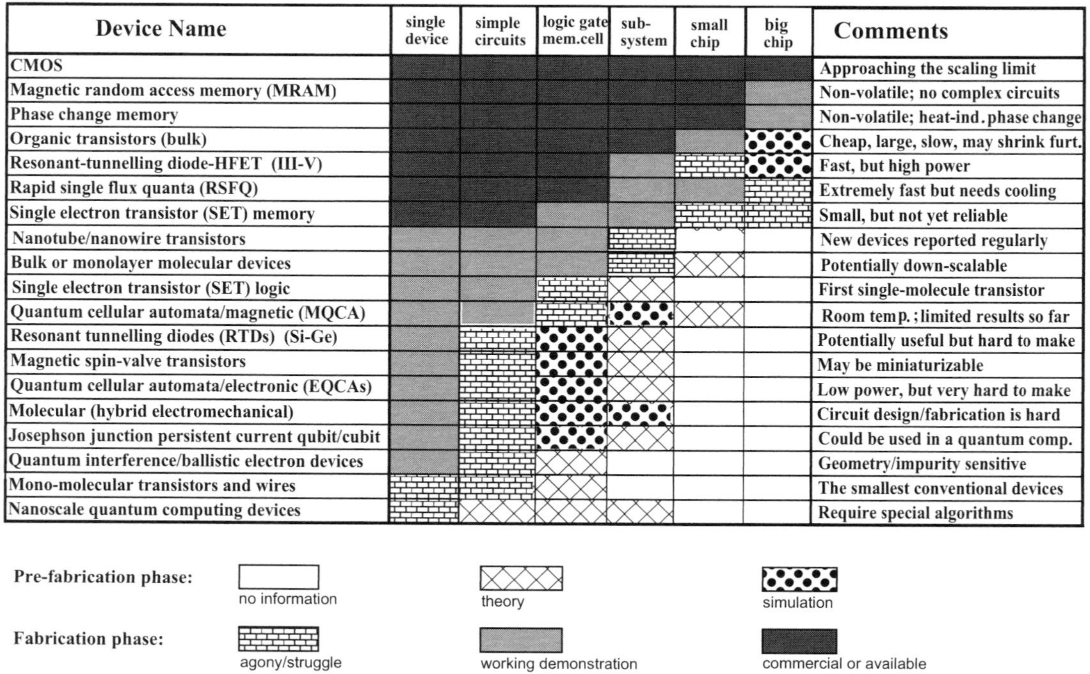

Device Name	single device	simple circuits	logic gate mem.cell	sub-system	small chip	big chip	Comments
CMOS							Approaching the scaling limit
Magnetic random access memory (MRAM)							Non-volatile; no complex circuits
Phase change memory							Non-volatile; heat-ind.,phase change
Organic transistors (bulk)							Cheap, large, slow, may shrink furt.
Resonant-tunnelling diode–HFET (III–V)							Fast, but high power
Rapid single flux quanta (RSFQ)							Extremely fast but needs cooling
Single electron transistor (SET) memory							Small, but not yet reliable
Nanotube/nanowire transistors							New devices reported regularly
Bulk or monolayer molecular devices							Potentially down-scalable
Single electron transistor (SET) logic							First single-molecule transistor
Quantum cellular automata/magnetic (MQCA)							Room temp.;limited results so far
Resonant tunnelling diodes (RTDs) (Si–Ge)							Potentially useful but hard to make
Magnetic spin-valve transistors							May be miniaturizable
Quantum cellular automata/electronic (EQCAs)							Low power, but very hard to make
Molecular (hybrid electromechanical)							Circuit design/fabrication is hard
Josephson junction persistent current qubit/cubit							Could be used in a quantum comp.
Quantum interference/ballistic electron devices							Geometry/impurity sensitive
Mono-molecular transistors and wires							The smallest conventional devices
Nanoscale quantum computing devices							Require special algorithms

Pre-fabrication phase: no information; theory; simulation

Fabrication phase: agony/struggle; working demonstration; commercial or available

Figure 4.15. The theoretical, experimental and commercial development status of some electronic devices and its dependence on the size of the circuits involved, after [154]. The labels 'single device' … 'big chip' describe units with approximately 1, 2, 4, 10–100, 1000–10 000, and greater than 10^6 devices respectively but they are intended to provide only a semi-quantitative guide and the actual numerical values may vary in any one column.

service. Nanodevices are at present hard to make reliably and fabrication problems are likely to increase as the devices get smaller and their numbers get larger. The nanodevice's performance on a chip can be significantly affected by fluctuations in the background charge (created by cosmic rays or some other high-energy particles), thermal fluctuations, quantum mechanical or classical electromagnetic coupling to other parts of the circuit, local heat dissipation, external electromagnetic radiation, etc. Some fault tolerant strategies that exist for overcoming the effects of inoperative devices are discussed in chapter 12.

References

[1] Peercy P S 2000 *Nature* **406** 1023
[2] Semiconductor Industry Association. International Technology Roadmap for Semiconductors (ITRS), 2002 edition, http://public/itrs.net
[3] Compañó R 2000 *Technology Roadmap for Nanoelectronics* European Commission, http://www.cordis.lu/ist/fetnid.htm
[4] Hutchby J A, Bourianoff G I, Zhirnov V V and Brewer J E 2002 *IEE Circuits Dev. Mag.* **18** 28
[5] Bourianoff G 2003 *Computer* **36** 44
[6] Herken R 1988 *The Universal Turing Machine: A Half-Century Survey* (Oxford: Oxford University Press)
[7] Deutsch D 1985 *Proc. R. Soc.* A **400** 96
[8] Nielsen M A and Chuang I L 2000 *Quantum Computation and Quantum Information* (Cambridge: Cambridge University Press)
[9] Steane A 1997 *Rep. Prog. Phys.* **61** 117
[10] Bennett C H 1995 *Physics Today* **October** 24
[11] Kilby J S 1976 *IEEE Trans. Electron. Devices* **23** 648
Noyce R N 1961 *US Patent* 2,981,887
[12] Moore G E 1965 *Electronics* **38** 114
[13] Hergenrother G D *et al* 2001 *IEDM Proceedings*
[14] *INTEL Press Releases* 2001 intel.com, 26 November 2001, web address: http://www.intel.com/pressroom/archive/releases/20011126tech.htm
[15] Critchlow D L 1999 *Proc. IEEE* **87** 659
Tompson S, Packan P and Bohr M 1998 *Intel Technol. J.* **Q3** 1
[16] Meindl J D, Chen Q and Davis J A 2001 *Science* **293** 2044
[17] Zhirnov V V, Cavin R K, Hutchby J A and Bourianoff G I 2003 *Proc. IEEE* **9** 1934
[18] Taur Y *et al* 1997 *Proc. IEEE* **85** 486
Wong H P, Frank D J, Solomon P M, Wann C H J and Welser J J 1999 *Proc. IEEE* **87** 537
[19] Schulz M 2000 *Physics World* **13** 22
[20] Joachim C, Gimzewski J K and Aviram A 2000 *Nature* **208** 541
[21] Heath J R and Ratner M A 2003 *Physics Today* **56** 43
[22] Reed M A, Zhou C, Muller C J, Burgin T P and Tour J M 1997 *Science* **278** 252
[23] Nitzan A and Ratner M A 2003 *Science* **300** 1384
[24] Pantelides S T, Di-Ventra M and Lang N D 2001 *Physica* B **296** 72
For a commercial example, see e.g. http://www.atomasoft.com/nanosim

[25] Martel R, Wong H-S P, Chan K and Avouris P 2001 IEDM
[26] Bachtold A, Hadley P, Nakanishi T and Dekker C 2001 *Science* **294** 1317
[27] Wind S J, Appenzeller J, Marte R, Derycke V and Avouris P 2002 *Appl. Phys. Lett.* **80** 3817
[28] Xiao K *et al* 2003 *Appl. Phys. Lett.* **83** 150
[29] Javey A, Wang Q, Ural A, Li Y and Dai H 2002 *Nano Lett.* **2** 929
[30] Avouris P, Martel R, Derycke V and Appenzeller J 2002 *Physica* B **323** 6
[31] Appenzeller J, Knoch J, Derycke V, Martel R, Wind S and Avouris P 2002 *Phys. Rev. Lett.* **89** 106801 and 126801
[32] Collins P G, Arnold M S and Avouris P 2001 *Science* **292** 706
[33] Cui J B, Sordan R, Burghard M and Kern K 2002 *Appl. Phys. Lett.* **81** 3260
[34] Ellenbogen J C and Love J C 2000 *Proc. IEEE* **88** 386
[35] Wada Y 2001 *Proc. IEEE* **89** 1147
[36] Zhitenev N B, Erbe A, Meng H and Bao Z 2003 *Nanotechnology* **14** 254
[37] Kagan C R *et al* 2003 *Nano Lett.* **3** 119
[38] Storm A J, van Noort J, de Vries S and Dekker C 2001 *Appl. Phys. Lett.* **79** 3881
[39] Stutzmann N, Friend R H and Sirringhaus H 2003 *Science* **299** 1881
[40] Cantatore E, Gelinck G H and de Leeuw D M 2002 *Proc. IEEE Bipolar/BiCMOS Circ. Tech.* p 167
[41] Huang Y, Duan X, Cui Y, Lauhon L J, Kim K-H and Lieber C M 2001 *Science* **294** 1313
[42] Yang P, Wu Y and Fan R 2002 *Int. J. Nanosci.* **1** 1
[43] Hu J and Bando Y 2003 *Ang. Chem.* at press
[44] Cui Y and Lieber C M 2001 *Science* **291** 851
[45] Cui Y, Zhong Z, Wang D, Wang W U and Lieber C M 2003 *Nano Lett.* **3** 149
[46] Duan X F, Niu C M, Sahi V, Chen J, Parce J W, Empedocles S and Goldman J L 2003 *Nature* **425** 274
[47] Chen J and Konenkamp R 2003 *Appl. Phys. Lett.* **82** 4782
[48] Ami S and Joachim C 2001 *Nanotechnology* **12** 44
[49] Stadler R, Ami S, Forshaw M and Joachim C 2001 *Nanotechnology* **12** 350
[50] Duerig U *et al* 2003 *Proc. Trends in Nanotechnology 2003 Conf.*
[51] Broekaert T P E, Lee W and Fonstad C G 1988 *Appl. Phys. Lett.* **53** 1545
[52] Paul D J *et al* 2000 *Appl. Phys. Lett.* **77** 1653
[53] Wei Y, Wallace R M and Seabaugh A C 1997 *J. Appl. Phys.* **81** 6415
[54] Slobodskyy A, Gould C, Slobodskyy T, Becker C R, Schmidt G and Molenkamp L W 2003 *Phys. Rev. Lett.* **90** 246601
[55] Stock J, Malindretos J, Indlekofer K M, Pottgens M, Forster A and Luth H 2001 *IEEE Trans. Electron. Devices* **48** 1028
[56] Reed M A 1999 *Proc. IEEE* **87** 652
[57] Spataru C D and Budau P 2002 *J. Phys.: Condens. Matter* **14** 4995
[58] Mazumder P, Kulkarni S, Bhattacharya M, Sun J P and Haddad G I 1988 *Proc. IEEE* **86** 664
[59] van der Wagt J P A, Seabaugh A C and Beam E A 1998 *IEEE Electron. Device Lett.* **EDL-19** 7
[60] van der Wagt J P A 1999 *Nanotechnology* **10** 174
[61] Levy H J and McGill T C 1993 *IEEE Trans. Neural Networks* **4** 427
[62] Bowen R C, Klimeck G, Lake R, Frensley W R and Moise T 1997 *J. Appl. Phys.* **81** 3207 http://hpc.jpl.nasa.gov/PEP/gekco/nemo/

[63] Schulman J N, De-Los-Santos J J and Chow D H 1996 *IEEE Electron. Device Lett.* **EDL-17** 220

[64] Broekaert T P E *et al* 1998 *IEEE J. Solid-State Circuits* **33** 1342

[65] Nikolic K and Forshaw M 2001 *Int. J. Electron.* **88** 453

[66] Pacha C *et al* 2000 *IEEE Trans. VLSI Systems* **8** 558

[67] Likharev K K 1999 *Proc. IEEE* **87** 606

[68] Neugebauer C A and Webb M B 1962 *J. Appl. Phys.* **33** 74
Lambe J and Jaklevic R C 1969 *Phys. Rev. Lett.* **22** 1371

[69] Fulton T A and Dolans G J 1987 *Phys. Rev. Lett.* **59** 109

[70] Kulik I O and Shekhter RI 1975 *Zh. Eksp. Teor. Fiz.* **62** 623
Kulik I O and Shekhter RI 1975 *Sov. Phys.–JETP* **41** 308

[71] Likharev K K 1988 *IBM J. Res. Devel.* **32** 144

[72] Wasshuber C, Kosina H and Selberherr S 1997 *IEEE Trans. CAD Integrated Circuits and Systems* **16** 937
Wasshuber C 1997 *PhD Thesis* TU Wien Österreichischer Kunst-und Kulturverlag. Web-address: http://home1.gte.net/kittypaw/simon.htm

[73] Fonseca L R C, Korotkov A N, Likharev K K and Odintsov A A 1995 *J. Appl. Phys.* **78** 3238

[74] Chen R H 1996 *Meeting Abstracts, The Electrochemical Society* **96-2** 576
http://hana.physics.sunysb.edu/set/software

[75] Yu Y S, Lee H S and Hwang S W 1998 *J. Korean Phys. Soc.* **33** 5269

[76] Thompson A and Wasshuber C 2000 *Proc. 2nd NASA/DoD Workshop on Evolvable Hardware* p 109

[77] Klunder R H and Hoekstra J 2000 *Proc. SAFE/IEEE Workshop* p 87

[78] Weimann T, Wolf H, Scherer H, Niemeyer J and Krupenin V A 1997 *Appl. Phys. Lett.* **71** 713 http://www.ptb.de/english/org/2/24/2401/hp.htm

[79] Yano K, Ishii T, Sano T, Murai F and Seki K 1996 *Proc. IEEE Int. Solid-State Circuits Conf.* p 266

[80] Smith R A and Ahmed H 1997 *J. Appl. Phys.* **81** 2699

[81] Chen W, Ahmet H and Nakazato K 1995 *Appl. Phys. Lett.* **66** 3383

[82] Meirav U, Kastner M A and Wind S J 1990 *Phys. Rev. Lett.* **65** 771

[83] Palasantzas G, Ilge B, Geerligs L J and de Nijs J M M 1998 *Surf. Sci.* **412–413** 509

[84] Shirakashi J, Matsumoto K, Miura N and Konagai M 1998 *Appl. Phys. Lett.* **72** 1893

[85] Ahlskog M, Tarkiainen R, Roschier L and Hakonen P 2000 *Appl. Phys. Lett.* **77** 4037

[86] Dresselhaus P, Li J, Han S, Ji L, Lukens J E and Likharev K K 1994 *Phys. Rev. Lett.* **72** 3226

[87] Geerligs L J *et al* 1990 *Phys. Rev. Lett.* **64** 2691

[88] Pothier H, Lafarge P, Orfila P F, Urbina C, Esteve D and Devoret M H 1991 *Physica B* **169** 1568

[89] Jensen H D and Martinis J M 1992 *Phys. Rev.* **46** 13407

[90] Heij C P, Dixon D C, Hadley P and Mooij J E 1999 *Appl. Phys. Lett.* **74** 1042
http://qt.tn.tudelft.nl/CHARGE/ppr2/PPR2.html

[91] Wasshuber C, Kosina H and Selberherr S 1998 *IEEE Trans. Electron. Devices* **45** 2365

[92] Durrani Z A K, Irvine A C and Ahmed H 2000 *IEEE Trans. Electron. Devices* **47** 2334

[93]   Nakajima F, Kumakura K, Motohisa J and Fukui T 1999 *Japan. J. Appl. Phys.* **38** 415

[94]   Heij C P, Hadley P and Mooij J E 2001 *Appl. Phys. Lett.* **78** 1140

[95]   Stone N J and Ahmed H 1999 *Electron. Lett.* **35** 1883

[96]   Ancona M G 1996 *J. Appl. Phys.* **79** 526

[97]   Ramcke T, Rösner W and Risch L 1998 *Proc. 3rd Workshop on Innovative Circuits and Systems for Nanoelectronics* Nano-EL98, p D4/1

[98]   Johnson M 2000 *IEEE Spectrum* **February** 33

[99]   Heinrich B 2000 *Can. J. Phys.* **8** 161

[100] Cowburn R P 2000 *Phil. Trans. R. Soc.* A **358** 281

[101] Jansen R *et al* 2001 *J. Appl. Phys.* **89** 7431–6

[102] Parkin S S P *et al* 1999 *J. Appl. Phys.* **85** 5828

[103] Boeve *et al* 1999 *IEEE Trans. Magnetics* **35** 2820

[104] http://e-www.motorola.com
       http://www.almaden.ibm.com/st/projects/magneto/index.html

[105] Landauer R 1988 *IBM J. Res. Dev.* **32** 306

[106] Büttiker M 1988 *IBM J. Res. Dev* **32** 317

[107] van Wees B J *et al* 1988 *Phys. Rev. Lett.* **60** 848

[108] Sols F, Macucci M, Ravaioli U and Hess K 1989 *Appl. Phys. Lett.* **54** 350

[109] Datta S 1997 *Electronic Transport in Mesoscopic Systems* (Cambridge: Cambridge University Press)

[110] Debray P, Raichev O E, Rahman M, Akis R and Mitchel W C 1999 *Appl. Phys. Lett.* **74** 768

[111] Sordan R and Nikolić K 1996 *Appl. Phys. Lett.* **68** 3599

[112] Del Alamo J A, Eugster C C, Hu Q, Melloch M R and Rooks M J 1998 *Supperlatt. Microstruct.* **23** 121

[113] Hieke K, Wesstrom J O, Palm T, Stalnacke B and Stoltz B 1998 *Solid State Electron.* **42** 1115

[114] Nikolic K and MacKinnon A 1994 *Phys. Rev.* B **50** 11008

[115] Collier C P *et al* 2000 *Science* **289** 1172

[116] Donhauser Z J *et al* 2001 *Science* **292** 2303

[117] Collier C P *et al* 1999 *Science* **285** 391
       Kuekes P J, Williams R and Heath J R 2000 *US Patent* 6,128,214
       Kuekes P J, Williams R and Heath J R 2001 *US Patent* 6,256,767 B1
       Kuekes P J, Williams R and Heath J R 2001 *US Patent* 6,314,019

[118] HP press release 9/9/02 http://www.hp.com/hpinfo/newsroom/press/09sep02a.htm

[119] Codd E F 1968 *Cellular Automata* (New York: Academic)

[120] Lent C S and Tougaw P D 1996 *J. Appl. Phys.* **80** 4722

[121] Lent C S and Tougaw P D 1997 *Proc. IEEE* **85** 541

[122] Porod W *et al* 1999 *Int. J. Electron.* **86** 549

[123] Cowburn R P and Welland M E 2000 *Science* **287** 1466

[124] Orlando T *et al* 1999 *Phys. Rev.* B **60** 15398

[125] Mooij J E, Orlando T P, Levitov L, Tian L, van der Waal C H and Lloyd S 1999 *Science* **285** 1036

[126] Likharev K K and Semenov V K 1991 *IEEE Trans. Appl. Supercond.* **1** 3

[127] Brock D K, Track E K and Rowell J M 2000 *IEEE Spectrum* **37** 4046

[128] Jonker P and Han J 2000 *Proc. CAMP2000, 5th IEEE Int. Workshop on Computer Architectures for Machine Perception* p 69

[129] Macucci M, Iannaccone G, Francaviglia S, Governale M, Girlanda M and Ungarelli C 2000 *Proc. NATO ARW Frontiers of Nano-Optoelectron. Sys.* invited

[130] Governale M, Macucci M, Iannaccone G, Ungarelli C and Martorell J 1999 *J. Appl. Phys.* **85** 2962

[131] Berzon D and Fountain T 1998 Internal report no 98/1, IPG, Department of Physics and Astronomy, UCL, at http://ipga.phys.ucl.ac.uk/reports/rep98-1.pdf

[132] Nikolic K, Berzon D and Forshaw M 2001 *Nanotechnology* **12** 38

[133] *A QUantum Interconnected Network Array Simulator (AQUINAS)* 1996–1997 http://www.nd.edu/ qcahome

[134] Berzon D and Fountain T 1999 *Proc. Great Lakes Symposium on VLSI*

[135] Niemier M T, Kontz M J and Kogge P M 2000 *Proc. Design Automation Conference* p 227

[136] Macucci M, Iannaccone G, Francaviglia S and Pellegrini B 2001 *Int. J. Circ. Theory Applic.* **29** 37

[137] Orlov A O, Amlani I, Bernstein G H, Lent C S and Snider G L 1997 *Science* **277** 928

[138] Orlov A O, Kummamuru R, Ramasubramaniam R, Lent C S, Bernstein G H and Snider G L 2003 *Surf. Sci.* **532** 1193

[139] Amlani I, Orlov A O, Toth G, Bernstein G H, Lent C S and Snider G L 1999 *Science* **284** 289

[140] Macucci M, Iannaccone G, Francaviglia S and Pellegrini B 2001 *Int. J. Circuit Theory Applic.* **29** 37

[141] Lent C S, Isaksen B and Lieberman M 2003 *J. Am. Chem. Soc.* **125** 1056

[142] Cowburn R P, Koltsov D K, Adeyeye A O, Welland M E and Tricker D M 1999 *Phys. Rev. Lett.* **83** 1042

[143] Parish M C B and Forshaw M 2003 *Appl. Phys. Lett.* **83** 2046

[144] Aharoni A 1996 *Introduction to Theory of Ferromagnetics* (Oxford: Clarendon)

[145] Aharoni A 2001 *Physica* B **306** 1

[146] Internet address: http://math.nist.gov/oommf/

[147] Csaba G, Imre A, Bernstein G H, Porod W and Metlushko V 2002 *IEEE Trans. Nanotechnol.* **1** 209

[148] Bandyopadhyay S, Das S and Miller A E 1994 *Nanotechnology* **8** 113

[149] Naveh Y, Averin D V and Likharev K K 2001 *IEEE Trans. Appl. Supercond.* **11** 1056

[150] Chen W, Rylyakov A V, Patel V, Lukens J E and Likharev K K 1999 *IEEE Trans. Appl. Supercond.* **9** 3212

[151] Polonsky S V, Kirichenko A F, Semenov V K and Likharev K K 1995 *IEEE Trans. Appl. Supercond.* **5** 3000

[152] Dorojevets M, Bunyk P and Zinoviev D 2001 *IEEE Trans. Appl. Supercond.* **11** 326

[153] Poundstone W 1987 *The Recursive Universe* (Oxford: Oxford University Press)

[154] Nikolić K, Forshaw M and Compañó R 2003 *Int. J. Nanosci.* **2** 7

Chapter 5

Molecular electronics

R Stadler and M Forshaw
University College London

Molecular electronics is one of the most promising candidates currently discussed for nanoelectronics. The big advantage of this field is that organic molecules are well-defined stable structures on an atomic scale, which, in principle, can be modified atom by atom in large quantities by means of chemical synthesis. This cannot be achieved for any inorganic or, more precisely, non-carbon-based material. Therefore, the use of molecules in electronic circuits might well represent the final frontier in miniaturization. There are, however, different approaches depending on the types and numbers of molecules used in an active device and on the way they are connected to form a circuit. These approaches vary widely in their near future feasibility, state of development and in their ultimate limits for miniaturization and performance and are usually difficult to categorize. For this chapter, recent research has been divided into (1) theory and experiments on electron transport through single small molecules, (2) design and fabrication of small circuits based on single macromolecules as active transistor elements and (3) integrated circuits with a few thousand molecules representing one diode in a regular metallic grid structure on the nanoscale. DNA computing has been deliberately omitted, since it has been recently established that DNA molecules are not suitable as elements in an electric circuit and the possible use of their recombination properties for self-assembly or chemical computing are beyond the scope of this book.

5.1 Introduction

In 1974, Aviram and Ratner introduced a concept for a molecular rectifier based on the use of a single organic molecule [1]. This molecular diode consisted of a donor π system and an acceptor π system separated by a sigma-bonded tunnelling bridge. In their theoretical work, the electron transport from the cathode to the anode through the molecule was divided into three steps, namely the hopping of electrons from the cathode to the acceptor part of the molecule, from the acceptor to the donor function inside the molecule and, finally, from the donor part of the molecule to the anode. The rectification effect was given by the partial polarization of the molecule due to their substituents, where electrons would have to overcome a larger energy barrier to hop on the donor function if the bias was reversed. Thirty years later, molecular electronics has become a very active field of research [2–4]. The reasons for this can partly be found in economic necessity, since despite continuous achievements in miniaturization of CMOS technology fundamental difficulties will be faced when the nanoscale is approached. Some of them, such as the irreproducibility of detailed atomic structures based on silicon, can be solved by replacing the components of electric circuits by organic molecules, where well-defined structures can be mass produced by means of chemical synthesis. In addition, the rich variety of different structures and functionalities in organic chemistry offers the potential of a huge tool box for the implementation of logic functions on the nanoscale. Another important reason for the increase in research in this field over the last ten years is that major breakthroughs in developing experimental techniques for studying systems consisting of a few or even single molecules have been achieved. Scanning probe techniques now allow the imaging, spectroscopical characterization and manipulation of single molecules adsorbed on crystal surfaces. By means of nanolithography (see chapter 6), metallic electrodes with a spatial gap of only a few nanometres between them can be fabricated. Advances in supra-molecular chemistry have enabled the synthesis of a variety of tailor-made complex molecules, which have been designed to meet different requirements such as a high conductivity, the possibility of switching between two states triggered either by photons or electrons, a stiff rod-like structure and good solubility.

Molecular electronics has now become a very vibrant and interdisciplinary field of science, where physicists, chemists and engineers in academia and industry work together on various aspects and levels in large-scale national and international collaborations. This inevitably led to a diversification into different approaches, where there are as many suggestions for a categorization of the field as there are reviews about it. One way is to distinguish between hybrid-molecular and mono-electronics [2]. In the former category, the molecule is just the active part of a diode or transistor and molecular units are interconnected by an electromechanical metallic grid, whereas the aim in the second category is to integrate an arbitrarily large number of logic functions including the wiring inside a single molecule. Other categorizations are based on classes of molecules, the physical process used for the implementation of 2-bit states (such as redox processes, configuration change, electronic excitations etc) or the type of the main circuit element (diodes, transistors, qubits, etc). For the current review, the authors decided to make the size of the molecular system under study the characteristic feature for the organization of this chapter. The reason for this is that 'larger' molecular systems have been integrated in rather complex circuits, which can be readily compared with CMOS-based electronics, whereas 'smaller' molecular systems are still on the level of basic research in theory and experiment, although in the long term they are showing more promise for performance enhancement. The next section gives an overview of theoretical calculations and scanning probe measurements of electron transport through single organic molecules. For such systems, quantum mechanical effects and atomistic details of the molecule/electrode interface play a decisive role. The following section focuses on small circuits, where the active elements are nanotubes or C_{60} molecules. Due to the larger size of these molecules and the range of conductivities they can adopt depending on their state of compression or charge polarization in an electric field, their applicability as circuit components is less sensitive with regard to details of the molecule/electrode interface on an atomic scale. In the next section, circuits based on organic molecular films are reviewed. Research in this category, where films of ~5000 molecules are used to implement a single diode, is so advanced that very complex circuits such as demultiplexers and large memory arrays have already been fabricated on very dense crossbar grids. Finally, a summary and outlook is given.

5.2 Electron transport through single organic molecules

The characterization of electron transport through single molecules is demanding, both theoretically [5] and experimentally [6]. For the theoretical description, quantum mechanical equations have to be solved numerically for a non-equilibrium system with open boundaries provided by long electrodes which maintain different chemical potentials due to an external bias. Within the framework of scattering theory, several approximations to this problem have been developed, where increased accuracy comes at the cost of increased computational expense. For experimental measurements the problem lies in the controlled and reproducible fabrication of nanojunctions, where two (or ideally three) metallic electrodes are positioned so close to each other, that a molecule can be placed in the gap between them and connected to both. For this task, scanning probe techniques are used, where several problems have to be faced and a simultaneous imaging of details of the molecule–electrode contacts and measurement of the current/voltage (I/V) curves can rarely be achieved. There have been many proposals in the literature for devices based on single molecule nanojunctions but none of them has reached a stage of development where small circuits could be designed in theory and reproducibly demonstrated in experiment. This section gives a short review of some of the latest advances in this research area.

5.2.1 Electron transport theory

For the theoretical modelling of an electrode/molecule/electrode nanojunction, the system has to be divided into separate parts: the finite contact region, for which Schrödinger's equation is solved electron by electron, and the semi-infinite leads where translational symmetry of a two-dimensional (2D) unit cell of the electrode material is assumed in the direction pointing away from the contact region. The problem can then be defined in terms of plane waves entering the system through the leads and scattered by the contact region. Since atomistic and electronic details of the electrode/molecule interface have an influence on the conductance through the electrode/molecule/electrode junction, a few layers of the electrode are often also defined as part of the contact region, which is then sometimes referred to as 'extended molecule'.

The great majority of theoretical work on the subject of electron transport through molecules deals with conditions very near to thermal equilibrium, particularly with very small voltage drop across the transporting system. These conditions are known as the linear-response regime, because the currents induced are linear in the applied voltage. A general approach to near-equilibrium transport is embodied in the Landauer formula [7], which expresses the conductance of a system at zero temperature in terms of the quantum mechanical transmission coefficients calculated from a steady-state scattering matrix. Despite the attention directed toward linear-response theories, they remain severely limited in the range of physical situations which they address and, in particular, do not include any inelastic effects such as dissipative scattering. The application of this approach for a 'small-signal analysis' in the engineering sense of this term is problematic. Such an analysis studies small departures from a steady-state but typically far from equilibrium situation where a significant voltage drop occurs. The linear-response theories study small departures from equilibrium, not from a non-equilibrium state.

A more sophisticated approach to electron transport theory, which also allows the description of systems where large voltages are applied, is supplied by the non-equilibrium Green's-function formulation of many-body theory [8], which had its origins in the development of the theory of quantum electrodynamics in the late 1940s and early 1950s. In such an approach, the subspaces defined by the wavefunctions representing the electrodes are projected into the subspace defined by the wavefunctions describing the contact region and enter the molecular Hamiltonian as self-energies. Techniques based on Green's functions have also been developed for a rigorous theoretical treatment of inelastic processes in molecular wires [9].

Apart from the general setup of the theoretical frame-work, in which the open-system scattering problem is defined, there is also the question of how accurate the contact region is described numerically. Semi-empirical methods, such as extended Hückel (EH) techniques [10–12] where the Hamiltonians are parametrized, allow the modelling of rather large molecules and have now been extended to circuit simulation software, where an arbitrarily large number of electrodes can be connected to a molecule [10] (see figure 5.1). However, these methods are implementations of a linear-response theory (with the exception of [12]) and additionally operate with parametrized Hamiltonians and, therefore,

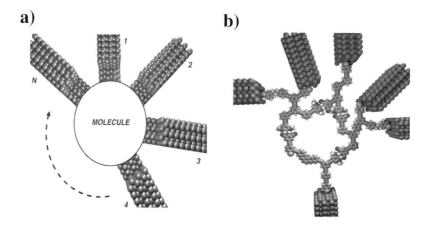

Figure 5.1. (*a*) Structure of a scattering circuit consisting of a central molecule interacting with *N* electrodes as developed recently based on an EH approach by Ami *et al* [10]. (*b*) An example of a complex molecule attached to six electrodes, which from its original design represents a full adder, where the *I*/*V* characteristics between each combination of electrodes can be assessed using the EH technique.

they cannot describe electrostatic effects such as the polarization of the electronic wavefunctions in the contact region, which might affect the potential distribution along the region.

For solving this problem, first-principle calculations using density functional theory based non-equilibrium Green's function (DFT-NEG) techniques have to be performed where the coupled Schrödinger and Poisson equation are solved self-consistently [13–15]. This has the advantage that the energetic position and shape of the molecular orbitals in the contact region depends on the voltage applied on the electrodes, which is necessary for a realistic non-equilibrium description of the system. Such calculations are computationally more expensive but also describe current-induced local forces acting on atoms in the contact region [16] and thereby allow predictions to be made about the junction's stability for a variety of applied voltages (see figure 5.2). In order to make the computational effort of such calculations more bearable, the electrodes have been modelled as jellium in early versions of this technique [13], where the positive charge of the nuclei is uniformly distributed in the region of the electrodes and the details of their atomic and electronic structure are not taken into account. The methods, where the electrodes are described atomistically [14, 15], differ in the way in which the semi-infinite boundary conditions are defined, where the matching between the 'extended molecule' and the leads presents a technical challenge. For a discussion of these issues, see [14].

EH methods have been used to scan the conductance properties of molecular wires for a variety of chemical compositions [17] and the influence of the way they

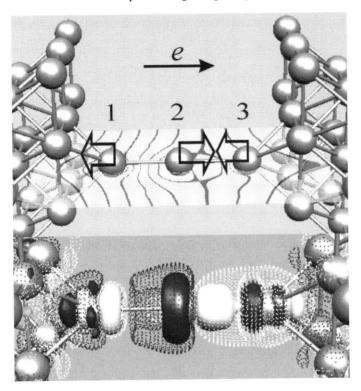

Figure 5.2. (Top) Direction of forces and voltage drop for a gold wire connecting (100) gold electrodes at 1 V bias [16] as calculated with a recently developed DFT-NEG method [14]. (Bottom) Isodensity surface for the change in electron density from 0 to 1 V, where dark is deficit and white is extra charge density. The full (dotted) surface correspond to $\pm 5 \times 10^{-4} e\,A^3 (\pm 2 \times 10^{-4} e\,A^3)$. Reprinted with permission from [16].

are interconnected to each other has also been addressed [18]. For the physically more complex issue of the influence of details of the molecule/electrode interface and related electron density polarization and charge transfer effects, DFT simulations have been performed for chemically rather simple test systems such as dithio-phenyl radicals [19] or the original proposal for the Aviram–Ratner diode [20]. For the research in this field in the future, it is expected that both methods will be further developed and are complementing rather than competing with each other, since semi-empirical techniques enable quick scans of whole classes of rather complex molecules, whereas first-principle methods provide a very accurate picture of all possible physical effects for small junctions.

5.2.2   Scanning probe measurements and mechanically controlled break junctions

Molecular junctions can be investigated by scanning probe techniques, where a layer of organic molecules is adsorbed on a single crystal surface and the adsorption sites are scanned with a scanning tunnelling microscope (STM) tip, which can act as the second electrode where the surface would be the first one [21]. This method has the advantage that it provides imaging and conductance measurements at the same time for the adsorbate system but there are a few shortcomings. The surface and the STM tip have different chemical potentials due to their difference in shape even if they are made of the same material and this creates an undesirable asymmetry in the junction, which would not be present in any application with a molecule connected to two metallic nano-wires.

Most molecules of interest are aromatic molecules, which are characterized by rings of conjugated double bonds. The resulting delocalized π electron system is responsible for the semiconducting properties of this class of molecules but this also makes a planar geometry their most stable equilibrium configuration. Consequently, such molecules have to be chemisorbed for STM studies of molecular conductance because, if physisorbed, they would lie flat on the surface due to Van der Waals interactions between its π system and the surface and current through the molecule from one of its atoms to another could not be measured. Usually such a chemisorption is achieved by using thiol-groups as 'alligator clips', since the sulphur atoms in these groups form strong bonds with gold surfaces, which are chosen as electrode material for most studies.

Even in the chemisorbed case, it is necessary to substitute the molecules with four thiol-legs or embed them in a carpet of insulating alkanethiols to make sure that they 'stand up' on the surface [22] (see figure 5.3(a)). Such an insulating matrix also allows, in principle, the isolation of single conducting molecules and the measurement of their conductance. To identify these single molecules, however, is not an easy task. It requires a distinction between the contributions to the conductance coming from the type and from the length of a molecule, which can be achieved by STM measurements of a surface covered with molecules using a combination of a conventional direct current–voltage STM with a microwave alternating current STM technique.

STM techniques have an even wider range of applications in molecular electronics than the imaging of the topography of a substrate/molecule system and measuring conductances through molecules. They can also be used for the controlled desorption of molecules, for placing atoms individually and as a spectroscopic tool which provides information about the electronic density of states and inelastic effects such as molecular vibrations [23]. In a recent study, copper(II) phthalocyanine (CuPc) molecules have been assembled on an NiAl(110) surface and have been bonded to two gold atomic chains by manipulation of single atoms [24] (see figure 5.4). The dependence of the electronic properties of this metal–molecule–metal junction on the length of

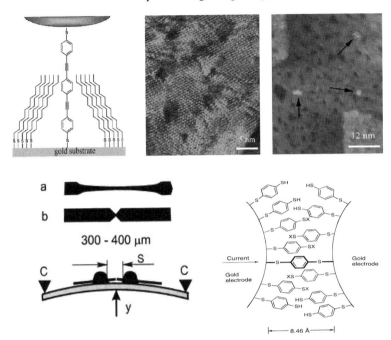

Figure 5.3. Different methods for measuring the conductance of single molecules: (Top) Conducting molecules are embedded in an insulating matrix [22]. The left-hand picture shows a schematic diagram of the setup. In the middle, a STM topographical image of a 25 nm × 25 nm alkane film with resolution of the molecular lattice is depicted. In the image on the right-hand side, 100 nm × 100 nm of the same film are presented, where inserted conducting molecules are marked by black arrows. (Bottom) The left-hand picture shows the design and principle of a mechanically controlled break-junction [26]. The right-hand image sketches the way in which molecules are adsorbed in such a junction as proposed in [25].

the gold chain have been systematically investigated by varying the number of atoms one by one and the influence of the detailed position of the molecule on the contacts could also be addressed. Such experimental studies with complete control over the atomic configurations in the junction are needed to complement the theoretical work described in the previous section for understanding the alignment of the molecular orbitals with respect to the wire's Fermi energy, the build-up of electrostatic barriers and other issues determined by the detailed electronic structure of the system.

As an alternative to STM measurements for single molecule conductance, metallic nanowires can be broken in a mechanically controlled way which allows for the formation of ~1 nm wide gaps and molecules deposited on top of this gap can be detected by changes in the I/V curves [25–27] (figure (5.3(b))). There is

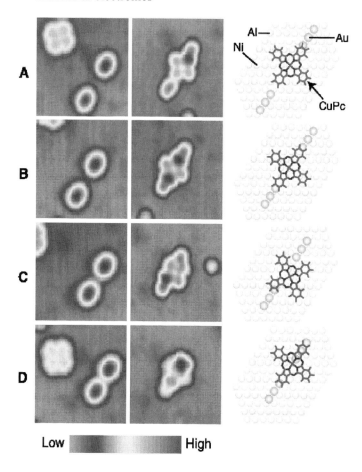

Figure 5.4. CuPc molecules on two Au$_3$ chains for different spacings between the chains [24]. Left-hand column: Bare Au$_3$ junctions before the molecules were added. Middle column: The molecules have been adsorbed. Right-hand column: Schematic diagrams attributed to each adsorption configuration.

no way to be certain that the increase in conductance in such a junction is due to the adsorption of single molecules rather than many of them, since no microscopy technique up to now is able to observe them directly. However, this method has the advantage that, in contrast to the STM measurements, the junction setup is symmetric with regard to the two electrodes and could, in principle, be used for the fabrication of integrated circuits if a reproducible, reliable and cost effective way for mass production could be found. The symmetry or lack of symmetry of the molecule on the other side has been used as an argument that conductances through single molecules have been measured in some experiments. Symmetric molecules give symmetric differential I/V curves, because there is no reason to assume that the conductance should change if the voltages on the two electrodes are reversed. For an asymmetric molecule, which has donor and acceptor groups at different positions, the I/V curves can be asymmetric. This effect, however, would be statistically averaged out if more than one molecule is adsorbed inside the junction, since there is no way of controlling or predicting which end of the molecule bonds to which electrode and a random behaviour must be assumed, which would make the measured conductance curves symmetric again.

Summarizing this section, it can be said that experimental manipulations and conductance measurements on single molecules are still big scientific challenges and a lot of the progress, which has been recently made, has been achieved for particular substrate/molecule systems and could not be easily transferred to other surface materials or types of molecules. The STM is certainly the most versatile instrument for manipulations and measurements on the nanoscale but it is not very suitable for integration into nano-electronic devices. New techniques such as mechanically controlled break-junctions will have to be developed for this purpose in the future.

5.2.3 Possible applications of single organic molecules as various components in electrical circuits

There have been several proposals for wires, diodes [1, 29], transistors [30], memory cells [31] or logic gates [32] based on the conductance properties of small single organic molecules. Recently, Kondo and Coulomb blockade effects have been observed, which could be the basis for the use of molecules as single electron transistors [33]. Similarly, observed negative differential resistance could enable resonant tunnelling transistors [34]. However, in some of the theoretical work, it is not clear whether the predicted device behaviour is an artifact of the level of approximation chosen for the calculations. The initial Aviram/Ratner proposal for a molecular rectifier for instance modelled originally by a three-step hopping process, has been recently re-investigated with DFT-NEG calculations where very poor diode characteristics were found [20]. However, there is a controversy about theoretical explanations for the device effects found in experiments, where the negative differential resistance found in molecular wires has been explained by such diverse concepts as charging of the molecule [34] and thermally activated rotation of functional groups [35]. On the molecular scale any device effect is not a property of a particular molecule but must be attributed to the electrode–molecule–electrode nano-junction, where details of the interface between the molecules and the electrodes can qualitatively change the characteristics of the system. Since the atomistic details of this contact can rarely be observed and much less controlled in conductance experiments, the big variety of theoretical explanations for the same measurement is hardly surprising.

It is rarely addressed in the literature that, for molecular versions of circuit elements to be useful, there has to be the possibility of connecting them together in a way, in which their electrical characteristics—measured individually between electrodes—would be preserved in the assembled circuit. However, it has been recently shown that such a downscaling of electrical circuits within classical network theory cannot be realized due to quantum effects, which introduce additional terms into Kirchhoff's laws and let the classical concept of circuit design collapse [18]. Circuit simulations on the basis of a topological scattering matrix approach have corroborated these results [36].

Recently, schemes for computational architectures, which make use of the quantum properties on the molecular scale rather than trying to avoid them, have been proposed, such as cascading CO molecules on metal surfaces [37] (figure 5.5), computing with optical excitations [38] or controlling the interference pattern of electron transport through aromatic molecules by modifying their chemical side-groups [39] (figure 5.6).

5.3   Nanotubes, nanowires and C_{60} molecules as active transistor elements

Single-walled carbon nanotubes are currently perceived as the main candidate as a material to replace silicon in information technology [40]. They can exhibit a large range of conductances depending on details of their structure ranging from semiconducting to metallic, where especially the semiconducting ones can be used as active elements in field effect transistors. The nearly dissipation-free ballistic electron transport along the tubes also suggests applications as high current density interconnects, these are reviewed in detail in chapter 7. The comparatively large size of these molecules make them less sensitive to details of their contact to the electrodes and conventional device schemes, e.g. field effect transistors (FETs), small logic gates and memory cells could be experimentally demonstrated and are theoretically well understood. Semiconducting nanowires made from GaP, GaN, InP or Si exhibit somewhat similar properties and have also been used for the fabrication of diodes, bipolar transistors and FETs as well as simple circuits such as complementary inverter-like structures, OR, AND and NOR gates. A 1-bit full-adder circuit has been demonstrated. A recent proposal for a single C_{60} molecule electromechanical transistor has been used as the basis for a performance evaluation study, where a memory/adder model has been designed and the signal response behaviour of the corresponding circuit has been tested with standard circuit simulation software. In this section, recent progress with such devices will be reviewed.

5.3.1   Carbon nanotube field effect transistors (CNTFETs)

Carbon nanotubes (CNTs) are rolled up sheets of graphene just a few nanometres in diameter that can behave as either metals or semiconductors depending on their atomic arrangement [40]. The mechanical robustness and long length of the nanotubes makes it easier to attach these macromolecules to electrodes. This can,

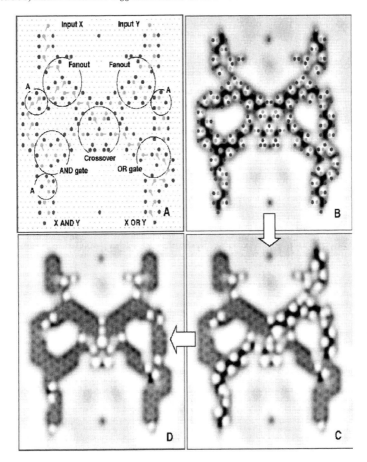

Figure 5.5. Two-input sorter in a CO molecule cascade scheme [37]: (A) Model of the sorter, which computes the logic AND and logic OR of the two inputs. (B)–(D) Succession of STM images. Starting from the initial setup (B), input X was triggered by manually moving the top CO molecule, which propagated a cascade to the OR output (C). Input Y was subsequently triggered, which propagated a cascade to the AND output as shown in (D).

for instance, be achieved by depositing nanotubes on source and drain electrodes, which are fabricated by lithographical means. If the substrate wafer is a material which can be capacitatively charged but is covered with an oxide layer, varying the voltage on the substrate, which then acts as a gate electrode, changes the number of charge carriers on the nanotube [41].

Measurements of semiconducting nanotubes reveal unusual electrical characteristics, such as an increase in resistance by several orders of magnitude with an increase of the gate voltage and nearly dissipation-free electron transport

Figure 5.6. A molecular data storage scheme based on an aromatic molecule (naphthalene) bonded to four gold electrodes by sulphur atoms and polyacetylene wires [39]. For the surface, an insulator has to be chosen to prevent crosstalk between the electrodes. The variables X, Y and Z could either be chemical substituents or alternatively connections to further electrodes. Some parts of the molecule, electrodes and variables are drawn light grey, which is meant to indicate an active state during a particular read-out. The darker parts are considered to be inactive.

in the ballistic regime. Recently, it has been shown that CNT-FETs can compete with the leading prototype silicon transistors currently available in terms of important transistor characteristics such as transconductance and maximum current drive [42].

One of the major challenges in nanotube electronics is to ensure that the coupling between the gate and the nanotube is strong enough to amplify a signal, so that the variations in the output voltage of one nanotube transistor can control the input of a second transistor, which is necessary for integrating nanotubes into circuits. Recently, a new layout that resembles a conventional MOSFET structure with the gate above the conduction channel has been developed, which is characterized by a very small gap between the nanotube, thus making the resistance much more sensitive to variations in the gate voltage [43]. One way to do so is to use electron beam lithography to pattern aluminium electrodes on top of an oxidized silicon wafer, depositing the nanotubes on top and adding gold electrodes by an evaporation technique [44].

Semiconducting nanotubes typically operate like p-type semiconductors (conducting holes rather than electrons), where the charge transfer from the electrodes acts as 'doping'. In this setup, molecules being adsorbed from the atmosphere onto the nanotube can affect the reproducibility of the device characteristics. This molecule adsorption can be controlled by embedding the nanotubes in a film and by applying thermal-annealing treatments in a vacuum [45]. The same techniques can be used to fabricate n-type devices as an alternative to the current method of doping nanotubes with alkali metals.

Recent progress in improving the intrinsic resistance of nanotubes as well as the contact resistance at the electrodes is due to advances in nanotube growth [46] and deposition [47] techniques and their ability to tailor the diameter (which determines the band gap) and chirality of the tubes as well as the Schottky barriers at the molecule/electrode interface. However, the synthesis of nanotubes, which can be done, for example, by electric arc discharge [48] or chemical vapour deposition [49], in general results in a mixture of metallic and semiconducting tubes, where only the latter can be used as a transistor element. A separation procedure has been recently proposed, where metallic tubes are 'burnt off' by high voltages and only the semiconducting ones remain on the substrate [50]. But to obtain transistors that can conduct the same amount of current as micrometre-wide silicon transistors, for example, 2D arrays of parallel tubes with exactly the right width are needed.

Different nanotubes have been assembled into basic logic circuits, such as a logic NOR, a static random-access-memory cell and a ring oscillator [44] (see figure 5.7). In these circuits, a simple inverter device consists of a nanotube FET and a large bias resistance and by adding an extra FET in parallel, the NOR gate could be realized. Any of the standard logic gates—AND, OR, NAND—can be created using different arrangements of these FETs. Using thermal annealing, which can transform p-type CNT-FETs into n-type tubes or just locally on part of a tube, an inverter with complementary p- and n-FET transistors on a single nanotube bundle could also be demonstrated [51] (see figure 5.8).

Nevertheless, current fabrication techniques fall far short of those needed for mass production. Most problematic, perhaps, is the lack of control when it comes to placing the tubes in predetermined positions during device fabrication, where different approaches such as controlled deposition from solution or lattice-directed growth are currently pursued [40]. A remarkable success has been recently achieved in using DNA molecules for the construction of a single CNTFET device [52]. However, more research is needed to assess whether such a biological approach can also be used for the assembly of a large number of devices.

5.3.2 Cross-junctions of nanowires or nanotubes

In order to create nanotube-based devices, researchers must carefully tailor their electronic properties. This is difficult to achieve as described in the previous section, since these properties depend on the diameter and crystal structure of the tubes and it is difficult to separate the semiconducting from the metallic ones. The composition of semiconducting nanowires, by contrast, where GaP, GaN,

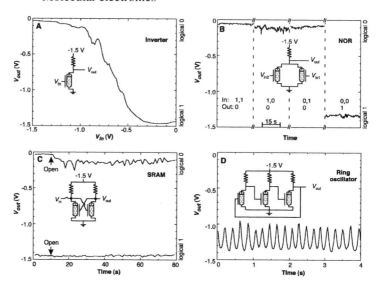

Figure 5.7. Experimental demonstration of one-, two-, and three-transistor logic circuits with CNTFETs [44]. Output voltages as a function of the input voltage are given for (A) an inverter, (B) a NOR gate, (C) a static random access memory cell (SRAM) and (D) a ring oscillator.

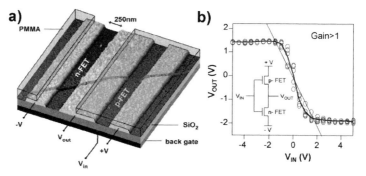

Figure 5.8. (*a*) An atomic force microscope (AFM) image of the design and (*b*) measured voltage output characteristics of a complementary CNTFET inverter [51]. In (*a*), a single nanotube bundle is positioned over the gold electrodes to produce two p-type CNTFETs in series. The device is covered by a resist and a window is opened by e-beam lithography to expose part of the nanotube. Potassium is then evaporated through this window to produce an n-CNTFET, while the other CNTFET remains p-type. In (*b*) open circles are raw data for five different measurements on the same device. The bold curve is the average of these five measurements. The thin straight line corresponds to an output/input gain of one. Reprinted with permission from [51], copyright 2001 American Chemical Society.

InP and Si are the most commonly used materials, is relatively easy to control. Silicon wires, for instance, can be produced by vapour–liquid–solid (VLS) growth combined with other techniques [53]. In a first step, a laser is used to vaporize silicon together with a metal catalyst, typically iron or gold. As the vapour cools, it forms nanosized liquid clusters containing both Si and some amount of the catalyst. Si wires then slowly grow out from the cluster, as more Si atoms condense.

If the nanowire is grown from a semiconductor, its electrical properties can be altered by adding small amounts of dopants. In conventional circuits, combinations of differently doped semiconductors are used to build electronic devices and the same can be done using doped nanowires. Recently, two differently doped Si nanowires have been synthesized and joined in crossbar structures [54]. The difference in the density of conducting electrons makes the junction where the wires meet behave as a diode, allowing the current to flow in only one direction across it. A crossing junction of nanowires made of different materials—one GaN, the other Si—has been used to create nanowire FETs [55]. Figure 5.9 shows how these nanowire diodes and transistors can be assembled to form logic gates [56], which has been experimentally demonstrated. But nanowires can also be used as the basis of a memory unit. In nanowires coated with charged molecules, the charge affects the nanowires' ability to conduct electricity and, therefore, by applying a voltage to change the charge, the wire's conductance can be switched between two different states, which could be used for storing information.

Cross-junction switches, where metallic nanotubes will bend toward a perpendicular semiconducting nanotube when electrically charged, have also been realized with nanotubes [57]. When a metallic tube is 1–2 nm away from a semiconducting one, the electrical resistance at the junction is low, creating an 'on' state. When the nanotubes are apart, the resistance is much higher, creating an 'off' state. In most recent research in this field, a Si wire/CNT hybrid memory array is proposed, where the Si wires are used for the semiconducting portion of the array. This will not produce the same density as a pure nanotube array but would help to avoid the problem of having to position metallic and semiconducting CNTs in the same circuit in a controlled way.

5.3.3 A memory/adder model based on an electromechanical single molecule C₆₀ transistor

The physical principle of a C_{60} electromechanical amplifier has been experimentally demonstrated with a single molecule on a gold surface compressed and, at the same time, electrically contacted by an STM tip [58] (figure 5.10). In such a setup, the substrate and STM tip act as source and drain electrodes, whereas the force on the tip compressing the molecule acts as the gate, where the state of compression changes the conductance through the molecule by a few orders of magnitude. This design facilitated the experimental demonstration, because the

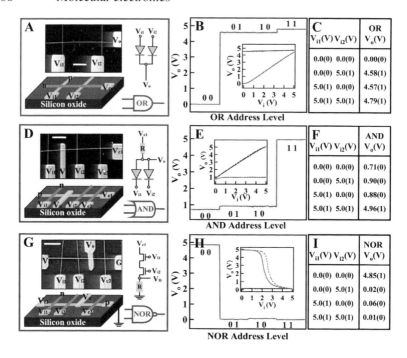

Figure 5.9. Schematic diagrams, AFM images, measured output voltages and experimental truth tables for OR- (A)–(C), AND- (D)–(F) and NOR-gates (G)–(I) based on crossed nanowires [56].

behaviour of a three-terminal device (source, drain and gate) could be mimicked with two terminals only. For devices assembled into circuits, however, this would be completely impractical and a physical separation of all terminals is needed.

In a later design [59] (figure 5.11a), which was the basis for a theoretical evaluation of the performance of C_{60} transistors in larger circuits [60, 61], a single C_{60} molecule is placed in a 1.24 nm wide gap between a source and a drain electrode in a co-planar arrangement. A tip at the end of a cantilever, which is equipped with a piezolayer to control its deflection, can be used to compress the C_{60} molecule, thereby changing its conductance which gives the transistor effect. The equivalent electrical circuit description (figure 5.11(b)) of the electromechanical conversion of the gate signal into a movement of the nanocantilever follows the theoretical framework outlined in [62] and has been adapted to C_{60} transistors by Ami and Joachim [59]. The C_{60} molecule comes into this circuit description twice. On the one hand, it acts as an electrical resistor where the value for the resistance is modulated by the compressing action of the cantilever tip. On the other hand, the molecule resists its deformation elastically, which has to be taken into account as a counter electromotive force in the model.

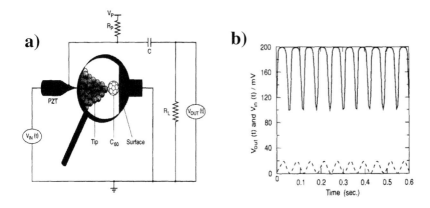

Figure 5.10. (*a*) Schematic diagram of an experimentally demonstrated electromechanical single molecule C$_{60}$ amplifier, where the molecule is connected between an STM tip and a surface, and the input voltage V$_{in}$ actuates a piezoelectric translator. (*b*) The output voltage V$_{out}$(t) (full curve) is determined from the experimentally measured conductance through the C$_{60}$ molecule and the characteristics of an external polarization circuit [58] and is compared with V$_{in}$(t) (dotted line).

Based on this design, a memory/adder model (figure 5.11(*c*)) using 464 transistors could be constructed and evaluated on grounds of SPICE circuit simulations. Four bits of information were read from four different memory cells. They were combined to give two 2-bit words by OR-gates, which passed the words through registers (clocked D-latches) to a serial adder (a full adder combined with an RS-latch). The output of the addition moved through another register, from where it could be written on a memory or used in a subsequent computation. Since, in a serial adder, the carry-out is used as carry-in when the next significant bits are added, the addition involves a clocked RS-latch to store the carry between two steps. It was found that, with improved two-input gates using about 12 C$_{60}$ transistors, an addition can be performed with very stable logical levels throughout the whole computation. It must be noted that, unlike suggestions for other nanoscale architectures [63], the described C$_{60}$ microprocessor does not use any CMOS components but consists of hybrid-molecular transistors only. This demonstrates that hybrid-molecular electronics can, in principle, be moved from the device to the computer architecture level without fundamental problems.

A performance evaluation, however, shows that such circuits are not likely to be a competitor for current CMOS devices. The first logic gate that was designed by Ami and Joachim [59] was a NOT-gate or inverter, where it was also demonstrated that a square-wave signal passing through a chain of eight such gates decays rapidly at a frequency of 1 MHz but moves undistorted at 1 kHz. The *RC* time delay ($\sim$0.1 μs) is dominated by the resistance of the C$_{60}$ molecule

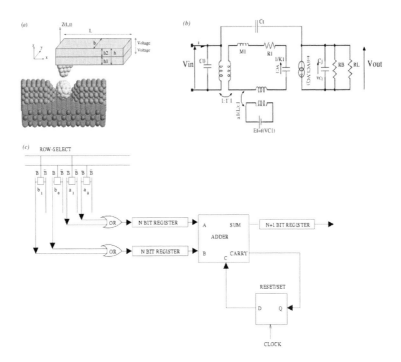

Figure 5.11. (*a*) Layout of a theoretically modelled modified C_{60} electromechanical transistor composed of a metal–C_{60}–metal nanojunction and of a small grid cantilever whose dimensions are specified.   (*b*) Electrical equivalent circuit model of the C_{60} transistor in (*a*) used for circuit simulations, where the mechanical properties of the cantilever have been described as a *RLC* cell plus a counterelastic force to describe the reaction of the molecule upon compression [62]. (*c*) Schematic diagram of a memory/adder model, where the arrows indicate the directions in which the signals move [59]. Reprinted from [59] and [62], copyright IOP Publishing Ltd.

in the compression range defined by the input gate voltages ($\sim$100 MΩ) and the capacitance of the cantilever for this case ($\sim$10^{-15} F). A value of $\sim$0.1 μs for the time delay was derived numerically using the electric circuit model developed in [59] for the simulation of a single pulse passing through a chain of inverters, as well as analytically by simply multiplying the resistance and capacitance given earlier. It could be shown that this circuit would only be effective at a frequency of 200 kHz or less.

A comparison with current CMOS devices ($R \approx 0.01$ MΩ, $C \approx 2 \times 10^{-16}$ F) shows that, although the capacitance is a bit higher in hybrid-molecular devices, it is the high resistance of the molecule that mainly limits the clock frequency on the device level. This resistance is not only defined by the conductance of the molecule as a passive device, as in a molecular wire, but also by the requirements of compatible input and output voltages as discussed in [59] and [60]. These restrictions on speed are mainly due to the delicate interplay between the molecule and the electromechanical grid or electric field of the device.

The dimensions of a C_{60} electromechanical transistor are limited by the length of the cantilever, which has to be larger than $\sim$200 nm in order to compress the C_{60} molecule efficiently and cannot, therefore, be further reduced. If this is compared with the minimum feature size of CMOS today ($\sim$150 nm) and extrapolated to 2012 ($\sim$50 nm) [64], the C_{60} architecture will not be able to compete. One might argue that these limits are due to the electromechanical nature of the device [58] and would not apply to transistors based on molecules in an electric field. Quite recently, a four-terminal CNTFET has been fabricated where two electrodes act as source and drain bridged by a nanotube and the other two electrodes are used to apply a local electric field [65]. By changing the field, the conductance of the molecule was varied, which gives the transistor effect. Although in [65] only device densities of only 0.1 Mbit cm^{-2} are realized, the authors claim that 100 Mbit cm^{-2} could be achieved. This would still not reach the densities that are expected for CMOS technology in the near future (475 Mbit cm^{-2} for SRAM memories in 2006 [64]). Wherever electric fields are considered as the 'non-molecular' part of a hybrid-molecular device, the field strength has to be large enough to change the conductance of the molecule inside the nano-junction. Since the field strength depends on the area of the electrodes, there is a natural limit to the minimum feature size of the device for each choice of a particular molecule.

From this comparison, it can be concluded that, on the device level, there are fundamental limits for the miniaturization of hybrid-molecular electronics, which are not only governed by the size of the molecule but also depend on the size that an electromechanical or capacitative grid has to have in order to influence its energy levels. These architectural issues regarding the interfacing of a molecular device to the outside world or the assembly of many devices into larger circuits will have to be addressed to a larger extent in the near future.

5.4 Molecular films as active elements in regular metallic grids

The most advanced concept in molecular electronics today makes use of recent progress in various different fields, such as the tailoring of chemical functions via supra-molecular synthesis, the subsequent imprinting of metallic nanowires and molecular monolayers using a combination of nanolithography and molecule deposition techniques and innovative circuit design on the basis of regular grid arrays. Molecular films are sandwiched between metallic wires of submicrometre size, where they act as molecular switches depending on their redox state, which can be manipulated by applying external voltages on the wires [66–71]. If such metallic wire/molecular film/metallic wire cross-junctions are fabricated in arrays, they can be used as memory cells, where the individual junctions can be addressed independently and each junction can store 1-bit of information. Alternatively, such crossbar arrays can be used for circuit design in the context of programmable gate logic arrays (PGLAs), where the junctions, which are 'on' and 'off' represent diodes or disconnected grid points, respectively. This scheme has received a lot of attention in recent years, since researchers have managed to fabricate complex circuits such as demultiplexers and memory cells, which are an order of magnitude smaller than anything which can be currently produced with CMOS technology. It must be noted, however, that this scheme is based on classical electron transport in the Boltzmann regime, where quantum effects are averaged out statistically and devices cannot be scaled down to the size of single molecules [36].

5.4.1   Molecular switches in the junctions of metallic crossbar arrays

Molecular switches are two-terminal devices which are electronically reconfigurable, where a relatively high voltage (e.g. ± 2 V for [2] catenane-based molecule [66], figure 5.12(a)) is used to close or open the switch but a relatively low voltage to read (~ 0.1 V). These molecular switches [67], a monolayer of so-called Rotaxane molecules, are not field-activated but can be described as small electrochemical cells, which are characterized by signature voltages at which current flows and the molecules switch. They are just a couple of nanometres wide and have a dumb-bell-shaped component with an interlocked ring. When the molecules sit between two wires, electrons flowing through one wire can hop onto them and across to the other wire. But if a voltage is applied to a particular crossbar junction, the interlocked ring of its molecules slides to a new position on the dumb-bell and blocks the electrical current. Once the ring slips to its new position, it cannot be moved back, making the rotaxane molecules of a particular junction a single-use switch. But since discovering the rotaxane switch, researchers have come up with a dozen more molecular switches, including reversible ones [68].

A crossbar architecture (figures 5.12(b) and (c)) has four advantages over other circuit design concepts in nanoelectronics:

(i)   the architecture is easily scalable at least down to the point where the dimensional limits of the Boltzmann electron transport regime begin;
(ii) it requires only $2N$ communication wires to address 2^N nanojunctions, which would allow a better addressability of the circuits with external CMOS circuits than a dedicated architecture;
(iii) a regular architecture is reconfigurable, since the position of diodes is not determined by fabrication but junctions are switched 'on' and 'off' by applying voltages to change their redox state, which allows to compensate for a high percentage of defects; and
(iv) the regular structure facilitates the fabrication process.

Figure 5.12(b) illustrates the concept with simple examples.

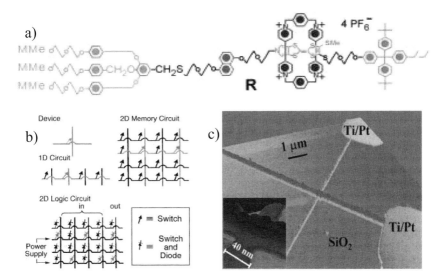

Figure 5.12. (*a*) The molecular structure of the bistable [2] rotaxane **R** used to form Langmuir–Blodgett monolayers [69]. (*b*) Crossbar circuit architecture at an increasing level of complexity in terms of fabrication and function [68]. Each junction is a switching device, with black arrows corresponding to open switches and grey arrows to closed switches. In the 1D circuit, the number 0100 is stored by addressing the second device and closing the switch. In the 2D memory circuit, the number 0101 is stored by writing the second row with the second and fourth column in parallel. In the 2D logic circuit, the switches are configured in such a way that six of them are closed, so that the circuit can perform as a half adder. (*c*) An atomic force micrograph of the cross-point molecular device with an insert showing the details of the cross-point [69]. Reprinted from (*a*) [69], (*b*) [68] and (*c*) [69].

5.4.2 High-density integration of memory cells and complex circuits

For the fabrication of the crossbar structures described in the previous section, an alternative to electron-beam lithography is needed, which does not have the same shortcomings in terms of the slow writing speed and the damage the high-energy electron beams can cause in molecular devices. For this purpose, an imprinting process [69] (figure 5.13) can be used, which produces sub-10-nm feature sizes and combines high throughput with low costs. In this process, an imprinting mould is fabricated into a 100-nm-thick, thermally grown SiO_2 on an Si substrate, which is patterned and etched to leave raised 40-nm-wide nanowires connected by 3-μm-wide wires on each end to 100-μm^2 pads (figure 5.14). The height of each wire is 80 nm above the etched surface of the mould and each mould has 400 such patterns laid out in a geometrical array to enable a large number of devices to be fabricated with a single imprinting step. To form the electrodes, a 100-nm-thick polymethylmethacrylate (PMMA) film is coated on another 100-nm-thick

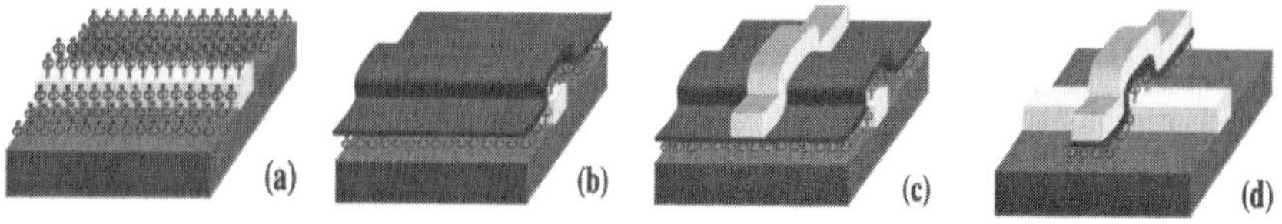

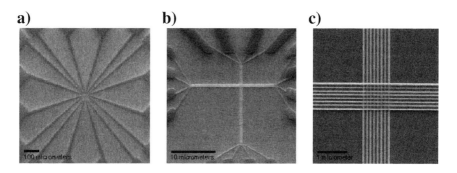

Figure 5.13. Schematic diagram of the procedure used for fabrication of nanoscale molecular-switch devices by imprint lithography [69]. (*a*) Deposition of a molecular film on Ti/Pt nanowires and their micrometre-scale connections to contact pads. (*b*) Blanket evaporation of a 7.5 nm Ti protective layer. (*c*) Imprinting 10 nm Pt layers with a mould that was oriented perpendicular to the bottom electrodes and aligned to ensure that the top and bottom nanowires crossed. (*d*) Reactive ion etching with CF_4 and O_2 (4:1) to remove the blanket Ti protective layer. Reprinted from [69].

a)   b)   c)

Figure 5.14. An electron microscope zooms in on an 8×8 nanowire grid [69]. Molecules between the grid junctions act as switches. Reprinted from [70], copyright IOP Publishing Ltd.

layer of SiO_2 on a Si substrate. The mould is then pressed onto the substrate coated with the PMMA resist in order to transfer the pattern. After imprinting, reactive ion-etching (RIE) with oxygen is used to remove the residual PMMA at the bottom of the imprinted trenches. Then, 5-nm Ti and 10-nm Pt metal layers are subsequently evaporated onto the substrate and an acetone liftoff of the unpatterned matrix reveals Ti/Pt nanowires and their micrometre-scale connection to the contact pads. A molecular monolayer of [2] rotaxane **R** (figure 5.12(*a*)) is then deposited over the entire device substrate including the bottom metal electrodes using a Langmuir–Blodgett (LB) method. The fabrication of the top electrodes begins with the evaporation of a Ti protective layer of 7.5 nm, which reacts with the top functional groups of the molecules to form a direct electrical contact. Patterned top-electrodes of Ti and Pt are then fabricated using the same imprinting process, where the mould is orientated and aligned perpendicular to the bottom electrodes. The removal of the Ti protective layer is then carried out with RIE.

This process has resulted in the highest density electronically addressable memory to date [70]. A 64-bit memory with an area of less than 1 μm^2 has been experimentally demonstrated. The researchers made the device by creating a master mould of eight 40 nm-wide parallel lines and pressed this mould into a polymer layer on a silicon wafer to make eight parallel trenches. After molecule deposition and fabrication of the top wires, the resulting device contains 64 points where the top and bottom wires cross. A bit of memory sits at each of these points in the roughly 1000 molecules sandwiched in a single junction between a higher and lower wire. To write a bit, a voltage pulse is applied to set the molecules' electrical resistance. Measuring the molecules' resistance at a lower voltage allows to read the bit. Using a combination of optical and electron-beam lithography, it takes about a day to create the master, which includes 625 separate memories connected to conventional wires so that they can be addressed with external circuits. After that, it takes just a few minutes to make an imprint. It is also possible to put logic in the same circuit by configuring molecular-switch junctions to make a demultiplexer—a logic circuit that uses a small number of wires to address memory, which is essential to make memories practical. This has been the first experimental demonstration that molecular logic and memory can work together on the same nanoscale circuits. The memories are also rewritable and non-volatile, i.e. unlike today's DRAM (dynamic random access memory) chips, they preserve information stored in them after the voltage is removed. Recently, the method for the fabrication of the nanowire pattern has been refined [71], where regular arrays of nanowires with a diameter of down to 8 nm can now be achieved.

In spite of the overall success of this particular scheme, there are some issues which need to be clarified, especially when it comes to details of the structure of the electrode/monolayer/electrode crossings on an atomistic level [4]. In a recent review on the vapour deposition of metal atoms on organic monolayers [72], the complexity of the process and the subtle effects of the surface structure and composition on the outcome are illustrated. A reflection–absorption infrared spectroscopy study [73] of a system, which is similar to the actual cross-junction, suggests undamaged organic monolayers with Ti coatings but further research is needed for the complete characterization of these complex structures.

A crossbar-array-based scheme is at the moment, perceived as the most realistic strategy to combine nanoscale components with CMOS circuits [74], where a concept for neuromorphic networks based on molecular single-electron transistors [75] has also been proposed.

5.5   Summary and outlook

Three different approaches to nanodevices in molecular electronics have been reviewed in this chapter. The use of single organic molecules in such devices is the least developed of those branches in terms of possible technological applications, although it is historically the oldest. Due to the challenges that circuit design and fabrication on an atomistic scale imply, it is still on a basic science level. Much more progress has been made with macromolecules such as C_{60} molecules and nanotubes, where the strictly speaking non-molecular but related semiconducting nanowires also have been discussed in this review. With these macromolecules, simple circuit elements have been demonstrated experimentally and circuit design on a system level has been done based on their experimentally determined device characteristics. The fabrication of larger circuits, however, still remains difficult due to the absence of a scheme for the fabrication of large arrays of identical devices. The most developed concept in molecular electronics so far is based on molecular films between the cross-junctions of metallic wires, which can act as switches in memories or configurable diodes in logic circuits. In this scheme, large memory cells and complex circuits have been experimentally demonstrated on the same chip in high density arrays.

All these approaches are essentially planar and can, therefore, be, in principle, envisioned as occupying stack layers in the 3D structure, which is the topic of this book. Since CNTs could also have an application as interconnect, an all-CNT-based 3D nanoelectronic system can be aimed at. Also single organic molecules with an internal 3D structure could possibly be used in molecular electronics in the future. It is usually assumed that the dissipation losses in molecular electronics are very low. However, it must be noted that these assumptions usually look at the internal molecular processes only and disregard the means needed to address them, such as the capacitance of metallic nanoelectrodes or the inevitably larger CMOS circuitry which is needed to connect the molecular circuit to the outside world.

References

[1]   Aviram A and Ratner M A 1974 *Chem. Phys. Lett.* **29** 277
[2]   Joachim C, Gimzewski J K and Aviram A 2000 *Nature* **408** 541
[3]   Joachim C 2002 *Nanotechnology* **13** R1
      Heath J R and Ratner M A 2003 *Physics Today* **56** 43
[4]   Mayor M, Weber H B and Waser R 2003 *Nanoelectronics and Information Technology* (Weinheim: Wiley-VCH) p 501
[5]   Nitzan A 2001 *Annu. Rev. Phys. Chem.* **52** 681
      Nitzan A and Ratner M A 2003 *Science* **300** 1384
[6]   Cui X D, Primak A, Zarata X, Tomfohr J, Sankey O F, Moore A L, Moore T A, Gust D, Harris G and Lindsay S M 2001 *Science* **294** 571
[7]   Büttiker M, Imry Y, Landauer R and Pinhas S 1985 *Phys. Rev.* B **31** 6207
[8]   Meir Y and Wingreen N S 1992 *Phys. Rev. Lett.* **68** 2512
[9]   Ness H and Fisher A J 1999 *Phys. Rev. Lett.* **83** 452
[10] Sautet P and Joachim C 1988 *Phys. Rev.* B **38** 12 238
      Ami S and Joachim C 2002 *Phys. Rev.* B **65** 155419
      Ami S, Hliwa M and Joachim C 2003 *Chem. Phys. Lett.* **367** 662
[11] Mujica V, Kemp M and Ratner M A 1994 *J. Chem. Phys.* **101** 6849

Samanta M P, Tian W, Datta S, Henderson J I and Kubiak C P 1996 *Phys. Rev.* B **53** R7626

[12] Emberly E G and Kirczenow G 1999 *Phys. Rev.* B **60** 6028
Emberly E G and Kirczenow G 2000 *Phys. Rev.* B **62** 10451

[13] Lang N D 1995 *Phys. Rev.* B **52** 5335
Hirose K and Tsukada M 1995 *Phys. Rev.* B **51** 5278

[14] Taylor J, Guo H and Wang J 2001 *Phys. Rev.* B **63** 245407
Brandbyge M, Mozos J L, Ordejon P, Taylor J and Stokbro K 2002 *Phys. Rev.* B **65** 165401

[15] Derosa P A and Seminario J M 2001 *J. Phys. Chem.* B **105** 471
Xue Y, Datta S and Ratner M A 2002 *Chem. Phys.* **281** 151

[16] Brandbyge M, Stokbro K, Taylor J, Mozos J L and Ordejon P 2003 *Phys. Rev.* B **67** 193104

[17] Magoga M and Joachim C 1997 *Phys. Rev.* B **56** 4722
Yaliraki S N, Kemp M and Ratner M A 1999 *J. Am. Chem. Soc.* **121** 3428

[18] Magoga M and Joachim C 1999 *Phys. Rev.* B **59** 16011

[19] Stokbro K, Taylor J, Brandbyge M, Mozos J L and Ordejon P 2003 *Comp. Mater. Sci.* **27** 151
Xue Y and Ratner M A 2003 *Phys. Rev.* B **68** 115406

[20] Stokbro K, Taylor J and Brandbyge M 2003 *J. Am. Chem. Soc.* **125** 3674

[21] Datta S, Tian W, Hong S, Reifenberger R, Henderson J I and Kubiak C P 1997 *Phys. Rev. Lett.* **79** 2530

[22] Blum A S, Yang J C, Shashidhar R and Ratna B 2003 *Appl. Phys. Lett.* **82** 3322

[23] Gimzewski J K, Jung T A, Cuberes M T and Schlittler R R 1997 *Surf. Sci.* **386** 101
Lopinski G P, Wayner D D M and Wolkow R A 2000 *Nature* **406** 48
Kuntze J, Berndt R, Jiang P, Tang H, Gourdon A and Joachim C 2002 *Phys. Rev.* B **65** 233405
Gross L, Moresco F, Alemani M, Tang H, Gourdon A, Joachim C and Rieder K H 2003 *Chem. Phys. Lett.* **371** 750
Rosei F, Schunack M, Naitoh Y, Jiang P, Gourdon A, Laegsgaard E, Stensgaard I, Joachim C and Besenbacher F 2003 *Prog. Surf. Sci.* **71** 95
Wallis T M, Nilius N and Ho W 2003 *J. Chem. Phys.* **119** 2296

[24] Nazin G V, Qiu X H and Ho W 2003 *Science* **302** 77

[25] Reed M A, Zhou C, Muller C J, Burgin T P and Tour J M 1997 *Science* **278** 252

[26] Kolesnychenko O Y, Toonen A J, Shklyarevskii O I and Van Kempen H 2001 *Appl. Phys. Lett.* **79** 2707

[27] Reichert J, Ochs R, Beckmann D, Weber H B, Mayor M and Von Löhneysen 2002 *Phys. Rev. Lett.* **88** 176804

[28] Tour J M, Jones L, Pearson D L, Lamba J J S, Burgin T P, Whitesides G M, Allara D L, Parikh A N and Atre S V 1995 *J. Am. Chem. Soc.* **117** 9529

[29] Metzger R M 2000 *Synthetic Met.* **109** 23

[30] Di Ventra M, Pantelides S T and Lang N D 2000 *Appl. Phys. Lett.* **76** 3448

[31] Seminario J M, Zacarias A G and Derosa P A 2001 *J. Phys. Chem.* A **105** 791

[32] Ellenbogen J C and Love J C 2000 *Proc. IEEE* **88** 386

[33] Park J, Pasupathy A N, Goldsmith J I, Chang C, Yaish Y, Petta J R, Rinkoski M, Sethna J P, Abruna H D, McEuen P L and Ralph D C 2002 *Nature* **417** 722

[34] Chen J, Reed M A, Rawlett A M and Tour J M 1999 *Science* **286**

[35] Pantelides S T, Di Ventra M, Lang N D and Rashkeev S N 2002 *IEEE Trans. Nanotechnol.* **1** 86

[36] Stadler R, Ami S, Forshaw M and Joachim C 2002 *Nanotechnology* **13** 424
Stadler R, Ami S, Forshaw M and Joachim C 2003 *Nanotechnology* **14** 722

[37] Heinrich A J, Lutz C P, Gupta J A and Eigler D M 2002 *Science* **298** 1381

[38] Remacle F, Speiser S and Levine R D 2001 *J. Phys. Chem.* B **105** 5589

[39] Stadler R, Forshaw M and Joachim C 2003 *Nanotechnology* **14** 138

[40] Dresselhaus M S, Dresselhaus G and Avouris P (ed) 2001 *Carbon Nanotubes: Synthesis, Structure, Properties, and Applications* (Berlin: Springer)
Appenzeller J, Joselevich E and Hönlein W 2003 *Nanoelectronics and Information Technology* (Weinheim: Wiley-VCH) p 473

[41] Tans S, Verschueren A and Dekker C 1998 *Nature* **393** 49
Martel R, Schmidt T, Shea H R, Hertel T and Avouris P 1998 *Appl. Phys. Lett.* **73** 2447

[42] Appenzeller J, Martel R, Derycke V, Radosavjevic M, Wind S, Neumayer D and Avouris P 2002 *Microelectron. Eng.* **64** 391

[43] Heinze S, Tersoff J, Martel R, Derycke V, Appenzeller J and Avouris P 2002 *Phys. Rev. Lett.* **89** 106801
Nakanishi T, Bachtold A and Dekker C 2002 *Phys. Rev.* B **66** 73307

[44] Bachtold A, Hadley P, Nakanishi T and Dekker C 2001 *Science* **294** 1317

[45] Derycke V, Martel R, Appenzeller J, Avouris P 2002 *Appl. Phys. Lett.* **80** 2773

[46] Cheung C L, Kurtz A, Park H and Lieber C M 2002 *J. Phys. Chem.* B **106** 2429

[47] Liu J, Casavant M J, Cox M, Walters D A, Boul P, Lu W, Rimberg A J, Smith K A, Colbert D T and Smalley R E 1999 *Chem. Phys. Lett.* **303** 125
Su M, Li Y, Maynor B, Buldum A, Lu J P and Liu J 2000 *J. Phys. Chem.* B **104** 6505
Zhang Y G, Chang A L, Cao J, Wang Q, Kim W, Li Y M, Morris N, Yenilmez E, Kong J and Dai H J 2001 *Appl. Phys. Lett.* **79** 3155

[48] Iijima S 1991 *Nature* **354** 56
Journet C, Maser W K, Bernier P, Loiseau A, delaChapelle M L, Lefrant S, Deniard P, Lee R and Fischer J E 1997 *Nature* **388** 756

[49] Kong J, Soh H T, Cassell A M, Quate C F and Dai H J 1998 *Nature* **395** 878

[50] Collins P C, Arnold M S and Avouris P 2001 *Science* **292** 706

[51] Derycke V, Martel R, Appenzeller J and Avouris P 2001 *Nano Lett.* **1** 453

[52] Keren K, Berman R S, Buchstab E, Sivan U and Braun E 2003 *Science* **302** 1380

[53] Morales A M and Lieber C M 1998 *Science* **279** 208

[54] Cui Y and Lieber C M 2001 *Science* **291** 851

[55] Huang Y, Duan X F, Cui Y and Lieber C M 2002 *Nano Lett.* **2** 101

[56] Huang Y, Duan X F, Cui Y, Lauhon L J, Kim K H and Lieber C M 2001 *Science* **294** 1313

[57] Rueckes T, Kim K, Joselevich E, Tseng G Y, Cheung C L and Lieber C M 2000 *Science* **289** 94

[58] Joachim C and Gimzewski J K 1997 *Chem. Phys. Lett.* **265** 353

[59] Ami S and Joachim C 2001 *Nanotechnology* **12** 44

[60] Stadler R, Ami S, Forshaw M and Joachim C 2001 *Nanotechnology* **12** 350

[61] Stadler R and Forshaw M 2002 *Physica* E **13** 930

[62] Tilmans H 1997 *J. Micromech. Microeng.* **7** 28

[63] Nikolic K, Berzon D and Forshaw M 2001 *Nanotechnology* **12** 38

[64] Compano R 2000 *Technology Roadmap for Nanoelectronics* (European Commission, IST Program)

[65] T. Ondarcuhu, C. Joachim and S. Gerdes 2000 *Europhys. Lett.* **52** 178

[66] Collier C P, Wong E W, Belohradsky M, Raymo F M, Stoddart J F, Kuekes P, Williams R S and Heath J R 1999 *Science* **285** 391

[67] Collier C P, Mattersteig G, Wong E W, Luo Y, Beverly K, Sampalo J, Raymo F M, Stoddart J F and Heath J R 2000 *Science* **289** 1172

[68] Luo Y, Collier C P, Jeppesen J O, Nielsen K A, Delonno E, Ho G, Perkins J, Tseng H R, Yamamoto T, Stoddart J F and Heath J R 2002 *Chem. Phys. Chem.* **3** 519

[69] Chen Y, Ohlberg D A A, Li X, Stewart D R, Williams R S, Jeppesen J O, Nielsen K A, Stoddart J F, Olynick D L and Anderson E 2003 *Appl. Phys. Lett.* **82** 1610

[70] Chen Y, Jung GY, Ohlberg D A A, Li X, Stewart D R, Jeppesen J O, Nielsen K A, Stoddart J F and Williams R S 2003 *Nanotechnology* **14** 462

[71] Melosh N A, Boukai A, Diana F, Gerardot B, Badotato A, Petroff P M and Heath J R 2003 *Science* **300** 112

[72] Fisher G L, Walker A V, Hooper A E, Tighe T B, Bahnck K B, Skriba H T, Reinard M D, Haynie B C, Opila R L, Winograd N and Allara D L 2002 *J. Am. Chem. Soc.* **124** 5528

[73] Chang S C, Li Z Y, Lau C N, Larade B and Williams R S 2003 *Appl. Phys. Lett.* **83** 3198

[74] Stan M R, Franzon P D, Goldstein S C, Lach J C and Ziegler M M 2003 *Proc. IEEE* **91** 1940

DeHon A 2003 *IEEE Trans. Nanotechnol.* **2** 23

[75] Turel O and Likharev K 2003 *Int. J. Circ. Theor. Appl.* **31** 37

Chapter 6

Nanoimprint lithography: A competitive fabrication technique towards nanodevices

Alicia P Kam and Clivia M Sotomayor Torres
Institute of Materials Science and Department of Electrical and
Information Engineering, University of Wuppertal, Germany

6.1   Introduction

The challenges of making the fabrication of both electronic and optical devices more competitive have led to an upsurge in research activity in numerous lithographic techniques, which include but are not limited to microcontact printing, photolithography and electron-beam lithography. Alternative lithographic techniques, in particular nanoimprint lithography (NIL), are emerging as a unique and viable novel nanofabrication approach towards the goal of submicrometre devices. A comparison of alternative lithographies are given in table 6.1. NIL has received considerable attention in recent years due to its potential as a parallel, high-resolution, high-throughput and, moreover, low-cost technique [1]. Moreover, small feature sizes [2], ∼10 nm, and reasonably high aspect ratios [3, 4], typically four, have been demonstrated by NIL. Such routinely attainable characteristics have undoubtedly contributed to the interest in this field as exemplified by the growing number of applications which employ this approach: molecular electronics [5, 6], photonics [7–9], biosensors [10, 11], dynamic and fundamental research in polymer characteristics [4, 12] and etch mask patterning [13] to list a few. Some of these devices will be discussed in-depth later. The aforementioned applications are indeed limited in breadth to one or two dimensions; however, the potential for imprinting in three dimensions or multilayer [14–19, 23] has also been explored.

The concept of 3D NIL is relatively new and is still being defined. It ranges from the fabrication of 3D nanostructured stamps, and subsequent imprints,

Table 6.1. Comparison of printing and related techniques. This table was presented during discussions amongst participants at the International Workshop on Nanoimprint Lithography held in Lund, Sweden, 16–17 January 2002 [3, 15, 20–22, 43, 44].

Process	Process type	Min. feature Min. pitch (nm)	Combined small and large features Largest area printed	Overlay accuracy and no. of times stamp used	Time for alignment, printing, release, cycle time
Imprint	Nano-imprint	6–40	10 nm– 100s μm 100 mm wafer	0.5 μm 50	Few mins 10 s Few mins 10–15 min
	Step and Stamp	300 600	300 nm– 100's μm 150 mm^2	1 μm 36	1 min 5 min Few seconds 6 min/step
	Step and Flash [a]	10 50	20 nm– ~ mm 1 inch wafer	1 μm >100	1 min 10–20 s 10 s 5 min/flash
Printing	Micro contact Printing	60–200	60 nm– 60 mm 12 inch wafer	0.5– 1 μm >100	1 min 1–30 s 10 s 2 min
Stencilling	Shadow	10	10 nm– 10 s μm 50 μm^2– 1 mm^2	— —	Few min 15 min Few min 30 min
Combined Techniques	UV Lith +NIL[b]	100	10 nm– contact pads 400 μm^2	1 μm	— —
	Laser assisted NIL	10	140 nm 10 nm– 10's μm	2.3 mm^2 — —	— 250 ns — —

[a] Kurz H, 2002, Private communication.
[b] Considering imprint first followed by UV photolithography.

to imprinting in multilayers. Though both 3D stamps and imprints, such as diffractive elements and T-gates, have been demonstrated, the realization of

3D NIL still remains a challenge, especially in the multilayer regime. There are several key issues which need to be addressed in multilayer imprinting: alignment, matching complementary polymers for successive layers and refining the glass transition temperatures of such polymers to allow for imprinting. More importantly, device design, which requires that the imprinted polymers be functional by incorporating either an electrically or an optically conductive element, is becoming more routine. It is hoped that, in the near future, the more important issue of alignment may be resolved with the implementation of optical controls and more research, such as the submicrometre alignment over a 4 in Si wafer demonstrated on a multilevel NIL technique by Zhang *et al* [15]. Nevertheless, in practical applications, such as packaging technology [10], microfluidic devices [24] and MEMS [25], imprinting has been limited to two dimensions and the micrometre scale. The ability to scale down to the submicrometre regime with NIL appears promising, while the possible applications of 3D nanostructuring remain an untapped area of research.

6.2   Nanoimprint lithography

Nanoimprint lithography (NIL) is based on the deformation of a thin polymer film with a rigid mould (figure 6.1). The master stamp, with sizes in the 2 cm^2 to 150 mm wafer [26] range, is typically fabricated by a combined process of optical and electron-beam lithographies followed by dry etching. The layer to be patterned is a thermoplastic polymer that is spun onto a solid substrate as a thin film. The NIL process consists of two main stages. In the first stage, the master stamp and printable polymer are placed in parallel stages and heated to approximately 100 °C above the glass transition temperature, T_g, of the polymer. Both the stamp and thin film are then brought into physical contact and pressure is applied. As the polymer is compressed, the now viscous polymer is forced to flow into the cavities of the mould, thereby imparting the features of the stamp onto the polymer. The applied pressure and time required for an imprint is dependent on several factors but primarily on the viscoelastic properties of the polymer being imprinted. Typical imprinting parametres are illustrated in table 6.2. They range from 100–200 °C for temperature, 1–4 MPa for pressure and imprinting times of a few minutes [9, 27–30]. On completion of the imprinting process, the polymer is cooled to below T_g, demounted and separated from the stamp. Assuming that the imprinted polymer contains some functionality, such as conductivity, optical linear or nonlinear response, it may be directly implemented as an organic device.

The second step of the NIL process entails the transfer of the imprinted features to a hard medium. This is achieved, first, by the removal of the residual thin polymer layer via an O_2 plasma, thus exposing the substrate. The pattern is then imparted to metallic, semiconducting or magnetic material depending on the application [31]. A further lift-off process, etching or electroplating renders the final pattern transfer. Devices fabricated in this manner are primarily suited

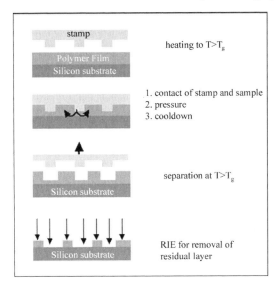

Figure 6.1. Illustration of the NIL process and residual layer removal.

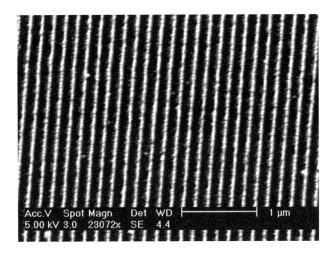

Figure 6.2. An SEM image showing the pattern transfer of a Cr/SiO_2 stamp with 160 nm period and 80 nm linewidth via NIL and RIE to PMMA.

for molecular electronic devices such as memory storage components, thin-film transistors and organic field effect transistors. A scanning electron microscope (SEM) image of a stamp with 160 nm period lines and its imprint are illustrated in figure 6.2.

Table 6.2. Typical imprinting parameters and glass transition temperatures (T_g) for printable polymers. The applied temperature, pressure and heat/cooling times are dependent on pattern density, viscosity and molecular weight of printed layer.

Polymer	T_g (°C)	T (°C)	P (MPa)
PMMA[ba]	105	200	13
PS[b]	100	180	4–10
mr-L 6000[c]	25–65	55–125	6
NEB 22[d]	80	120	5
PPM[e]	115	180	10
PBM[e]	56	146	10
PDAP[e]	63	160	10
Hybrane HS3000[f]	70	140	3
Hybrane HS2550[f]	0–10	50	3
PVC[g]	87	140	3

[a] Chou S Y *et al* 1995 *Appl. Phys. Lett.* **67** 3114
[b] Seekamp J *et al* 2002 *Nanotechnology* **13** 581
[c] Pfeiffer K *et al* 2002 *Microelectron. Eng.* **62** 393
[d] Gourgon C *et al* 2000 *Microelectron. Eng.* **53** 411
[f] Lebib A *et al* 2002 *Microelectron. Eng.* **62** 3717
[g] Scheer H-C *et al* 2002 *Handbook of Thin Film Materials* vol 5 (New York: Academic)

6.2.1 Fabrication issues

Limitations in the resolution of the NIL technique are solely attributed to the minimum feature size of the master stamp, with the minimum feature achieved being 6 nm [32], and the inherent viscoelastic properties of the polymer [30]. As a result, particular attention must be given to the fabrication of the master stamp and the type of printable polymer used. Furthermore, there are two major issues which contribute to the viability of this technique towards commercial applications: overlay accuracy or alignment and critical dimensions (CDs).

The issue of controlling CDs has recently been experimentally addressed [33–35]. In conventional lithographies, the CD is managed by adjusting the exposure dose, mask feature sizes and the post-lithographic processes [34]. The replication process in the NIL technique is thought to be of high fidelity and, therefore, the control of CDs is not warranted [34]. However, though this may be true for larger feature sizes, in the range of hundreds of nanometres, for more demanding applications, such as photonic crystal waveguides and quantum magnetic dots, dimensions are often in the sub-10 nm regime and the CD plays a significant role. Perret *et al* have studied the influence of the CD on the uniformity of the residual layer and the etching process after imprinting in two polymers: mrl-I-8030 and NEB 22 [36]. A uniform CD over the imprinted area is obtained only when both the residual layer and oxygen-based plasma are uniform. An in-depth investigation into imprinting parameters—temperature, pressure and imprinting time, as related to the uniformity of the residual layers for these polymers—was also undertaken and found to be limited to the viscoelasticity of the polymer [36]. Gourgon *et al* studied the influence of the residual layer uniformity as a function of pattern density [33, 37]. Here, 500 nm trenches with varying periods between 0.65 and 10 μm were investigated. With the proper attenuation of imprinting time and polymer viscosity, the residual thickness uniformity was demonstrated to be independent of pattern density [33]. Moreover, Lebib *et al* have demonstrated good control of CD accuracy, sub-10 nm, for imprinting and etching of a 100 nm dots array in a trilayer process [34], while Matsui *et al* [38] have investigated room-temperature NIL with hydrogen silsequioxane in attempt to manage the CD and pattern displacement.

Several factors influence the alignment accuracy during an imprint: stability of the press, relative thermal expansion of polymer and stamp, uniformity of the printing polymer and the overlay accuracy of the aligner [15]. Zhang *et al* have successfully obtained submicrometre alignment over a 100 μm wafer and suggest that this alignment is fully scalable to 150 mm wafer processes [15]. The average alignment accuracy demonstrated was 1 μm with a standard deviation of 0.4 μm in both the X and Y directions. These results reflect the limited alignment capabilities inherent to the press employed, 0.5 μm, and were not due to the properties of the imprinted polymer [15]. The alignment accuracy as applied to multilevel transistors fabricated by NIL was also investigated and found to be 0.5 μm [39]. Avenues to increase the alignment accuracy through lower imprinting temperatures [40, 41], 3D nanoimprinting [23] and optical capacitive coupling between alignment markers on stamp and substrate [42] have also been explored.

6.2.2 Instrumentation

There exist two concepts for the method of imprinting [43]: a flatbed approach and a roller technique. In flatbed geometry, the stamp is supported on a rigid platform which is brought into contact with the thin layer to be imprinted. Both single-print and step-and-repeat imprints can be facilitated by this method [44]. Alternatively, the roller geometry enables the stamp to be rolled over the printable polymer. This technique offers a few advantages over the flatbed approach, such as better uniformity, simple construction and smaller applied imprinting pressures [45]. Conventionally, imprinting machines or presses were custom built; however, with the research demands for such high-resolution presses growing, these machines are now commercially available from a few companies, e.g. Suss MicroTec AG, Jenoptik and the EV Group and are analogous to current wafer-bonding equipment.

The method of fabrication of the master stamp employed in NIL is, in general, dictated by the target application. For high-resolution stamps, primarily electron-beam lithography and pattern transfer processes such as dry etching and metal liftoff are employed. Alternatively, other lithographic and pattern transfer techniques, such as optical lithography and reactive ion etching, may be used to pattern silicon or silicon dioxide, which in turn can be electroplated. NIL itself may also be utilized in the replication of stamps, as has been demonstrated with a thermoset mr-L-prepolymer by Pfeiffer *et al* [46].

Though no well-defined protocol on stamp size or pattern density exists, it is clear, from the view of CD and alignment issues, that both these parameters should be controlled. As the stamp size dictates the imprinting area, issues of parallelity and homogeneity of the printable polymer, as well as the uniformity of the residual layer, need to be addressed. Conceptually, larger stamps increase the difficulty in controlling the parallelization problem. This dilemma may be combated in step-and-repeat processes, such as the step-and-stamp imprint lithography (SSIL) offered by Suss MicroTec AG, in which small stamps ($\sim$1 cm^2) are used to imprint larger areas [44] such as 200 mm diameter wafers [45]. The step-and-repeat process can also be applied to solve the problem of polymer flow on imprinting with a high pattern density.

Two major concerns with the ability for stamp regeneration after successive imprints are contamination and damage. These may be due in part to the cleanliness of stamp and/or wafer, adhesion of stamp to wafer or inhomogeneity of the pressure applied. Thus, an anti-adhesion layer is normally applied to the stamp prior to imprinting to combat these problems. Recipes for coating the master stamp abound and range from the deposition of thin metal films [47], plasma deposition of a fluorocarbon thin film [48] to the casting of polymeric silanes [49] and self-assembled anti-adhesion layers [50].

The imprinting parameters, such as pressure and temperature, are dependent on the viscoelasticity of the printable polymer. This viscoelasticity has been found to be related to the molecular weight of the printable polymer and their glass transition temperatures [29]. During the imprinting process, pattern transfer is realized via the displacement of the printing polymer material. The rate of flow of the polymer is a function of the rates and values of applied temperature and pressure [29]. As a short processing time is desirable, this requires that the polymer must be under viscous flow when heat and pressure are applied. Thus, the viability of the NIL process is higher when lower imprinting temperatures and pressures enable the flow.

Various printable polymers exist and new ones continue to be developed. The more common thermoplastic polymers used in NIL are the polymethyl-methacrylate (PMMA) analogues, with the smallest feature size obtained on imprinting being 10 nm. However, other polymers have shown to be very promising [41, 46, 51, 52]. Lebib *et al* have developed and employed a hyperbranched polymer, Hybrane, which demonstrated better critical dimension control than PMMA in a trilayer imprinting process [34, 41, 53]. Crosslinking polymers devel-

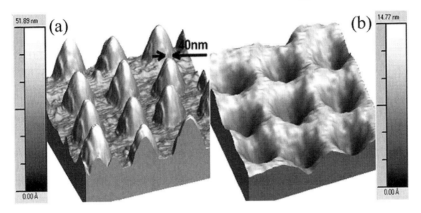

Figure 6.3. An AFM image of (*a*) an mr-L6000 polymer stamp with 40 nm features and (*b*) its respective imprint into mr-L6000.

oped specifically for NIL, such as the mrI-9000 and mr-L6000, have recently been made commercially available. The mr-L and mrI series of prepolymers, offered by Micro Resist Technology GmbH [54], have lower glass transition temperatures and molecular weights than PMMA analogues, thus they may be imprinted at much lower temperatures [46,51,55–57]. In particular, the Duroplast mrI-9000 and Thermoplast mrI-8000 are thermosetting polymers which are crosslinked on imprinting, whereas the mr-L6000 is a photochemical crossing linking polymer that is set in two phases: imprinting followed by a UV flash [46]. Moreover, the mr-L polymers offer higher dry etch resistance than PMMA, with the advantage that after crosslinking the polymer is robust enough to be utilized as a stamp [46,52]. The smallest feature size obtained after imprinting in these polymers is 40 nm [56], figure 6.3.

6.3 Device applications

The demands on semiconductor devices have been relentless. The drive to increase processing times and device density, while reducing the overall dimensions of the circuitry, has resulted in an interconnection bottleneck. Three-dimensional (3D) architectures enable shorter interconnect wires, which translates to a minimization in heat and electrical losses and increase in operating speeds. This new dimension has alleviated the bottleneck, though more research is required in the area of making the fabrication of 3D architectures viable.

The fabrication of 3D microelectronic architectures can be realized in several ways. However, only two processes have garnered any attention as a viable fabrication technique: (1) wafer bonding and (2) low-temperature silicon epitaxy [58]. The wafer bonding process consists of joining sufficiently thin layers of 2D (planar, 2D) interconnected integrated circuits via high-quality interconnect bonds, such as copper–copper metal bonds. The planar layers should be thinned, peeled and prepared prior to bonding. This method affords the parallel production of several planar layers which can then be assembled. In contrast, the fabrication of 3D architectures by silicon epitaxy is accomplished by growing crystalline silicon at low temperatures on top of each functional 2D layer, thereby forming the gate for the devices [59]. There are three significant disadvantages with this fabrication approach. First, deposition temperatures must be significantly low as so not to compromise the integrity of the underlying structures. Typical deposition temperatures for crystalline silicon are around 900–1500 °C, while the melting temperatures of commonly employed metals in circuitry, such as gold, are in the range of 1000–2000 °C. Furthermore, caution must be taken in ensuring that the evaporated silicon layer does not penetrate the insulating layer or the active device layer. Finally, the processing time for this method is longer than that of wafer bonding due to the sequential nature of the technique, where each layer is created in sequence.

The similarities of both the processing steps and instrumentation (technologies) for NIL and wafer bonding have created a role for imprinting in the realization of 3D architectures. It is this shared platform that allows for the transfer of technological successes from NIL to wafer bonding, such as the issue of wafer alignment [60, 61]. In fact, nanoimprinting has allowed for the cheaper fabrication of both 2D and 3D [16] nano- and micrometre functional structures, interconnects and circuitry on a wafer scale, albeit in a planar configuration. These layers can then be easily assembled by conventional wafer-bonding techniques to fabricate 3D architectures. Though such architectures are being actively pursued to alleviate the system-performance bottleneck, methods such as electronic-to-optical interconnects [62] and the development of tri-gate CMOS transistors [63] are also being investigated. Moreover, architectural templating, whereby successive imprints are done in different polymers and functionalized, has been successfully demonstrated for 60 nm transistors by Zhang *et al* [39]. In this research, the patterning of four lithographic layers in a standard four-mask MOSFET process were all done by NIL. Such templating has also been considered in metal-dielectric/light-emitting systems. Certainly the issues of polymer glass transition temperature in multilayer imprinting processes must be considered and a mix-and-match technique for complementary polymers explored.

As the limits of inorganic-based circuitry are approached, especially in respect of dimensionality, the merits of organic electronics as a low cost substitute have been realized. Though not expected to replace their inorganic counterparts in high-performance or high-density devices, the interest lies in applications where ease of fabrication, mechanical flexibility and avoidance of high temperatures are of major consideration [64]. As the transport phenomena in organics are confined to length scales in the nanometre range, downscaling the dimensionality of devices to this range is envisioned as a promising route to unleash the potential of molecular devices. These aspects motivate the investigation of NIL for organic electronics [30], especially in aiming to devices with lateral feature sizes below 100 nm and with tolerance below 10 nm. Moreover, the patterning of organic semiconductors by NIL has been demonstrated to show a lack of degradation in both electrical [65] and electro-optical [66] properties of such devices. Thus, it is within this scope that the fabrication of devices, such as organic field effect transistors and polymeric photovoltaic cells, via NIL has been undertaken.

Device applications can, for all intensive purposes with regards to NIL, be divided into three main branches: nanomagnetism, optoelectronics (organic light-emitting diodes, electroluminescent devices) and organic semiconductors.

6.3.1 Magnetism

Magnetic nanostructures, as applied to the field of data storage, have a variety of applications ranging from quantum magnetic discs, high-density magnetic dot structures, microrings for magnetization reversal and magnetic memory cell

The application of nanoimprinted surfaces, containing nanostructures of metal dots or pillars which can be magnetized, in quantized magnetic discs (QMD) has been pioneered by Chou *et al* [67].   QMDs consisting of both perpendicular [68, 69] and longitudinal [70, 71] nanoimprinted single-domain nickel or chromium bars with densities ranging from 3 Gbits in^{-2} to 45 Gbits in^{-2}, depending on orientation of the bars, have been fabricated and characterized. The disc area imprinted was nominally 4 cm $\times$ 4 cm for perpendicular quantized magnetic discs. The adaptation of this nanostructuring technique to imprinting in multilayers has led to the development of spin-valve magnetic memory cells [72, 73] where the multilayers consisted of thin films of elemental metals and metal alloys, to high-density magnetic dot structures [74] in which a trilayer process is implemented for pattern transfer. The smallest feature size obtained for the imprinted spin valves was 70 nm [73], while that of 30 nm [74] for the imprinted dot structures.

The employment of high-density arrays as magnetic recording material requires that the dots be single domain and there is no direct magnetic exchange coupling [75]. Thus, investigations were performed to determine the influence of dot size, thickness and periodicity in the high-density dot arrays on the single domain state and coupling. The single domain state was demonstrated to depend on the thickness of the dots [76], while magnetic dipole interactions, which are deemed important in the magnetic reversal process, were shown as a function of both dot size and periodicity [53, 77].

Furthermore, a successful variant approach to NIL involving a photocuring process has been applied to the patterning of high-density magnetic dots consisting of cobalt and platinum multilayers [77]. In this process, the minimum feature size attained was 55 nm diameter dots [78]. Chen *et al* [79] have fabricated alternative nanoimprinted structures for data storage (figure 6.4). Magnetic microrings were produced with varying inner and outer diameters and similar to the high-density dot arrays, the magnetism reversal was demonstrated to be

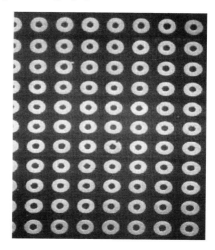

Figure 6.4. An SEM image of Cr magnetic rings after imprinting at low temperature and subsequent lift-off. Rings were 20 nm in thickness and had an outer diameter of 5 μm (Courtesy of Y Chen.)

a function of the size and thickness of the rings [79, 80]. Thus, NIL therefore provides the means via which magnetic structures can be cost-effectively procured with high resolution.

6.3.2   Optoelectronics

The term 'optoelectronics' has broad implications. In the most general sense, any device that functions as an optical-to-electrical or electrical-to-optical transducer or incorporates such components may be deemed optoelectronic. Optoelectronic devices can be divided into two categories: inorganic (typically silicon and GaAs based) and organic. The field of organic optoelectronic devices encompasses organic light-emitting diodes (OLED) and electroluminescent devices [81–89], organic photovoltaics [90–92], organic field effect transistors (OFET) [64, 93], molecular switches [94], organic lasers [95–98] and integrated optical circuits [99, 100].

The direct structuring of organic materials, whether semiconductor or light-emitting media, via NIL is an area of untapped potential. The ability to imprint functional organic materials with submicrometre resolution has been explored by Wang *et al* NIL was employed to pattern transfer gratings with periods of 200 or 300 nm to a quinoline compound, Alq3, impregnated with DCMII dye molecules [66]. The degradation of the light emission efficiency on exposure to an external source, such as water, oxygen or incident radiation, which induces a change in the light-emitting media, has been well documented [88, 101–107]. However, studies of light emission efficiency both prior and post-imprinting with

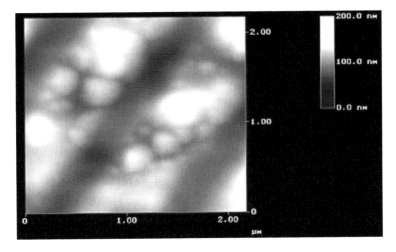

Figure 6.5. An AFM image of 500-nm period lines imprinted into PF/PANI. The image shows that the phase separation is influenced by imprinting. (Courtesy of F Bulut.)

the NIL technique demonstrated no such degradation [66, 108], thereby making NIL a competitive high-resolution alternative for patterning of organic materials and controlling their light emission [95]. Polymer matrices that support functional organic media have also been investigated [109], with particular interest into the molecular alignment of the molecules upon imprinting. Reversible spontaneous molecular alignment of a DCMII dye molecule impregnated in a PMMA matrix was illustrated on imprinting with gratings of 200 and 300 nm periods [109].

The application of NIL to control the molecular orientation and, thus, morphology in an organic layer has widespread implications in polymeric solar cells [110, 111]. Here, the role of the imprinting technique is twofold: (1) patterning in which the lateral sizes of the electrodes are commensurable with the diffusion length of excitons in polymers regardless of material specific ordering properties; and (2) control the network morphology of the phase separated functional material for enhanced transport and carrier generation [110]. An AFM image of an imprinted polymer demonstrating phase separation induced by printing is shown in figure 6.5 [112].

6.3.3 Organic semiconductors

The direct patterning of organic semiconductor material, such as a-sexithiophene [27] and polyaniline [113], via NIL has been reported. The development of a novel fabrication extension to the NIL technique for low-temperature imprinting has been achieved by Behl *et al* [5]. In this research, UV curable polymers based on triphenylamine analogues were patterned with nanoelectrodes of 100 nm feature sizes. However, due to the sensitivity of organic semiconductors to

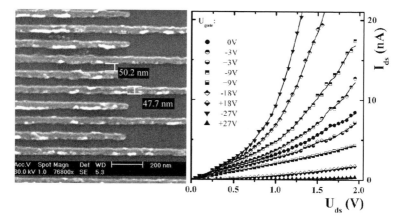

Figure 6.6. (*a*) An SEM image of interdigitated Au electrodes with 50 nm channel length. (*b*) I/V measurements of α-sexithiophene OFETs with 100 nm electrode spacing, with varying gate voltage.

the pressures and temperatures used in the imprinting technique, nano-electrode patterning has been limited to fabrication of the metal electrodes prior to the deposition of functional organic materials or the bottom-up approach. The goal of such an approach would be to provide a platform on which the scaling behaviour of charge mobility in conjugated thin organic films may be investigated. This can be viewed as a strategy to increase the mobility of the device and the switching speed, by bringing electrode separation close to the size of a single domain of the material and thereby reducing scattering caused by grain boundaries (Moore's law). The implications are that nanoelectrodes with a separation of around a few tens up to a few hundred nanometres, commensurable to the size of the molecular domains being employed, will be required. Within this respect, the fabrication of metal nanoelectrodes separated by sub-100 nm has been successfully attempted [57, 114], as illustrated (figure 6.6(*a*)). The optimization of nanoelectrode size to match that of the molecule investigated, such as α, ω-dihexylquaterthiophene, is an iterative process and electrical characterization of FET devices produced have become routine (figure 6.6(*b*)). However, the characterization of such small devices usually involves a combined fabrication approach, NIL for the fabrication of the nanoelectrodes while photolithography is often used to define the submillimetre contact pads.

Alternative methods employing NIL to pattern nanoelectrodes for molecular device applications have emerged. Austin and Chou [115] have fabricated gold nanocontacts with sub-10 nm channel lengths using a combination of NIL and electromigration techniques. While the development of a cold welding technique for both the addition [116] and subtraction [65] of a metal layer has been achieved. Furthermore, Kim *et al* [116] have successfully illustrated the application of an additive cold welding NIL process to a top-down approach to the fabrication of pentacene thin film transistors with channel lengths approaching 1 μm.

6.4 Polymer photonic devices

Polymer-based photonic devices propose another unique system from which the applicability of NIL can be extracted. Here the feature size is not the limitation, as for molecular electronics, but rather the parallelization of the technique. With present alignment capabilities, structuring optical devices with NIL is of interest for single-step processes which have to deliver a surface roughness compatible with the demands of light guiding. The tolerance limits for such imprints being on the order of 5 nm. Thus, the imprinting of optical systems, which contain waveguides, filters, splitters, and more complex functions on a low refractive index contrast polymer platform, is a realistic goal.

6.4.1 Integrated passive optical devices

A rapid parallel fabrication technique has been realized in NIL towards the goal of passive optical devices. The most obvious application of NIL would be in the production of diffraction [117–120] or Bragg gratings [95] due to the demand of high-resolution and relatively large dimensions, approximately a few square centimetres, of the device. Various types of submicrometre gratings have been successfully fabricated for specific applications: polarizers with 190 nm periods [121, 122], microspectrometer arrays [123] with 1 μm periods, subwavelength resonant grating filters [124], Fresnel zone plates [125] with circular gratings of 20 nm linewidths and antireflective surfaces [126]. Although the potential of NIL to manufacture periodic structures for large-area applications has been investigated [15, 127], optical characterization has been limited [127]. Recently, Seekamp *et al* [9] have fabricated a 25 mm^2 grating with a period of 800 nm by NIL. The diffraction grating demonstrated homogenous scattering of light and illustrates the viability of NIL to produce practical optical devices. It should be noted that, though there has been some successes in fabricating polymeric optical devices, it is imperative that the optical properties remain unchanged by the imprinting process and the device's lifetime as to avoid changes in absorption or refractive index. To this end, further studies into polymer relaxation and deformation are required.

The drive to develop platforms for integrated organic optical circuits has opened the door to the possibility of NIL to be utilized in the fabrication of microoptical components with submicrometre resolution. Thus, polymeric optical waveguides and microring resonators have been successfully fabricated via NIL [128–131], with losses under 0.4 dB cm^{-1} [132] and Q-factors as high as 5800 [129] respectively. Multimode waveguides with 100 μm^2 cross sections have been realized and successfully integrated into printed circuit boards by Lechmacher and Neyer [133]. Moreover, two types of waveguides, slab and

Y-branch, were fabricated by the nanoimprinting of polystyrene on SiO_2 and characterized [4, 134]. Overall measured losses for the slab waveguide after fibre coupling for 514 nm light, taking into account coupling and detection losses, gave values of 7 dB mm^{-1} [134]. The design of the imprinted Y-branched waveguide was based on applications in the 1.3 mm communication window [134]. With decreasing optical losses and greater Q-factors achieved in optical circuitry fabricated by imprinting, NIL is poised to be a premier fabrication technique to enable integration in both electronic–optical and purely optical systems.

6.4.2 Organic photonic crystals

The NIL technique inherently limits the type of photonic crystals (PC) to 1D (figure 6.7) or 2D planar system. As a result, research in this area has so far been very challenging and limited. The direct patterning of organic light-emitting structures, Alq3, and its potential applicability in the production of photonic devices was suggested by Wang *et al* [66] While the fabrication of photonic structures, which were reactive ion etched into silicon using NIL-imprinted polymer as the reactive ion etching (RIE) mask, was achieved by Grigaliunas *et al* [7], PC fabricated via NIL imprinting of a thin aluminium foil with a SiC stamp has also been demonstrated [135]. However, the first known NIL-imprinted polystyrene PC structures on silicon oxide were realized by Seekamp *et al* [9]. Two PC structures were imprinted, one consisting of a triangular lattice with 300 nm diameter columns, the other with a hexagonal geometry and 100 nm diameter columns. Calculations performed for given PC geometries suggest that photonic band gaps may be expected for the polystyrene PC, with no influence of the residual layer thickness, typical of the NIL process, on the quality of the PC bandgap [9]. Moreover, NTT Basic Research Laboratories have patterned microstructures, which form the basis of a photonic crystal laser, in an organic thin film layer via nanoimprinting [136]. With the variation of the pitch of the silicon carbide master stamp employed, the laser could be tuned to emit light at a given wavelength. The height of the imprinted columns ranged from 200 to 1000 nm [136]. It stands to reason that with research in this area being limited to the second dimension, progress may be realized mainly through developments in varying the refractive index of the imprinted polymer.

6.5   Conclusion

The fabrication of various organic optoelectronic and photonic devices in the submicrometre regime have been realized with NIL and revised in this article. This approach offers a competitive edge over conventional lithographic techniques due to its low cost and fast turnover. Moreover, NIL has demonstrated the capability of achieving feature sizes in the sub-10 nm regime, sizes which are required for the fabrication of novel devices such as quantum magnetic discs, organic field transistors, and photonic crystal waveguides. Such applications are

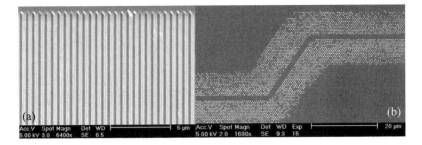

Figure 6.7. SEM images of an imprinted 1D PC (*a*) and a polymeric stamp of a waveguide fabricated by imprinting in mr-L6000 (*b*).(Courtesy of A Goldschmidt.)

not limited to the aforementioned devices but are rapidly expanding to other areas of nanotechnology. One such area, which has garnered much interest and promises to alleviate the bottleneck of interconnects as well as opening the avenue to novel applications, is 3D templating and structuring. Several technical issues, however, still need to be addressed in order to enhance the viability of this technique, the most pressing being the question of critical dimensions and alignment. Though some solutions to these problems have been demonstrated, such as anti-adhesive measures, utilizing and developing low-T_g polymers and step-and-stamp imprint lithography, there is still much to be done to ensure uptake by industrial stake-holders.

Acknowledgments

The authors would like to acknowledge the German Research Council (DFG), the European Union (IST-1999-13415-CHANIL, IST-2000-29321-APPTECH, and GRD1-2000-2559-MONALISA) and the Volkswagen Stiftung for their financial support.

References

[1] Chou S Y, Krauss P R, and Renstrom P J 1996 *J. Vac. Sci. Technol.* B **14** 4129–33
[2] Chou S Y and Krauss P R 1997 *Microelectron. Eng.* **35** 237–40
[3] Chou S Y, Krauss P R, and Renstrom P J 1995 *Appl. Phys. Lett.* **67** 3114–16
[4] Sotomayor Torres C M, Hoffmann T, Sidiki T P, Bruchhaus L, Bruch L U, Ahopelto J, Schulz H, Scheer H-C, Cardinaud C, and Pfeiffer K 2000 *Conf. Lasers and Electro-Optics (CLEO 2000) (San Francisco, CA)* p 211
[5] Behl M, Seekamp J, Zankovych S, Sotomayor Torres C M, Zentel R and Ahopelto J 2002 *Adv. Mater.* **14** 588–91
[6] McAlpine M C, Friedman R S and Lieber C M 2003 *Nano Lett.* **3** 443
[7] Grigaliunas V, Kopustinskas V, Meskinis S, Margelevicius M, Mikulskas I and Tomasiunas R 2001 *Opt. Mater.* **17** 15–8

[8] Jian W, Xiaoyun S, Lei C and Chou S Y 1999 *Appl. Phys. Lett.* **75** 2767–9

[9] Seekamp J, Zankovych S, Helfer A H, Maury P, Sotomayor Torres C M, Boettger G, Liguda C, Eich M, Heidari B, Montelius L and Ahopelto J 2002 *Nanotechnology* **13** 581–6

[10] Han A, Wang O, Mohanty S K, Graff M and Frazier B 2002 *IEEE EMBS Conf. Microtechnologies in Medicine and Biology (Madison, WI)* pp 66–70

[11] Montelius L, Heidari B, Graczyk M, Maximov I, Sarwe E-L and Ling T G I 2000 *Microelectron. Eng.* **53** 521–4

[12] Hirai Y, Fujiwara M, Okuno T, Tanaka Y, Endo M, Irie S, Nakagawa K and Sasago M 2001 *J. Vac. Sci. Technol.* B **19** 2811–15

[13] Gaboriau F, Peignon M C, Turban G, Cardinaud C, Pfeiffer K and Bleidiesel G 2000 *Microelectron. Eng.* **53** 501–5

[14] Xiaoyun S, Lei Z, Wei Z and Chou S Y 1998 *J. Vac. Sci. Technol.* B **16** 3922–5

[15] Zhang W and Chou S Y 2001 *Appl. Phys. Lett.* **79** 845–7

[16] Li M T, Chen L and Chou S Y 2001 *Appl. Phys. Lett.* **78** 3322–4

[17] Macintyre D S, Chen Y, Lim D and Thoms S 2001 *J. Vac. Sci. Technol.* B **19** 2797–800

[18] Kim Y S, Lee H H and Hammond P T 2003 *Nanotechnology* **14** 1140

[19] Morita T, Kometani R, Watanabe K, Kanda K, Haruyama Y, Kaito T, Fujita J, Ishida M, Ochiai Y and Matsui S 2002 *Int. Microprocesses and Nanotechnology Conf.* p 156

[20] Deshmukh M M, Ralph D C, Thomas M and Silcox J 1999 *Appl. Phys. Lett.* **75** 1631–3

[21] Brugger J *et al* 2000 *Microelectron. Eng.* **53** 403

[22] Chou S Y, Kaimel C and Gu J 2002 *Nature* **417** 835–7

[23] Jayatissa W, Alkaisi M M and Blaikie R J 2002 *Int. Microprocesses and Nanotechnology Conf. (Tokyo, Japan)*

[24] Zhang J, Tan K L and Gong H Q 2001 *Polym. Test.* **20** 693–701

[25] Zhang J, Tan K L, Hong G D, Yang L J and Gong H Q 2001 *J. Micromech. Microeng.* **11** 20–6

[26] Heidari B, Maximov I and Montelius L 2000 *J. Vac. Sci. Technol.* B **18** 3557–60

[27] Clavijo Cedeno C *et al* 2002 *Microelectron. Eng.* **62** 25–31

[28] Faircloth B, Rohrs H, Tiberio R, Ruoff R and Krchnavek R R 2000 *J. Vac. Sci. Technol.* B **18** 1866–73

[29] Scheer H-C, Schulz H and Lyebyedyev D 2001 *Proc. SPIE* **4349** 86–9

[30] Scheer H-C, Schulz H, Hoffmann T and Sotomayor Torres C M 2002 *Handbook of Thin Film Materials* vol 5, ed H-C Scheer *et al* (New York: Academic)

[31] Heyderman L J, Schift H, David C, Gobrecht J and Schweizer T 2000 *Microelectron. Eng.* **54** 229–45

[32] Chou S Y, Krauss P R, Zhang W, Guo L and Zhuang L 1997 *J. Vac. Sci. Technol.* B **15** 2897–904

[33] Gourgon C, Perret C and Micouin G 2002 *Microelectron. Eng.* **62** 385–92

[34] Lebib A, Natali M, Li S P, Cambril E, Manin L, Chen Y, Janssen H M and Sijbesma R P 2001 *Microelectron. Eng.* **58** 411–16

[35] Perret C, Gourgon C, Micouin G and Grolier J P 2002 *Japan. J. Appl. Phys.* **41** 4203–7

[36] Perret C, Gourgon C, Micouin G and Grolier J P E 2001 *Int. Microprocesses and Nanotechnology Conf. IEEE* pp 98–9

[37] Gourgon C, Perret C, Micouin G, Lazzarino F, Tortai J H, Joubert O and Grolier J P E 2003 *J. Vac. Sci. Technol.* B **21** 98

[38] Matsui S, Igaku Y, Ishigaki H, Fujita J, Ishida M, Ochiai Y, Namatsu H and Komuro M 2003 *J. Vac. Sci. Technol.* B **21** 688

[39] Zhang W and Chou S Y 2003 *Appl. Phys. Lett.* **83** 1632

[40] Alkaisi M M, Blaikie R J and McNab S J 2001 *Microelectron. Eng.* **58** 367–73

[41] Lebib A, Chen Y, Cambril E, Youinou P, Studer V, Natali M, Pepin A, Janssen H M and Sijbesma R P 2002 *Microelectron. Eng.* **62** 371–7

[42] Islam R, Wieder B, Lindner P, Glinsner T and Schaefer C 2002 *Int. Conf. on Sensors* **2** 931

[43] Michel B *et al* 2001 *IBM J. Res. Dev.* **45** 697–719

[44] Haatainen T and Ahopelto J 2003 *Phys. Scr.* **67** 357–60

[45] Chou S Y 2001 *Mater. Res. Soc. Bull.* **26** 512–17

[46] Pfeiffer K *et al* 2002 *Microelectron. Eng.* **62** 393–8

[47] Heidari B, Maximov I, Sarwe E-L and Montelius L 1999 *J. Vac. Sci. Technol.* B **17** 2961–4

[48] Schulz H, Osenberg F, Engemann J and Scheer H-C 2000 *Proc. SPIE* **3996** 244–9

[49] Borzenko T, Tormen M, Schmidt G, Molenkamp L W and Janssen H 2001 *Appl. Phys. Lett.* **79** 2246–8

[50] Martini I, Kuhn S, Kamp M, Worschech L, Forchel A, Eisert D, Koeth J and Sijbesma R 2000 *J. Vac. Sci. Technol.* B **18** 3561–3

[51] Pfeiffer K, Fink M, Bleidiessel G, Gruetzner G, Schulz H, Scheer H-C, Hoffmann T, Sotomayor Torres C M, Gaboriau F and Cardinaud C 2000 *Microelectron. Eng.* **53** 411–14

[52] Schulz H, Lyebyedyev D, Scheer H-C, Pfeiffer K, Bleidiessel G, Grutzner G and Ahopelto J 2000 *J. Vac. Sci. Technol.* B **18** 3582–5

[53] Natali M, Lebib A, Cambril E, Chen Y, Prejbeanu I L and Ounadjela K 2001 *J. Vac. Sci. Technol.* B **19** 2779–83

[54] www.microresist.de Microresist Technology GmbH (Köpenicker Strasse 325, D-12555 Berlin, Germany)

[55] Pfeiffer K, Fink A, Gruetzner G, Bleidiessel G, Schulz H and Scheer H-C 2001 *Microelectron. Eng.* **58** 381–7

[56] Pfeiffer K *et al* 2003 *Microelect. Eng.* **67-8** 266–73

[57] Sotomayor Torres C M *et al* 2003 *Mater. Sci. Eng.* C **23** 23

[58] Reif R, Fan A, Chen K-N and Das S 2002 *Proc. Int. Symp. Quality Electronic Design* p 33

[59] Friedrich J A and Neudeck G W 1989 *Electron. Device Lett.* **10** 144

[60] Matsumoto T, Satoh M, Sakuma K, Kurino H, Miyakawa N, Itani H and Koyanagi M 1998 *Japan. J. Appl. Phys.* **37** 1217

[61] Dragoi V, Lindner P, Tischler M and Schaefer C 2003 *Mater. Sci. Semicond. Process.* **5** 425

[62] Himmler A *et al* 2001 *Proc. SPIE* **4455** 221–30

[63] Doyle B, Boyanov B, Datta S, Doczy M, Hareland S, Jin B, Kavalieros J, Linton T, Rios R and Chau R 2003 *Symp. VLSI Technology and Circuits*

[64] Katz H E 1997 *J. Mater. Chem.* **7** 369–76

[65] Kim C, Burrows P E and Forrest S R 2000 *Science* **288** 831

[66] Wang J, Sun X, Chen L and Chou S Y 1999 *Appl. Phys. Lett.* **75** 2767

[67] Krauss P R and Chou S Y 1995 *J. Vac. Sci. Technol.* B **13** 2850–2

[68] Wei W, Bo C, Xiao Yun S, Wei Z, Lei Z, Linshu K and Chou S Y 1998 *J. Vac. Sci. Technol.* B **16** 3825–9

[69] Bo C, Wei W, Linshu K, Xiaoyun S and Chou S Y 1999 *J. Appl. Phys.* **85** 5534–6

[70] Linshu K, Lei Z and Chou S Y 1997 *IEEE Trans. Magnet.* **33** 3019–21

[71] Linshu K, Lei Z, Mingtao L, Bo C and Chou S Y 1998 *Japan. J. Appl. Phys.* **37** 5973–5

[72] Kong L, Pan Q, Li M, Cui B and Chou S Y 1998 *Annual Device Res. Conf.* pp 50–1

[73] Kong L, Pan Q, Cui B, Li M and Chou S Y 1999 *J. Appl. Phys.* **85** 5492–4

[74] Chen Y, Lebib A, Li S, Pepin A, Peyrade D, Natali M and Cambril E 2000 *Eur. Phys. J. Appl. Phys.* **12** 223–9

[75] Moritz J, Dieny B, Nozieres J P, Landis S, Lebib A and Chen Y 2002 *J. Appl. Phys.* **91** 7314–16

[76] Lebib A, Li S P, Natali M and Chen Y 2001 *J. Appl. Phys.* **89** 3892–6

[77] Moritz J, Landis S, Toussaint J C, Bayle Guillemaud P, Rodmacq B, Casali G, Lebib A, Chen Y, Nozieres J P and Dieny B 2002 *IEEE Trans. Magnet.* **38** 1731–6

[78] McClelland G M, Hart M W, Rettner C T, Best M E, Carter K R and Terris B D 2002 *Appl. Phys. Lett.* **81** 1483–5

[79] Chen Y, Lebib A, Li S P, Natali M, Peyrade D and Cambril E 2001 *Microelectron. Eng.* **58** 405–10

[80] Li S P, Lew W S, Bland J A C, Natali M, Lebib A and Chen Y 2002 *J. Appl. Phys.* **92** 7397

[81] Kido J, Harada G, Komada M, Shionoya H and Nagai K 1997 *Photonic and Optoelectronic Polymers, ACS Symposium Series* vol 672, ed J Kido J *et al* (Washington, DC: American Chemical Society) pp 381–94

[82] Kaneto K, Kudo K, Ohmori Y, Onoda M and Iwamoto M 1998 *IEEE Trans. Electron.* **E81C** 1009–19

[83] Roth S, Burghard M and Leising G 1998 *Curr. Opin. Solid State Mater. Sci.* **3** 209–15

[84] Shim H K and Jin J I 2002 *Polymers for Photonics Applications I* vol 158, ed H K Shim and J I Jin (Berlin: Springer) pp 193–243

[85] Thelakkat M 2002 *Macromol. Mater. Eng.* **287** 442–61

[86] Otsubo T, Aso Y and Takimiya K 2002 *J. Mater. Chem.* **12** 2565–75

[87] Mitschke U and Bauerle P 2000 *J. Mater. Chem.* **10** 1471–507

[88] Rothberg L J and Lovinger A J 1996 *J. Mater. Res.* **11** 3174–87

[89] Cheng X, Hong Y T, Kanicki J and Guo L J 2002 *J. Vac. Sci. Technol.* B **20** 2877

[90] Brabec C, Shaheen S, Fromherz T, Padinger F, Hummelen J, Dhanabalan A, Janssen R and Sariciftci N 2001 *Synthetic Met.* **121** 1517–20

[91] Shaheen S, Brabec C, Sariciftci N, Padinger F, Fromherz T and Hummelen J 2001 *Appl. Phys. Lett.* **78** 841-3

[92] Petritsch K and Friend R 1999 *Synthetic Met.* **102** 976

[93] Katz H E, Bao Z N and Gilat S L 2001 *Acc. Chem. Res.* **34** 359–69

[94] Feringa B 2001 *Acc. Chem. Res.* **34** 504–13

[95] Lawrence J R, Andrew P, Barnes W L, Buck M, Turnbull G A and Samuel I D W 2002 *Appl. Phys. Lett.* **81** 1955–7

[96] McGehee M D and Heeger A J 2000 *Adv. Mater.* **12** 1655–68

[97] Kranzelbinder G and Leising G 2000 *Rep. Prog. Phys.* **63** 729–62

[98] Scherf U, Riechel S, Lemmer U and Mahrt R F 2001 *Curr. Opin. Solid State Mater. Sci.* **5** 143–54

[99] Yang P *et al* 2000 *Science* **287** 465–7

[100] Ma H, Jen A K-Y and Dalton L 2002 *Adv. Mater.* **14** 1339–65
[101] Nguyen T P, Jolinat P, Destruel P, Clergereaux R and Farenc J 1998 *Thin Solid Films* **325** 175–80
[102] Sampietro M, Ferrari G, Natali D, Scherf U, Annan K O, Wenzl F P and Leising G 2001 *Appl. Phys. Lett.* **78** 3262–4
[103] Schaer M, Nuesch F, Berner D, Leo W and Zuppiroli L 2001 *Adv. Funct. Mater.* **11** 116–21
[104] Zou D C, Yahiro M and Tsutsui T 1997 *Synthetic Met.* **91** 191–3
[105] Kam A P, Aroca R, Duff J and Tripp C P 2000 *Langmuir* **16** 1185–8
[106] Aziz H, Popovic Z, Tripp C P, Hu N X, Hor A M and Xu G 1998 *Appl. Phys. Lett.* **72** 2642–4
[107] Aziz H, Popovic Z, Xie S, Hor A M, Hu N X, Tripp C P and Xu G 1998 *Appl. Phys. Lett.* **72** 756–8
[108] Pisignano D, Persano L, Visconti P, Cingolani R, Gigli G, Barbarella G and Favaretto L 2003 *Appl. Phys. Lett.* **83** 2545
[109] Wang J, Sun X Y, Chen L, Zhuang L and Chou S Y 2000 *Appl. Phys. Lett.* **77** 166–8
[110] Brabec C J, Sariciftci N S and Hummelen J C 2001 *Adv. Funct. Mater.* **11** 15–26
[111] Brabec C J and Sariciftci N S 2000 *Semiconducting Polymers* ed C J Brabec and N S Sariciftci (Wiley: Weinheim) pp 515–60
[112] Bulut F and Seekamp J 2002 unpublished results
[113] Maekela T, Haatainen T, Ahopelto J and Isolato H 2001 *J. Vac. Sci. Technol.* B **19** 487
[114] Austin M D and Chou S Y 2002 *Appl. Phys. Lett.* **81** 4431
[115] Austin M and Chou S Y 2002 *J. Vac. Sci. Technol.* B **20** 665–7
[116] Kim C, Shtein M and Forrest S R 2002 *Appl. Phys. Lett.* **80** 4051–3
[117] Li M T, Tan H, Chen L, Wang J and Chou S Y 2003 *J. Vac. Sci. Technol.* B **21** 660
[118] Engelbrecht R S, Gale M T and Knop K 1976 *Conf. Laser and Electrooptical Systems* p 14
[119] Knop K 1976 *Opt. Commun.* **18** 298–303
[120] Knop K and Gale M T 1978 *J. Photogr. Sci.* **26** 120–3
[121] Wang J, Schablitsky S, Yu Z N, Wu W and Chou S Y 1999 *J. Vac. Sci. Technol.* B **17** 2957–60
[122] Yu Z N, Deshpande P, Wu W, Wang J and Chou S Y 2000 *Appl. Phys. Lett.* **77** 927–9
[123] Traut S, Rossi M and Herzig H P 2000 *J. Mod. Opt.* **47** 2391–7
[124] Chang A S P, Wei W and Chou S Y 2001 *LEOS-2001-IEEE* **2** 584–5
[125] Li M T, Wang J A, Zhuang L and Chou S Y 2000 *Appl. Phys. Lett.* **76** 673–5
[126] David C, Haberling P, Schnieper M, Sochtig J and Zschokke C 2002 *Microelectron. Eng.* **62** 435–40
[127] Yu Z N, Wu W, Chen L and Chou S Y 2001 *J. Vac. Sci. Technol.* B **19** 2816–19
[128] Park S J, Cho K S and Choi C G 2003 *J. Colloid and Interface Sci.* **258** 424
[129] Chao C Y and Guo L J 2002 *J. Vac. Sci. Technol.* B **20** 2862
[130] Kalveram S and Neyer A 1997 *Proc. SPIE* **3135** 2–11
[131] Paatzsch T, Smaglinski I, Bauer H D and Ehrfeld W 1998 *Proc. SPIE* **3276** 16–27
[132] Gale M T, Baraldi L G and Kunz R E 1994 *Proc. SPIE* **2213** 2–10
[133] Lehmacher S and Neyer A 2000 *Electron. Lett.* **36** 1052–3
[134] Zankovych S, Seekamp J, Maka T, Sotomayor Torres C M, Böttger G, Liguda C, Eich M, Heidari B, Montelius L and Ahopelto J 2001 *Workshop on Nanoimprint Lithography (Lund, Sweden)*

[135] Pang S W, Tamamura T, Nakao M, Ozawa A and Masuda H 1998 *J. Vac. Sci. Technol.* B **16** 1145–9

[136] Hewett J 2002 *Optics.org-news* http://optics.org/article/news/8/5/14

Chapter 7

Carbon nanotubes interconnects

B O Bošković and J Robertson
Department of Engineering, Cambridge University

7.1   Introduction

Carbon nanotubes (CNTs) can behave like metals or semiconductors, can conduct electricity better than copper, can transmit heat better than diamond and they rank among the strongest materials known. The electronic properties of the resulting nanotube depend on the direction in which the sheet was rolled up. Some nanotubes are metals with high electrical conductivity, while others are semiconductors with relatively large band gaps. In several decades from now, we may see integrated circuits with components and wires made from nanotubes.

A nanotube can be considered as a single sheet of graphite that has been rolled into a tube and capped on each end with half of a fullerene molecule. This is termed a single-wall carbon nanotube (SWNT). If nanotubes are nested inside each other like Russian dolls, researchers then call them multi-walled CNTs (MWNTs). Some MWNTs grown by the chemical vapour deposition (CVD) method have a very similar structure to carbon nanofibres (CNFs), also known as filaments, which usually have catalyst particles attached to one end and graphene layers adopting a herringbone or stacked cup arrangement.

The discovery of the C_{60} molecule in 1985 (also known as fullerene or buckminsterfullerene) marked the beginning of a new era in carbon science [1]. In 1991, Iijima reported the first observation of structures that consisted of several concentric tubes of carbon, nested inside each other like Russian dolls (figure 7.1) [2]. He called them microtubules of graphitic carbon but, from 1992, Iijima and other researchers began to refer to them as CNTs. He observed MWNTs in the soot produced by an electric arc discharge between graphite electrodes in a helium atmosphere in a very similar apparatus to the one used for the production of large quantities of C_{60} by Kratshmer *et al* [3]. Subsequently, in 1993, Iijima and Ichihashi [4] and Bethune *et al* [5] independently discovered SWNTs. The

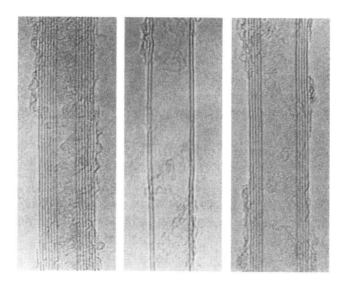

Figure 7.1. The first image of multiwalled carbon nanotubes published by Iijima in 1991 [2]. Copyright Nature Publishing Group.

diameters of SWNTs are 1 or 2 nm, compared to MWNT diameters of the order of tens of nanometres. In the following years, a considerable effort was invested in finding efficient means for producing large quantities of nanotubes. In 1992, Ebbesen and Ajayan [6] developed a method of producing larger quantities (a few grams) of high-quality MWNTs by vaporizing carbon electrodes. Since then other CNT synthesis methods have been developed. CNTs have been produced by vaporizing a graphite target using a laser [7, 8] and electron beam [9]. Catalytic CVD of hydrocarbons [10] is now widely used for CNT synthesis as a simple and efficient method. The catalytic CVD method is now regarded in both the scientific and business community as the best method for the low-cost production of large quantities of CNTs. The plasma-enhanced CVD (PECVD) process allows low-temperature growth of vertically aligned carbon nanofibres on a patterned catalyst [11]. CNTs produced using the CVD method may have more structural defects compared to nanotubes produced by arc-discharge [2, 6, 12] or laser vaporization [7, 8]. However, a perfect CNT structure is not important for many applications.

By the end of the 1990s, applications of CNTs and nanofibres were attracting extensive interest. CNTs have proven to be good field emitters [13, 14]. In the past few years, research aimed at obtaining large-area films of nanotubes or nanofibres producing a uniform field emission has been carried out [13–17]. CNTs have now been used for STM tips [18] as conducting fillers in polymer composite materials [19], and in fuel cells [20]. It has been proven that CNTs can be used in nanoelectronics as diodes and transistors [21], in supercapacitors [22], as electromechanical actuators [23] and as chemical sensors [24].

Experimental observation of the CNT with diameter 0.8 nm, corresponding to the C_{60} molecule was reported in 1992 by Ajayan and Iijima [25]. A single-wall C_{60}-derived CNT consists of a bisected C_{60} molecule with the two resulting hemispheres joined together by a cylindrical tube one monolayer thick and with the same diameter as C_{60}. If the C_{60} molecule is bisected normal to the fivefold axes, an 'armchair' nanotube, is formed and if the C_{60} molecule is bisected normal to the threefold axis, a 'zigzag' nanotube is formed [26]. In addition to armchair and zigzag nanotubes, a large number of chiral nanotubes can be formed with a screw axis along the axis of the nanotube and a variety of 'hemispherical'-like caps. Armchair and zigzag CNTs with larger diameters have correspondingly larger caps. Structures with smaller caps, for example C_{36}, with a diameter of 0.5 nm [27], and C_{20}, with a diameter of 0.4 nm, have been reported [28, 29].

7.2   Synthesis of CNTs

The first CNTs identified by Iijima in 1991 were synthesized using a carbon arc-discharge [2]. Since then a number of other methods for CNT growth have been developed including vaporizing a graphite target using a laser [7, 8], and electron-beam [9] sources. In addition, the production of relatively imperfect CNTs by catalytic CVD methods at high temperatures has been known for decades [30].

The arc-evaporation apparatus used by Iijima for CNT synthesis in 1991 [2] was very similar to that used in the Kratschmer–Huffman experiment for the mass production of C_{60} one year earlier [3]. A variety of different arc-evaporation reactors have been used for nanotube synthesis since Iijima's report. In the original method used by Iijima, graphite electrodes were held a short distance apart, rather than being kept in contact as in the Kratschmer–Huffman experiment. Under these conditions, some of the carbon that evaporated from the anode re-condensed as a hard deposit on the cathodic rod. CNTs were found to grow on the negative end of the carbon electrode, plentifully on only certain regions of the electrode. The carbon deposits on the electrode contained CNTs, ranging from 4 to 30 nm in diameter and up to 1 μm in length, and also polyhedral particles with a spherical shell structure, which were 5–20 nm in diameter.

Subsequent modifications to the procedure by Ebbesen and Ajayan [6] have enabled greatly improved yields to be obtained by arc-evaporation. They reported the synthesis of graphitic nanotubes in gram quantities, demonstrating that purity and yield depend on the helium pressure in the chamber. A striking increase in the number of tubes was observed as the pressure was increased and at 500 Torr, the total yield of nanotubes as a proportion of starting graphitic material was optimal. Under these conditions, approximately 75% of the consumed graphite rod was converted into a deposit of cylindrical shape and similar diameter. Journet *et al* [12] reported 80% estimate of the yield using an electric-arc technique.

CNTs and nanofibres can be synthesized efficiently by the catalytic decomposition of a reactant gas that contains carbon, when made to flow over a transition metal catalyst, CVD. This process has two main advantages. First, the nanotubes are obtained at a much lower temperature; however, this can be at the cost of lower structural quality. Second, the catalyst can be grown on a substrate, which allows the growth of aligned nanotubes and control of their growth. Nanotubes can be grown on surfaces with a degree of control that is unmatched by arc-discharge or laser-ablation techniques. The defective nature of CVD-grown MWNTs remains poorly understood. This may be due to the relatively low temperatures that are used, which do not allow the nanotube walls to crystallize fully.

CNT production using a thermal CVD process is very similar to the method used for CNF production developed and studied in the 1970s and 1980s [31, 32]. Catalytically produced CNTs using the CVD process are similar to the CARBON FIBRILS^{TM} patented by Hyperion Catalysis International in 1989 [33].

The first researchers who reported carbon nanofibre formation in the early 1950s [34, 35] established that CNFs had metal particles associated with them. Catalytic 'ice-cream-cone' shaped metal particles can usually be seen at the end of the nanofibre. Alternatively, 'diamond' shaped metal catalyst particle can sometimes be seen in the middle of the nanofibre. However, the structure of the CNFs could not be determined until more powerful transmission electron microscopes (TEMs) were in use. High-resolution TEM (HRTEM) microscopy studies by Rodriguez *et al* [36] have revealed that the nanofibres consist of a well-ordered graphite platelet structure, the arrangement of which can be engineered to the desired geometry by choice of the correct catalyst system. Depending on the chemical nature of the catalyst and the conditions of the reaction, assorted nanofibre structures with various morphologies and different degrees of crystallinity can be produced (figure 7.2). Nanofibres consist of graphite platelets, with 0.34 nm distance between the planes, that can be oriented in various directions with respect to the fibre axis.

CNTs and nanofibres can be synthesized using plasma enhanced CVD (PECVD) where the hydrocarbon gas is in an ionized state over the transition metal catalyst (nickel, iron, cobalt, etc). The electrical self-bias field from plasma is sufficient to induce aligned CNT growth perpendicular to the substrate. The PECVD method can use different energy sources for the creation of the plasma state and allows lower deposition temperatures than thermal CVD. Hot-filament PECVD uses thermal energy for plasma creation and has been used successfully for CNT production by Ren and co-workers [11]. Microwave PECVD, widely used for the preparation of diamond films, has also been successfully used in the production of CNTs and CNFs [37–41]. Synthesis of vertically aligned CNTs and CNFs requires an electric field normal to the substrate and dc PECVD is the most suitable method for achieving this [42, 43]. Inductively coupled plasma PECVD [44, 45] and radiofrequency PECVD [46, 47] methods have also been used successfully for CNTs and CNFs synthesis.

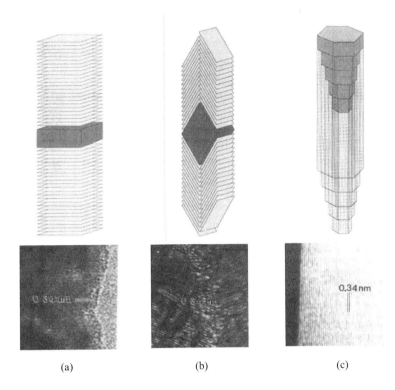

Figure 7.2.   Schematic diagram showing the structural relationship between the catalyst particle and precipitated graphite platelets, on the top row and corresponding high-resolution micrograph: (*a*) iron catalyst with CO/H_2 (4:1) at 600 °C; (*b*) iron-copper (7:3) catalyst with C_2H_2/H_2 (4:1) at 600 °C; (*c*) silica supported iron catalyst with CO/H_2 at 600 °C (after Rodriguez *et al* [36]). Reprinted with permission from [36].

Ren *et al* in 1998 [11] reported the first successful growth of large-scale well-aligned CNFs on nickel foils and nickel-coated glass at temperatures below 666 °C. Acetylene gas was used as the carbon source and mixed with ammonia (ratio from 1:2 to 1:10) in a plasma-enhanced hot-filament CVD (PE-HFCVD) system. CNF films were grown at pressures of 1–20 Torr maintained by flowing acetylene and ammonia gases with total flow rate of 120–200 sccm. They concluded that plasma intensity, acetylene-to-ammonia gas ratio and their flow rates affect the diameter and uniformity of the CNFs. A spherical ball of Ni was found on the tip of each nanofibre.

Bower *et al* [40] have grown well-aligned CNTs using microwave PECVD with an additional radiofrequency graphite heater. They found that switching the plasma source off effectively turns the alignment mechanism off, leading to the thermal growth of curly nanotubes.

Merkulov *et al* [42] reported the synthesis of a vertically aligned CNFs on a patterned catalyst using dc PECVD. The catalyst patterns were fabricated using conventional electron-beam lithography. Acetylene (C_2H_2) and a mixture of 10% ammonia (NH_3) and 90% helium (He), with gas flows of 15 and 200 sccm, respectively, were used as the gas source. They used ammonia to etch away amorphous carbon (a-C) film that was continuously formed during the growth. The shape of the CNFs depends on how much growth occurs at the tip by catalysis and how much by deposition of a-C from the plasma along the sidewalls [48]. This ratio is controlled by the catalyst activity and by the balance of deposition and etching of a-C. The balance between deposition and etching depends on the plasma and the etchant (NH_3) and hydrocarbon gas (C_2H_2). This balance has been studied by Merkulov *et al* [42] and Teo *et al* [49].

The PECVD method allows the growth of CNTs and CNFs at low temperatures suitable for use with temperature-sensitive substrates. A radiofrequency PECVD CNFs synthesis at room temperature has been reported by Boskovic *et al* [47]. Using dc PECVD, Hofmann *et al* [50] demonstrated the synthesis of aligned CNFs at temperatures as low as 120 °C and on plastic substrates [51].

The key parameters involved in growing nanotubes using CVD are the types of hydrocarbons and catalysts used and the temperature at which the reaction takes place. Most of the CVD methods used to grow MWNTs and CNFs use ethylene or acetylene gas as the carbon feedstock and iron, nickel or cobalt nanoparticles as the catalyst. The growth temperature is typically in the range 650–750 °C for thermal CVD and much lower for PECVD. Hydrocarbon dissociates on the metal nanoparticles, the carbon diffuses across the metal particles, which eventually become saturated. The carbon then precipitates to form solid carbon tubes, the diameters of which are determined by the size of the metal particles in the catalyst. CNTs can grow via the 'tip' growth mechanism where the catalyst particle is lifted off the support and on the tip of the CNT/CNF or via a 'base' or 'root' growth mechanism where the catalyst particle remains attached to the support. Tip growth occurs when the catalyst-support interaction is weak. The activation energy for thermal CVD growth of CNFs on nickel catalyst, reported to be 1.5 eV [52] and 1.21 eV [53], is similar to the activation energy of carbon bulk diffusion in nickel, 1.5 eV [54]. Hofmann *et al* [50] have found that PECVD growth of CNFs at low temperatures on nickel catalyst is described by a much lower activation energy of 0.23 eV, corresponding to that for surface diffusion of carbon on nickel (0.3 eV) [55]. These results suggested that carbon surface diffusion is the dominant process for low-temperature CNF growth using PECVD.

Carbon filament (nanofibre) growth was studied by many researchers in the 1970s and several growth models were proposed, Baker *et al* [31, 56], Oberlin *et al* [57], Rostrup-Neilsen and Trimm [58]. These models have been adopted for CNT CVD growth. However, no complete mechanism which can explain the various, and often conflicting data has yet been proposed. Baker *et al* [56] developed the most influential model for CNF growth as a result of direct growth observations using controlled atmosphere microscopy assuming that the diffusion flow was primarily driven by the temperature gradient. Rostup-Nielsen and Trimm [58], introduced a concentration-driven mechanism for the diffusion of carbon through the metal instead of the temperature-driven process proposed by Baker *et al* [56]. Evidence for the CNF growth mechanism was obtained recently using time-resolved, high-resolution *in situ* TEM observation of the formation of CNFs from methane decomposition over a supported nickel nanocrystral [59]. CNFs were observed to develop through a reaction-induced reshaping of the nickel nanocrystals. The nucleation and growth of graphene layers was found to be assisted by the dynamic formation and restructuring of mono-atomic step edges at the nickel surface.

Crystalline ropes of metallic SWNTs were synthesized for the first time in 1996 by the condensation of a laser-vaporized carbon–nickel–cobalt mixtures at 1200 °C [60]. The nanotubes were nearly uniform in diameter and self-organized into 'ropes', which consist of 100–500 SWNTs. The ropes were metallic, with a single-rope resistivity of $<10^{-4}$ Ω cm at room temperature. They estimated that typically 70–90% of the material was SWNT ropes and the ends of the ropes could not be found in the TEM images due to their long length. This 'self-orientation' of the nanotubes into ropes is due to strong Van der Waals interactions between the tubes, which cause the nanotubes to bundle together and form rigid rope structures.

In 1998, Cheng *et al* [61], Kong *et al* [62] and Hafner *et al* [63] reported production of SWNTs using the CVD of hydrocarbons. Kong *et al* [62] used methane gas as the carbon source and temperatures in the range of 900–1000 °C. Such high temperatures are needed to form SWNTs that have small diameters and high strain energies, and to produce virtually defect-free tubes. Long and wide ropes of SWNT bundles with rope diameters of 100 μm and lengths up to 3 cm, synthesized by the catalytic decomposition of hydrocarbons, were described by Cheng *et al* [61]. A process of SWNT production by flowing CO with a small amount of Fe(CO)$_5$ through a reactor at pressures 1–10 atm and temperatures 800–1200 °C, called 'HiPco', is now commercially exploited by Carbon Nanotechnologies Inc [64, 65].

A nucleation mechanism for SWNT has recently been proposed by Fan *et al* [66] where, in the presence of the metal surface, nucleation of a SWNT cap is favoured in order to minimize the number of the dangling bonds. Using first-principle calculations, they show that the introduction of the pentagons, on the surface of the Ni metal, to form the closed tube is energetically proffered over a graphine sheet, fullerene or an open tube. They also showed that nanosized clusters are not essential for the nucleation of SWNTs and that nucleation occurs on the larger metal particles.

The size of the metal catalyst nanoparticle generally dictates the diameter and the growth mechanism of SWNTs. A differently sized nanocluster catalysts has been used to demonstrate diameter control of the CVD-grown SWNTs [67, 68]. When the size of the nanoclusters is below 5 nm, particles corresponding

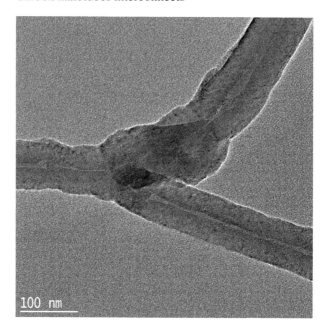

Figure 7.3. A TEM image of a Y-shaped CNF interconnecting junction synthesized using radiofrequency-supported MW PECVD at room temperature.

to the nanotube diameter were found in the tip of the SWNTs [69] suggesting a tip-growth mechanism similar to the one proposed for CNFs and MWNTs. Gavillet *et al* [70] investigated the nucleation and growth of SWNTs from catalyst particles larger than the tube diameter synthesized using different techniques. They concluded that, in this case, SWNTs growth occurs via root growth mechanism.

The synthesis of connections between two or more different CNTs or CNFs (figure 7.3) is important for the development of CNT-based electronic devices and conducting structures [71–75]. Branched CNTs have been synthesized using pyrolysis at temperatures of around 1000 °C [76, 77]. Multijunction CNTs networks have also been synthesized using pyrolysis method at 1100 °C [78]. Using an alumina template with branched channels produced by etching [72], Y-shaped junctions of CNTs were synthesized by CVD at 650 °C. For CNTs synthesized using radiofrequency PECVD at around 650 °C, nanojunction formation was observed after the application of a high electric field [79].

7.3   Carbon nanotube properties

The remarkable electrical and mechanical properties of CNTs make them excellent candidates for the range of electrical, mechanical and electromechanical applications. A review of the properties of CNTs is given in [80].

7.3.1   Electrical properties

The unusual electronic structure of a graphene layer made up of sp^2 bonded carbon atoms is the origin of the remarkable electrical properties of CNTs. The electronic properties of nanotubes depend on the direction in which the graphene sheet is rolled up, the so-called chirality.

Before measuring the conducting properties of nanotubes, it is necessary to attach metal electrodes to the tube. The electrodes, which can be connected to either a single tube or a bundle of up to several hundred tubes, are usually made using electron-beam lithography. The tubes can be attached to the electrodes in a number of different ways. One way is to make an electrode on a substrate and then drop a dilute dispersion of the tubes onto the electrode. Another is to deposit the tubes on a substrate, locate them with a scanning electron microscope (SEM) or atomic force microscope (AFM) and then deposit leads to the tubes using lithography. More advanced techniques are also being developed to make device fabrication more reproducible and controllable. These include the possibility of growing the tubes between the electrodes.

In semiconducting materials, a small amount of an impurity added as a dopant can make it n- or p-type. A junction between a p- and n-type semiconductor acts as a diode. A junction can also be formed between a semiconductor and a metal. In the case of CNTs, they can be either metallic or semiconducting, depending on the chirality. Hence, experimentally observed diode properties can be explained by the presence of the junction between two electronically different nanotubes [81].

CNTs have shown the behaviour of p-type semiconductors. One might expect an isolated semiconducting nanotube to be an 'intrinsic' semiconductor, where only excess electrons would be created by thermal fluctuations alone. However, it is now believed that the metal electrodes, as well as chemical species adsorbed on the tube, 'dope' the tube p-type. In other words, they remove electrons from the tube, leaving the remaining mobile holes responsible for conduction.

Lieber *et al* [82] carried out one of the first conductivity measurements of individual nanotubes. To measure the electrical properties of individual CNTs, they deposited a drop of a nanotube suspension on a flat insulating surface and covered it with a uniform layer of gold. A pattern of open slots was produced in the gold layer by conventional lithography procedures to expose the nanotube for measurement. AFM studies showed that, after this procedure, many of the single nanotubes have one end covered by the gold pattern and the other end

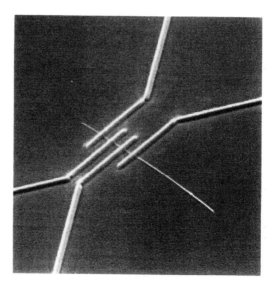

Figure 7.4. Focused-ion-beam image of four tungsten wires, each 80 nm wide, connected to an individual nanotube for four-probe measurement, from the work of Ebbesen *et al* [83]. Copyright Nature Publishing Group.

extended into an open slot. With a conduction cantilever in the AFM, it was possible to establish electric contacts and measure the axial conduction through a single nanotube to the gold contacts while simultaneously recording the nanotube structure. Depositing a layer of NbN onto commercial Si_3N_4 cantilevers made the tips conducting. The nanotubes used in this study were prepared catalytically and samples consisted primarily of MWNT with diameters between 7 and 20 nm. Measurements were made on nanotubes with diameters of 8.5 and 13.9 nm and their resistance per unit length was 0.41 and 0.06 MΩ μm^{-1}, respectively.

In order to avoid possible ambiguities due to poor sample contacts, Ebbesen *et al* [83] have developed a four-probe measurement method on single nanotubes by lithographic deposition of tungsten leads across the nanotubes (figure 7.4). They found that each MWNT has unique conductivity properties and that the differences between the electrical properties of different nanotubes were far greater than expected. Both metallic and semiconducting behaviour was observed on arc-grown nanotubes. However, they found that the tubes were essentially metallic. By analogy with carbon fibres and filaments [84], the variations in resistivities and their temperature dependence are possible due to the interplay of changes in carrier concentration and mobility in the metallic tubes.

Resistance measurements were reported for eight different nanotubes and the results differed widely. The highest resistance per unit length was 10^8 Ω μm^{-1} with a 0.8 Ω cm resistivity for the 10 nm diameter nanotube, and lowest was

$2 \times 10^2 \ \Omega \ \mu m^{-1}$ and $4 \times 10^{-3} \ \Omega$ cm resistivity for the 18.2 nm diameter nanotube. Most of the nanotubes had a resistance per unit length of around $10^4 \ \Omega \ \mu m^{-1}$ and a resistivity of around $10^{-4} \ \Omega$ cm. They also investigated the temperature dependence of the conductivity of various nanotubes and the results were widely different. Finally, they examined the effect of applying a magnetic field perpendicular to the axis of the tubes but this effect was found to be negligible.

The conductivity of an individual SWNT was studied by Tans *et al* [85]. Because of the structural symmetry and stiffness of SWNT, their molecular wavefunctions may extend over the entire tube. These researchers have found that electrical transport measurements on individual SWNTs confirm theoretical predictions and that they appear to behave as coherent quantum wires. Electrical conduction seems to occur through well-separated discrete electron states that are quantum mechanically coherent over long distances that are at least 140 nm from contact to contact. The two-point resistance at room temperature of a single tube was generally found to be around 1 MΩ.

Traditional electronic devices are based on classical electron diffusion. The size of the devices, at nanometre level, becomes comparable to the electron coherence length and quantum interference between electron waves starts to influence the device properties. CNTs are promising candidates for exploiting quantum effects for the benefit of future nanoelectronic devices.

Experimental [85–95] and theoretical studies [81, 86, 96] have indicated that CNTs behave as 1D ballistic conductor with quantized conductance. When the length of the conductor is smaller than the electronic mean path, then electron transport is ballistic, in which case each transverse wave guide mode or conducting channel has a quantum resistance of $h/2e^2 \approx 4 \ k\Omega$ [96,97], where h is Planck's constant. In the case of ballistic transport, there is no energy dissipation in the conductor and the Joule heat is dissipated in the electrical leads, which connect the ballistic conductor to the elements of the circuit. A theoretical study done by Chico *et al* [81] has shown that SWNTs have two conducting channels, from which a conduction of $2G_0 = 4e^2/h$, where G_0 is the unit quantum conductance $G_0 = 2e^2/h$, is predicted

Frank *et al* [85] have found that MWNTs are ballistic conductors. Each layer of the MWNT is closed and disconnected from every other layer. Hence, the top layer is the only one contributing to the conductance. McEuen *et al* [87] suggested that a simple diffusive conductor model can be used to explain experimental single-nanotube conductance results. Transport through the nanotube is limited by scatterers spaced at the distance l, each with transmission probability of $\frac{1}{2}$. The conductance of a tube length L is then $G \approx (4e^2/h)(l/L)$. Liang *et al* [94] made more than 100 devices from metallic and semiconducting SWNTs. Most of the metallic SWNT devices have shown room-temperature resistance below 100 kΩ, and more than 20 devices exhibited a resistance below 15 kΩ. The lowest resistance values observed in some metallic SWNT devices were 7 kΩ, approaching the theoretical limit for nanotube device with perfect ohmic contacts and indicating that electron transport is ballistic.

The first report of SWNT rope synthesis by Colbert *et al* [98] also reported the first measurement of the electrical properties of ropes of SWNT. They measured the electrical resistivity of a single rope using a four-point technique. They calculated resistivity values for different ropes ranged from 0.34×10^{-4} to 1×10^{-4} Ω cm. Bockrath *et al* [88] observed dramatic peaks in the conductance as a function of the gate voltage that modulated the number of electrons in the rope and interpreted these results as the effect of single-electron charging and resonant tunnelling through the quantized levels of the nanotubes composing the rope. Inter-tube conductance inside the rope was small compared to the conductance along the tube, inhibiting inter-tube transport.

7.3.2 Mechanical properties

In a sheet of graphite, each carbon atom is strongly bonded to three other atoms, which makes graphite very strong in certain directions. However, adjacent sheets are only weakly bonded by Van der Waals forces. Layers of graphite can be easily peeled apart, as happens when writing with a pencil but it is not easy to peel a carbon layer from a MWNT. Carbon fibres have already been used to strengthen a wide range of materials and the special properties of CNTs mean that they could be the ultimate-strength fibre. CNTs' remarkable mechanical properties have already been exploited as 'tips' in scanning probe microscopes. Since they are composed entirely of carbon, nanotubes also have a low specific weight. Nanotubes also offer great promise as the active elements in 'nano-electromechanical' systems. Their remarkable mechanical and electronic properties make them excellent candidates for applications such as high-frequency oscillators and filters.

In 1996, Treacy *et al* [99] measured the Young's modulus of MWNTs. Young's modulus of a material is a measure of its elastic stiffness. They arranged MWNTs vertically on a surface so that the tubes were fixed at the bottom and free to move at the top, and then used a TEM to measure the thermal vibration frequency of the free ends. The measured vibration amplitude revealed an exceptionally high Young's modulus of 10^{12} N m^{-2}, about five times the value for steel. The carbon–carbon bonds within the individual layers mainly determine Young's modulus. Salvetat *et al* [100] found that MWNTs grown by arc discharge had a modulus of one or two orders of magnitude greater than those grown by catalytic CVD of hydrocarbons. These results demonstrate that only highly ordered and well-graphitized nanotubes have a stiffness comparable to graphite, whereas those grown by catalytic decomposition have many more defects and, hence, a lower Young's modulus. The strong carbon–carbon bonds within each layer characterize well-graphitized nanotubes, while the interactions between layers are weak. TEM images of CVD-grown nanotubes indeed reveal that the carbon sheets are neither continuous nor parallel to the tube axis.

Ruoff *et al* [101] have developed and built a manipulation tool that can also be used as a mechanical loading device, which operates inside an SEM. Individual MWNTs were picked up and then attached at each end of opposing tips of AFM cantilever probes. Nanotubes were stress-loaded *in situ* in the SEM and observations were recorded on video. Measurements of tensile strength for individual MWNT has revealed a 'sword-in-sheath' breaking mechanism, similar to that observed for carbon fibres [102]. Tensile strengths of up to 20 GPa were reported for graphite whiskers [103], which were stated to have a scroll-like structure rather then the 'Russian doll' structure (nested cylinders) observed in MWNTs. The tensile strength of the outermost layer ranged from 11 to 63 GPa for the set of 19 MWNTs. Analysis of the stress–strain curves indicated that Young's modulus for this layer varied from 270 to 950 GPa. This work was further extended to measure the strength of individual MWNT [103, 104] to SWNT ropes [105]. They measured the mechanical response of 15 SWNT ropes under a tensile load and concluded that the force–strain data fit well to a model that assumes the load is carried by the SWNT on the perimeter of each rope. The average breaking strength of the SWNT in the perimeter of each rope ranged from 13 to 53 GPa, with a mean value of 30 GPa. Based on the same model, a Young's modulus ranging from 320 to 1470 GPa, with mean value of 1002 GPa was calculated.

MWNTs can be bent repeatedly through large angles using the AFM tip, without undergoing catastrophic failure suggesting that nanotubes are remarkably flexible and resilient. Folvo *et al* [106] used the Nanomanipulator AFM system to produce and record nanotube translations and bends by applying lateral stress at location along the tube. These fascinating mechanical properties of CNTs can be exploited in applications, which might include lightweight bullet-proof vests and earthquake-resistant buildings, while nanotube tips for scanning probe microscopes are already commercially available. There are a number of problems to overcome before starting to utilize nanotubes for reinforcement applications. The properties of nanotubes need to be optimized. Tubes must be efficiently bonded to the material they are reinforcing (the matrix), in order to maximize load transfer. New composite materials have been made from various polymers and CNTs. Some of these materials have shown improved mechanical or electrical properties when compared to polymers not containing nanotubes.

7.4   Electronic applications of carbon nanotubes

The age of semiconductor technology started in 1947, just half a century ago, when the first semiconductor device, a germanium-based transistor, was invented [107]. In 1965, Moore [108] observed an exponential growth of number of transistors per integrated circuits and predicted that this trend will continue. Since then, the miniaturization of devices has been continuous with exponential growth in the number of transistors and computers have become faster and smaller.

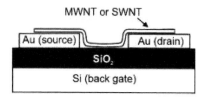

Figure 7.5. Schematic cross section of the field emission transistor device. A single nanotube which bridges the gap between the gold electrodes could be either multi-walled CNT (MWNT) or single-walled CNT (SWNT). The silicon substrate is used as a back gate [111]. Reprinted with permission from [111].

At the present pace of miniaturization, it is expected that the end of the path could be reached within a decade. In order to overcome this technological limit, several types of devices that make use of quantum effects rather than trying to overcome them are being investigated. For this reason, the nanometre-scale carbon materials, namely the fullerenes and nanotubes, have attracted great interest not only in the scientific fields but also in the fields of semiconductor technology.

In general, there are two types of elemental device structures: two-terminal and three-terminal devices. The transistor is a three-terminal device with a variety of structures, materials and basic functional mechanisms. A typical two-terminal device is the diode, which also has a variety of structures and applications, such as switching, rectification and solar cells. The first 'nanotube nanodevice' reported by Collins *et al* [109] was a type of nanodiode.

Semiconducting nanotubes can work as transistors. The tube can be made to conduct (turned ON) by applying a negative bias to the gate and turned OFF with positive bias. A negative bias induces holes on the tube and allows it to conduct. Positive biases, in contrast, deplete the holes and decrease the conductance. The resistance of the OFF state can be more than a million times greater than the ON state. This behaviour is analogous to that of a p-type metal–oxide–silicon field effect transistor (MOSFET), except that the nanotube replaces silicon as the material that hosts the charge carriers. It has been shown that an FET can be made from individual semiconducting SWNTs by Tans *et al* [110] and later by Martel *et al* [111] (figure 7.5).

Individual SWNTs could be jointed together to form multi-terminal devices. Junctions of two SWNT, one with metallic and one with semiconduction properties, were fabricated with an electrical contact on each end by Fuhrer *et al* [112]. A semiconducting nanotube was depleted at the junction by the metallic nanotube, forming an Schottky barrier. Rueckes *et al* [113] fabricated non-volatile random access memory based on CNTs. Device elements formed from suspended crossed nanotubes were switchable between ON and OFF states and addressable in large arrays by the CNT molecular wires making up the device. Logical circuits containing one, two and three single nanotube FETs were demonstrated performing a range of digital operations, such as inverter, a logic NOR, a static random-access memory cell and an ac ring oscillator by Bachtold *et al* [114]. Deryche *et al* [115] have shown that an n-type FET from CNTs could be changed into a p-type FET by simple annealing in vacuum. By bonding together a p- and n-FET, they built a logic gate based on CNT transistors, a NOT gate or voltage inverter. After these advances in CNT device fabrication, IBM announced that nanotube electronics would be realized in about decade [116]. However, further advances reported recently could provide additional reassurance about CNT-based electronics.

The presence of a Schottky barrier at the nanotube–metal contact junction is the common feature for all SWNT FETs. Javey *et al* [117] reported that with Pd contacts, the ON states of semiconducting nanotubes could behave like ohmically contacted ballistic metallic nanotubes, exhibiting room-temperature conductance near the ballistic transport limit of $4e^2/h$. For a long semiconducting SWNT device (length $L = 3$ μm, diameter $d = 3$ nm) with back gates (SiO$_2$ thickness $t_{ox} = 500$ nm) the room-temperature ON state resistance through the valance band is 60 kΩ. For short channel nanotubes at room temperature ($L = 300$ nm), the resistance was 13 kΩ.

Metallic SWNTs could function as interconnects in nanoscale circuits and semiconducting ones could perform as nanoscale Schottky-type FETs. Separation of metallic and semiconducting CNTs is an enabling step for electronic applications. Recently, two methods have been reported. The first method, reported by Krupke *et al* [118], used ac dielectrophoresis to separate metallic and semiconductiong nanotubes. The second method was based on molecular recognition between DNA and SWNT [119]. Zheng *et al* [120] found that wrapping CNTs with single-stranded DNA is sequence-depended. They reported the discovery of an oligonucleotide sequence that self-assembles into a highly ordered structure on CNTs, allowing not only the separation of semiconducting and metallic nanotubes but also a diameter-depended separation.

The precise localization and interconnection of CNTs are requirements for further advancements in CNT-based electronics. Keren *et al* [121] reported the realization of a self-assembled CNT FET using DNA to provide the address for the precise localization of the semiconducting SWNT as well as the template for the extended metalic wires contacting it. Rao *et al* [122] have created chemically functionalized patterns on the surface that can attract and align the SWNT along pre-determined lines. Self-assembly based on molecular recognition is a promising approach for bypassing the need for precise nanofabrication and mechanical manipulation of SWNTs. This could allow large-scale fabrication of CNT circuits with single-nanotube precision.

7.5   Carbon nanotube interconnects

Future development in interconnections is driven by the demand for the high-speed chip transmission was emphasized in the International Technology Roadmap for Semiconductors (ITRS) [123]. Material requirements and difficulties in processing are the challenges in interconnecting technology. While demands for cross sections of interconnects current density increases and electromigration at high current densities ($>10^6$ A cm^{-2}) is the problem for common interconnecting materials like copper. It would be extremely difficult to achieve the required high aspect ratio in future device interconnections using only conventional semiconducting processing technologies. CNTs and less crystalline CNFs with diameters from few nanometres up to 100 nm and several micrometres in length could be an ideal material for nanoelectronics 3D interconnections due to their high aspect ratio and high electrical, thermal conductivity and good mechanical properties [80,124]. It was shown that the current-carrying capacity of MWNTs did not degrade after 350 h at a current density of 10^{10} A cm^{-2} at 250 °C [125], which is 10^3 times more than copper before electromigration occurs. The scattering-free, ballistic transport of electrons in defect-free CNTs is the most attractive property for possible use of CNTs as interconnects. Unfortunately, large-contact resistance limits ballistic current-carrying capability of CNTs, where Pd contacts could be the most promising candidates to establish near ballistic conductance [117]. The thermal conductivity of the individual MWNT was measured to be greater than 3000 W mK^{-1}, which is greater than diamond and the basal plane of the graphite (both 2000 W mK^{-1}) [126]. However, in order to enable the use of the unique properties of CNTs for future nanodevices interconnections, conventional semiconductor processing technologies could be exploited.

Controlled and reproducible CNT synthesis is the first step towards CNT-based nanodevice production. The synthesis of vertically aligned CNTs on lithographically predefined locations as a step towards the parallel interconnections in nanotube-based devices has been already achieved. Synthesis of such CNT structures is not possible using only thermal CVD and PECVD synthesis is required when the electric field is normal to the substrate. PECVD methods have been used to achieve vertical alignment of CNTs and CNFs using growth temperatures as low as 120 °C that enables synthesis on plastic in addition to conventional silicon substrates [50, 51] (figure 7.6(a)). The next step is to reassure that vertically aligned CNT channels are electrical insulated and mechanically supported (figure 7.6(b)). This could be achieved by direct synthesis of CNTs inside the channels or creation of the insulated channel structure by gap filling after vertical CNT synthesis.

The first attempt to use CNTs for interconnections was conducted by researchers from Infineon [127, 128]. They deposited CNT inside 400 nm vias (figure 7.7) and 5 μm × 5 μm contact holes. Vias are interconnections between the different wiring layers in a chip. The ability to grow CNTs with a high

(a)

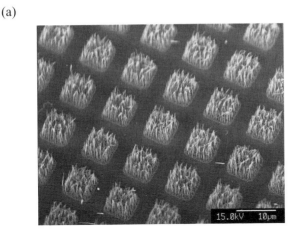

(b)

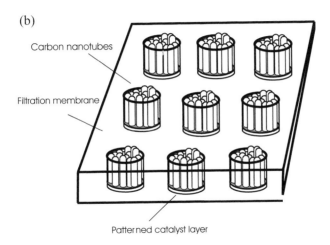

Figure 7.6. (*a*) SEM image of aligned carbon nanotubes grown on patterned metal catalyst. (*b*) Schematic diagram of proposed carbon nanotubes interconnections grown inside pores of plastic membranes.

current density at specific sites has opened the possibility for a nanotube via. They started with a conventional 6-inch wafer and, after the etching process, the resist serve as a shadow mask for a subsequent 5 nm-thick iron-based catalyst deposition. After catalyst deposition, a lift-off is performed leaving the catalyst material solely in the vias or in the bottom of the contact holes. MWNTs grown from the bottom of the vias and contact holes were contacted with tungsten pads by a dual-beam focused-ion-beam tool. With only about 3% of the vias available volume filled with CNTs, with an estimated density of 70 CNTs/μm^{-2},

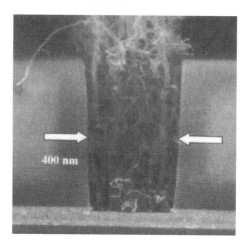

Figure 7.7. Selective growth of MWNTs in a via, when the top contact has not yet been formed [128]. Reprinted with permission from [128], copyright 2003 Elsevier.

the corresponding average resistance was 10 kΩ/CNT. Electrical measurements were performed on two contact holes of 5 μm × 5 μm in size and 1.25 μm in-depth gave a zero-bias resistance of 339 and 370 Ω. The measured resistance of two contact holes in series gave a resistance of 710 Ω. Assuming a CNT density of 70/μm^2, relatively high resistance of 600 kΩ per nanotube was obtained. This value could be improved with more control in CNT alignment and growth density and improvement in the CNT–metal contact resistance.

CNTs could be grown at a specific location using the catalyst patterning method. However, this method could not be used for the synthesis of individual CNTs. During growth, the catalysts later break up into clusters giving untangled or spaghetti-like CNT growth. Using a dc field during the growth process, the CNT could be aligned in the direction of the field. Exact positioning of the catalyst particle is essential for individual CNT growth. This could be achieved using ion implantation of the catalyst dots. Patterned growth of freestanding CNFs on submicrometre nickel dots on silicon was first reported by Ren *et al* [129]. A thin-film nickel grid was fabricated on a silicon wafer by standard microlithographic techniques and PE-HFCVD was performed using an acetylene and ammonia mixture. Growth was performed at a pressure of 1–10 Torr at temperatures below 660 °C for 5 min. An alternative method has been demonstrated by Duesberg *et al* [130]. Individual CNTs have been grown out of nanoholes in silicon dioxide created by optical lithography (figure 7.8).

Researchers from NASA argued that the Infineon approach simply replaces Cu and Al with CNTs and relies on the traditional etch-deposition-planarization path leaving all problems of the high aspect ratio of etching vias and holes, and opening a new problem of seeding at the bottom of a deep trench with

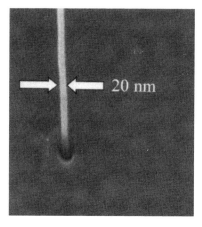

Figure 7.8. A single MWNT growing out of a 20 nm hole, produced in Infineon. Reprinted with permission from [128], copyright 2003 Elsevier.

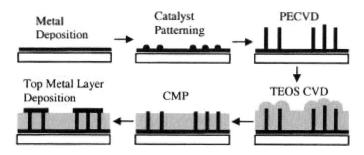

Figure 7.9. Schematic of the process sequence used by NASA researchers for production of CNT arrays embedded in SiO_2. Reprinted with permission from [131].

the catalyst [131]. They proposed an alternative bottom-up approach in which MWNTs are first grown at specified locations and then the gap is filled with SiO_2 and finally planarized (figure 7.9). A Si (100) wafer was first covered with a 500 nm thermal oxide and 200 nm Cr (or Ta) lines before a 20 nm-thick Ni catalyst layer deposition. For local wiring or contact applications, ion-beam sputtering was used to deposit Ni. Vertically aligned CNTs were grown using PECVD [132–134]. Then, the free space between the CNTs is filled with SiO_2 by CVD using tetraethylorthosilicate (TEOS) [134]. The top metal line was deposited. To produce a CNT array embedded in SiO_2 with only one end exposed over the planarized solid surface, chemical mechanical polishing (CMP) was applied.

7.6   Conclusions

The combination of CNT electronic properties and dimensions makes them ideal building blocks for future electronic devices and circuits. Integration of techniques for CNT growth on specified locations using PECVD in conventional semiconducting processing could provide the first step towards the exploitation of CNT properties. However, this 'top-down' approach could be replaced with a 'bottom-up' approach where the molecular properties of CNT will be exploited as the necessary requirement for high device density architectures. Self-assembly of CNTs into functional devices based on molecular recognition could be a promising approach for future large-scale integrated circuits. The nanometre dimensions of CNT and their high current-carrying capacity, with possibility for ballistic electron transport, are ideal properties for the interconnect applications in the electronics. The main challenge of the large-scale production is a good electrical coupling of the nanotubes to the contacts. Patterned catalyst PECVD growth of aligned CNTs could be the most suitable for 3D interconnections. Controlled, reproducible and low-cost production of SWNTs and the separation of metallic and semiconducting tubes is the main requirement for further advances towards the large-scale production of CNT-based electronics. Metallic SWNTs could function as interconnects and semiconducting could perform as nanoscale transistors. A CNT-FET based on semiconducting SWNT, with superb characteristics compared to the silicon MOSFET could be a promising candidate for a silicon-free 3D integration. In future 3D nanoelectronic architectures, CNT-based devices could be one of the most promising candidates for building blocks.

References

[1]   Kroto H W, Heathe J H, O'Brien S C, Curl R F and Smalley R E 1985 *Nature* **318** 162
[2]   Iijima S 1991 *Nature* **354** 56
[3]   Kratshmer W, Lamb D L, Fostiropolos K and Huffman D R 1990 *Nature* **347** 354
[4]   Iijima S and Ichihashi T 1993 *Nature* **363** 603
[5]   Bethune D S, Kiang C H, de Vries M S, Gorman G, Savoy R, Vazquez J and Beyers R 1993 *Nature* **363** 605
[6]   Ebbesen T W and Ajayan P M 1992 *Nature* **358** 220
[7]   Thess A *et al* 1996 *Science* **273** 483
[8]   Rao A M *et al* 1997 *Science* **275** 187
[9]   Kozavskaja Z Ja, Chernozatonskii L A and Federov E A 1992 *Nature* **359** 670
[10] Amelinck S *et al* 1994 *Science* **256** 678
[11] Ren Z F, Huang Z P, Xu J W, Wang J H, Bush P, Siegal M P and Provencio P N 1998 *Science* **282** 1105
[12] Journet C *et al* 1997 *Nature* **388** 756
[13] de Heer W A, Chatelain A and Ugerate D 1995 *Science* **270** 1179
[14] Rinzler A G *et al* 1995 *Science* **269** 1550
[15] Xu X and Brandes G R 1999 *Appl. Phys. Lett.* **74** 2549

[16] Xu D *et al* 1999 *Appl. Phys. Lett.* **75** 481
[17] Lee C J *et al* 1999 *Appl. Phys. Lett.* **75** 1721
[18] Dai H, Hafner J H, Rinzler A G, Colbert D T and Smalley R E 1996 *Nature* **384** 147
[19] Curran S A *et al* 1998 *Adv. Mater.* **10** 1091
[20] Dillon A C, Jones K M, Bekkedahl T A, Kiang C H, Bethune D S and Heben M J
        1997 *Nature* **386** 377
[21] Tans S J, Verschueren A R M and Dekker C 1998 *Nature* **393** 49
[22] Niu C, Sichel E K, Hoch R, Moy D and Tennent H 1997 *Appl. Phys. Lett.* **70** 1480
[23] Baughman R H, *et al* 1999 *Science* **284** 1340
[24] Kong J, Franklin N R, Zhou C, Chaplene M G, Peng S, Cho K and Dai H 2000
        Science **287** 622
[25] Ajayan P M and Iijima S 1992 *Nature* **358** 23
[26] Dresselhaus M S, Dresselhouse G and Eklund P C 1996 *Science of Fillerenes and
        Carbon Nanotubes* (New York: Academic)
[27] Sun L F *et al* 2000 *Nature* **403** 384
[28] Qin L-C, Zhao X, Hirahara K, Miyamoto Y, Ando Y and Iijima S 2000 *Nature* **408**
        50
[29] Wang N, Tang Z K, Li G D and Chen J S 2000 *Nature* **408** 50
[30] Davis W R, Slawson R J and Rigby G R 1953 *Nature* **171** 756
[31] Baker R T K and Barber M A 1978 *Chemistry and Physics of Carbon* vol 14, ed
        P L Walker and P A Throver (New York: Dekker) p 83
[32] Rodriguez N M 1993 *J. Mater. Res.* **8** 3233
[33] Snyder C E, Mandeville M W, Tennent H G, Truesdale L K and Barber J J 1989
        Carbon Fibrils, Hyperion Catalysis International WO8907163A1
[34] Radushkevich L V and Luk'yanovich V M 1952 *Zh. Fiz. Khim.* **26** 88
[35] Davis W R, Slawson R J and Rigby G R 1953 *Nature* **171** 756
[36] Rodriguez N M, Chambres A and Baker R T K 1995 *Langmuir* **11** 3862
[37] Qin L C, Zhou D, Krauss A R and Gruen D M 1998 *Appl. Phys. Lett.* **72** 3437
[38] Tsai S H, Chao C W, Lee C L and Shih H C 1999 *Appl. Phys. Lett.* **74** 3462
[39] Choi Y C *et al* 2000 *Appl. Phys. Lett.* **76** 2367
[40] Bower C, Zhu W, Jin S and Zhou O 2000 *Appl. Phys. Lett.* **77** 830
[41] Okai M, Muneyoshi T, Yaguchi T and Sasaki S 2000 *Appl. Phys. Lett.* **77** 3468
[42] Merkulov V I, Lowndes D H, Wei Y Y, Eres G and Voelkl E 2000 *Appl. Phys. Lett.*
        76 1534
[43] Chhowalla M, *et al* 2001 *J. Appl. Phys.* **90** 5308
[44] Li J, Stevens R, Delzeit L, Ng H T, Cassell A, Han J and Meyyappan M 2002 *Appl.
        Phys. Lett.* **81** 910
[45] Delzeit L, McAninch I, Cruden B A, Hash D, Chen B, Han J and Meyyappan M
        2002 *J. Appl. Phys.* **91** 6027
[46] Wang Y H, Lin J, Huan C H A and Chen G S 2001 *Appl. Phys. Lett.* **79** 680
[47] Boskovic B O, Stolojan V, Khan R U, Haq S and Silva S R P 2002 *Nature Mater.* **1**
        165
[48] Lee O-J and Lee K-H 2003 *Appl. Phys. Lett.* **82** 3770
[49] Teo K B K *et al* 2001 *Appl. Phys. Lett.* **79** 1534
[50] Hofmann S, Ducati C, Kleinsorge B and Robertson J 2003 *Appl. Phys. Lett.* **83** 135
[51] Hofmann S, Ducati C, Kleinsorge B and Robertson J 2003 *Appl. Phys. Lett.* **83** 4661
[52] Baker R T K and Barber M A 1978 *Chemistry and Physics of Carbon* vol 14, ed
        P L Walker and P A Throver (New York: Dekker) p 83

[53] Ducati C, Alexandrou I, Chhowalla M, Amaratunga G A J and Robertson J 2002 *J. Appl. Phys.* **92** 3299

[54] Diamond S and Wert C 1967 *Trans. AIME* **239** 705

[55] Mojica J F and Levensen L L 1976 *Surf. Sci.* **59** 447

[56] Baker R T K, Barber M A, Harris P S, Feates F S and Waite R J 1972 *J. Catal.* **26** 51

[57] Oberlin A, Endo M and Koyama T 1976 *J. Cryst. Growth* **32** 335

[58] Rostrup-Nielsen J and Trimm D 1977 *J. Catal.* **48** 155

[59] Helveg S *et al* 2004 *Nature* **427** 426

[60] Thess A *et al* 1996 *Science* **273** 483

[61] Cheng H M *et al* 1998 *Chem. Phys. Lett.* **289** 602

[62] Kong H, Cassell A M and Dai H 1998 *Chem. Phys. Lett.* **292** 567

[63] Hafner J H *et al* 1998 *Chem. Phys. Lett.* **296** 195

[64] Nikolaev P, Bronikowski M J, Bradley P K, Rohmund F, Colbert D T, Smith K A and Smalley R E 1999 *Chem. Phys. Lett.* **313** 91

[65] Bronikowski M J, Willis P A, Colbert D T, Smith K A and Smalley R E 2001 *J. Vac. Sci. Technol.* A **19** 1800

[66] Fan X, Buczko R, Puretzky A A, Geohegan D B, Howe J Y, Pantelides S T and Pennycook S J 2003 *Phys. Rev. Lett.* **90** 145501

[67] Li Y, Kim W, Zhang Y, Rolandi M, Wang D and Dai H 2001 *J. Phys. Chem.* B **105** 11424

[68] Cheung C L, Kurtz A, Park H and Lieber C M 2002 *J. Phys. Chem.* B **106** 2429

[69] Dai H, Rinzler A G, Nikolaev P, Thess A, Colbert D T and Smalley R E 1996 *Chem. Phys. Lett.* **260** 471

[70] Gavillet J, Loiseau A, Journet C, Willaime F, Ducastelle F and Charlier J-C 2001 *Phys. Rev. Lett.* **87** 275504

[71] Menon M and Sristava D 1997 *Phys. Rev. Lett.* **79** 4453

[72] Li J, Papadopoulos C and Xu J M 1999 *Nature* **402** 253

[73] McEuen P L 1998 *Nature* **393** 15

[74] Kouwenhoven L 1997 *Science* **275** 1896

[75] Chico L, Crespi V H, Benedict L X, Louie S G and Cohen M L 1996 *Phys. Rev. Lett.* **76** 971

[76] Satishkumar B C, Thomas P J, Govindaraj A and Rao C N R 2000 *Appl. Phys. Lett.* **77** 2350

[77] Li W Z, Wen J G and Ren Z F 2001 *Appl. Phys. Lett.* **79** 1879

[78] Ting J-M and Chang C-C 2002 *Appl. Phys. Lett.* **80** 324

[79] Ho G W, Wee A T S and Lin J 2001 *Appl. Phys. Lett.* **79** 260

[80] Baughman R H, Zakhidov A A and de Heer W A 2002 *Science* **297** 787

[81] Chico L *et al* 1996 *Phys. Rev. Lett.* **76** 971

[82] Dai H, Wong E W and Lieber C M 1996 *Science* **272** 523

[83] Ebbesen T W, Lezec H J, Hiura H, Bennett J W, Ghaemi H F and Thio T 1996 *Nature* **382** 54

[84] Dreselhous M S, Dresselhaus G, Sugihara K, Spain I L and Goldberg H A 1988 *Graphite Fibres and Filaments* ed U Gonser *et al* (New York: Springer) p 173

[85] Tans S J, Devoret M H, Dai H, Thess A, Smalley R E, Geerlings L J and Dekker C 1997 *Nature* **386** 474

[86] Frank S, Poncharal P, Wang Z L and de Heer W 1998 *Science* **280** 1744

[87]   McEuen P L, Bockrath M, Cobden D H, Yoon Y-G and Louie S G 1999 *Phys. Rev. Lett.* **83** 5098

[88]   Bockrath M, Cobden D H, McEuen P L, Chopra N G, Zenttl A, Thess A and Smalley R E 1997 *Science* **275** 1922

[89]   Tans S J, Devoret M H, Groeneveld R J A and Dekker C 1998 *Nature* **394** 761

[90]   Venema L C *et al* 1999 *Science* **283** 52

[91]   Bachtold A, Strunk C, Salvetat J-P, Bonard J-M, Forro L, Nussbaumer T and Schonenberger C 1999 *Nature* **397** 673

[92]   Tsukagoshi K, Alphenaar B W and Ago H 1999 *Nature* **401** 572

[93]   Nygard J, Cobden D H and Lindelof P E 2000 *Nature* **408** 342

[94]   Liang W, Bockrath M, Bozovic D, Hafner J H, Tinkham M and Park H 2001 *Nature* **411** 665

[95]   Javey A, Guo J, Wang Q, Lundstrom M and Dai H 2003 *Nature* **424** 654

[96]   White C T and Todorov T N 1998 *Nature* **393** 240

[97]   Thouless D J 1977 *Phys. Rev. Lett.* **39** 1167

[98]   Colbert D T *et al* 1994 *Science* **266** 1218

[99]   Treacy M M J, Ebbesen T W and Gibson J M 1996 *Nature* **381** 678

[100] Salvetat J-P *et al* 1999 *Adv. Mater.* **11** 161
       Salvetat J-P *et al* 1999 *Phys. Rev. Lett.* **82** 944

[101] Yu M, Lourie O, Dryer M J, Molini K, Kelly T F and Ruoff R S 2000 *Science* **278** 637

[102] Tibbets G G and Beetz C P 1987 *J. Phys. D: Appl. Phys.* **20** 292

[103] Bacon R 1959 *J. Appl. Phys.* **31** 283

[104] Yu M, Yakobson B I and Ruoff R S 2000 *J. Phys. Chem.* B **104** 8764

[105] Yu M, Files B S, Arepalli S and Ruoff R S 2000 *Phys. Rev. Lett.* **84** 5552

[106] Folvo M R, Clary G J, Taylor II R M, Chi V, Brooks Jr V P, Washburn S and Superfine R 1997 *Nature* **389** 582

[107] Saito S 1997 *Science* **278** 77

[108] Moore G E 1965 *Electronics* **38**

[109] Collins P G, Zehl A, Brando H, Thess A and Smalley R E 1997 *Science* **278** 100

[110] Tans S J, Verschueren A R M and Dekker C 1998 *Nature* **393** 49

[111] Martel R, Schmidt T, Shea H R, Hertel T and Avouris Ph 1998 *Appl. Phys. Lett.* **73** 2447

[112] Fuhrer M S *et al* 2000 *Science* **288** 494

[113] Rueckes T, Kim K, Joselevich E, Tseng G Y, Cheung C-L and Lieber C M 2000 *Science* **289** 94

[114] Bachtold A, Handley P, Nakanishi T and Dekker C 2001 *Science* **294** 1317

[115] Derycke V, Martel R, Appenzeller J and Avouris Ph 2001 *Nano Lett.* **1** 453

[116] 2002 *At IBM A Thinner Transistor Outperforms its Silicon Cousins* New York Times, 20 May 2002, p C4

[117] Javey A, Guo J, Wang Q, Lundstrom M and Dai H 2003 *Nature* **424** 654

[118] Krupke R, Hennrich F, Lohneysen H V and Kappes M M 2003 *Science* **301** 344

[119] Williams K A, Veenhuizen P T M, de la Torre B G, Eritja R and Dekker C 2000 *Nature* **420** 761

[120] Zheng M *et al* 2003 *Science* **302** 1545

[121] Keren K, Berman R S, Buchstab E, Sivan U and Braun E 2003 *Science* **302** 1380

[122] Rao S G, Huang L, Setyawan W and Hong S 2003 *Nature* **425** 36

[123] International Technology Roadmap for Semiconductors (Semiconductor Industry Association, San Jose, CA, 2003) http://public.itrs.net/

[124] Dreselhaus M S, Dresselhouse G and Avouris P (ed) 2001 *Carbon Nanotubes: Synthesis, Structure, Properties and Applications* (New York: Springer)

[125] Wei B Q, Vajtai R and Ajayan P M 2001 *Appl. Phys. Lett.* **79** 1172

[126] Kim P, Shi L, Majumdar A and McEuen P L 2001 *Phys. Rev. Lett.* **87** 215502

[127] Kreupl F, Graham A P, Duesberg G S, Steinhogl W, Liebau M, Unger E and Hoenlein W 2002 *Microelectron. Eng.* **64** 399

[128] Hoenlein W, Kreupl F, Duesberg G S, Graham A P, Liebau M, Seidel R and Unger E 2003 *Mater. Sci. Eng.* C **23** 663

[129] Huang Z P, Xu J W, Ren Z F, Wang J H, Siegal M P and Provencio P N 1998 *Appl. Phys. Lett.* **73** 3846

[130] Duesberg G S, Graham A P, Liebau M, Seidel R and Unger E, Kreupl F and Hoenlein W 2003 *Nano Lett.* **3** 257

[131] Li J, Ye Q, Cassell A, Ng H T, Stevens R, Han J and Meyyappan M 2003 *Appl. Phys. Lett.* **82** 2491

[132] Delzeit L, McAninich I, Cruden B A, Hash D, Chen B, Han J and Meyyappan M 2002 *J. Appl. Phys.* **91** 6027

[133] Matthews K, Cruden B A, Chen B, Meyyappan M and Delzeit L 2002 *J. Nanosci. Nanotech.* **2** 475

[134] Li J, Stevens R, Delzeit L, Ng H T, Cassell A, Han J and Meyyappan M 2002 *Appl. Phys. Lett.* **89** 910

Chapter 8

Polymer-based wires

J Ackermann and C Videlot
Université Aix-Marseille II/Faculté des Sciences de Luminy,
Marseille, France

8.1   Introduction

The specific task of the CNRS/Université des Sciences de Luminy in the CORTEX project was to develop a self-supporting insulating micro- and/or nanomolecular-based 'charge transfer' layer, containing very many narrow conducting channels from one surface of the layer to the other (figure 8.1). This permits electrical connection between pads on two circuit layers. In this context, we investigated the development of organic fibrils based on micro- and nano-wires in self-supporting insulating layers, which could be used as a chip interconnection matrix. In order to send signals between two chips across the matrix of conducting fibrils, the wires had to be isolated from each other. Therefore, the conducting fibrils were not allowed to be connected to each other, either inside the insulating material or along the surface.

The first stage in this technology development was to demonstrate the feasibility of a two-chip stack with vertical interconnections formed by organic conducting fibrils. In our approach, the organic conducting fibrils were produced by polymer wires grown vertically using polymerization techniques. Figure 8.2 shows a schematic representation of the transmitted electrical signals, via polymer conducting wires, between an upper chip and a lower chip. Both stacked chips were equipped with contact pads used for the vertical interconnections with sizes in the range of tens of micrometres. We defined the self-supporting insulating 'charge transfer' layer and realized the growth of polymer wires to form many narrow conducting channels from one surface of the layer to the other.

The choice of polymer fibrils was based on their advantageous properties of being easy to process, compactness, flexibility and good conductivity. In particular, electrically conducting polymers have received much attention since

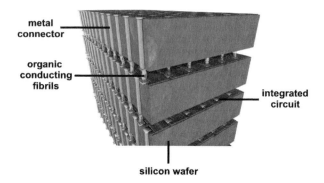

Figure 8.1. CORTEX logo describing the components of the 3D chip stack using organic conducting fibrils.

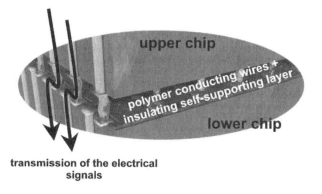

Figure 8.2. Schematic representation of a two-chip stack with vertical interconnections between the two chips based on polymer conducting wires.

their discovery as they promise to combine the optoelectronic properties of metals and semiconductors with the processability of polymers [1, 2]. For example, polypyrrole, first synthesized in 1916 [3], is one of the most widely investigated conducting polymers [4–6] because of its good thermal and environmental stability and good electrical conductivity. Further attractive features of the pyrrole system were the degree of freedom available to modify the electrical and physical properties by preparing derivatives, i.e. monomers containing a substituent on the pyrrole ring, in order to achieve any desired polymer properties.

Polythiophene and its derivatives represent a second example of a polymer family which has been widely studied. This new class of monomers appeared later in the 1970s to provide several critical characteristics of polymers such as chemical and electrochemical stability in air and moisture in both doped and undoped states, a high reversible doping level and a highly regular polymeric backbone [7]. By chemical or electrochemical doping (oxidation or reduction of the polymer), the electrical conductivity films of polythiophene-based polymers can be varied over 12 orders of magnitude, ranging from an insulator ($\sim 10^{-10}$ Ω^{-1} cm^{-1}), through a semiconductor, to a metal ($\sim 10^2$ Ω^{-1} cm^{-1}).

We focused our work on pyrrole and thiophene-based polymers which incorporated all the necessary elements mentioned earlier with an especially high electrical conductivity along the polymer chain. In addition, new functionalized pyrrole and thiophene monomers were developed in order to produce polymers with properties adapted to the demands of the CORTEX project.

As self-supporting insulating 'charge transfer' layers, we first used commercial filtration membranes that were demonstrated to be versatile templates for the generation of conducting polymer nanowires and nanotubules in combination with chemical and electrochemical polymerization techniques [8,9]. Our study first focused on the influence of the pore size on the polymer wire growth inside the polycarbonate membranes, while polymerization in liquid support layers such as gels were studied as an alternative for more flexible supporting layers. The investigation of classical polymerization techniques such as chemical and electrochemical polymerization (CP and ECP) on porous membranes demonstrated that the polymer formation through the pores with these techniques is very inhomogeneous and leads to serious short circuits between the wires via uncontrolled polymer deposition on the surface of the membranes [10–12].

Therefore, the main task in the CORTEX project was to control the polymer growth and develop techniques in order to produce isolated highly conducting polymer wires [13, 14]. The most promising approach to controlling the growth was the use of structured anodes in combination with a well-adapted polymerization process. This technique could be optimized to produce micropatterning of materials like porous membranes and gels with conducting polymers. The microscopic polymer wires thus formed could be grown vertically through porous membranes or gels with a high conductivity without any crosstalk. The feasibility of using molecular-wire-based interconnection layers for 3D chip-stacking technologies has been investigated. We could demonstrate a successful 3D chip stack, using polymer wires embedded in an insulated layer as vertical interconnections between two chips. Furthermore, the possibility of downscaling the directional electropolymerization (DEP) technique to form molecular wires for eventual use in nano-electronic systems was investigated.

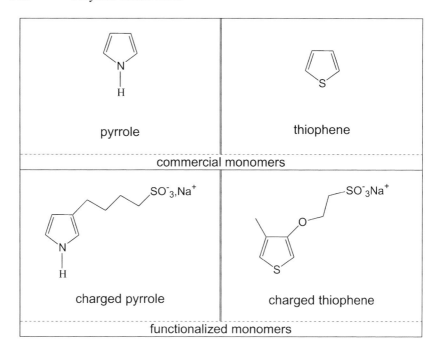

Figure 8.3. Chemical structure of pyrrole and thiophene monomers.

8.2   Experimental part

8.2.1   Monomers and polymers

Molecular self-assembled interconnections are based on two different polymer families: polypyrrole and polythiophene. The pyrrole and thiophene, non-functionalized, monomers are commercial products. The top part of figure 8.3 shows the chemical structure of the commercial monomers.

The conducting polymer polypyrrole (PPy), represented in figure 8.4, has attracted much interest in the polymer field due to its electrical conductivity and its potential application in biosensors, conductive paints, batteries and electronic devices [4–6]. It is highly stable in air, intensely black in colour in its oxidized form (conducting form) and transparent green in its undoped form. Its capability for good charge transport was used in the present work [10–14].

The synthesis of the polypyrrole can be realized either by CP or by ECP in several solvents such as acetonitrile or water, whereas the thiophene monomer only forms a polymer by electrochemical polymerization. Thus, to compare the two polymerization techniques, described in the following section, we mainly chose polypyrrole as the polymer to form our conducting fibrils.

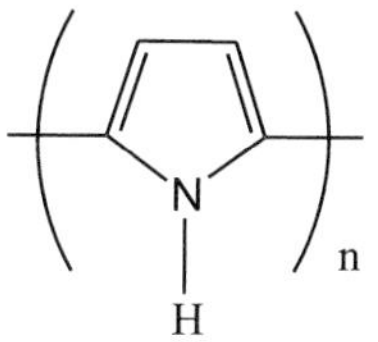

Figure 8.4. Chemical structure of polypyrrole (PPy).

Nevertheless during the project, both monomers were chemically modified in order to add functional groups to influence the growth and/or conductivity characteristics. The bottom part of figure 8.3 shows the chemical structure of the synthesized functionalized monomers. The substitution in form of a charged alkyl chain has to feature in the polymer formation process. First of all, the substituted charge plays the rôle of a fixed dopant inside the polymer. This has the advantage of a higher stability of the polymer for direct current (dc) applications in electronics, avoiding any charge migration problems. Second, in solution the functionalized monomer is charged negatively with the SO_3^- and it is, therefore, responsive to the electric field applied during the polymerization process. This gives an additional parameter in the control of growth in order to induce a directional growth of the polymer from one electrode to the other. The synthesis routes were adapted following chemical synthesis described in the literature [15, 16]. Figures 8.5 and 8.6 give a schematic representation of the synthesis.

From the literature, it is known that polymers based on charged pyrrole monomer units have a lower conductivity by a factor of 1000. However, a combination of these special polymers with the incorporation of metallic nanoparticles as the dopant in the polymer matrix could lead to an interesting combination of high conductivity and good control of the growth of the polymer. Therefore, these charged polymers are used in the present work to study the directional growth of polymers in more detail.

8.3 Self-supporting layers

8.3.1 Commercial filtration membranes

The first self-supporting polymer-based wiring layers were realized in commercial polycarbonate filtration membranes. Polycarbonate membranes, purchased from Millopore, were isopore membranes with different pore diameters (0.05–10 μm). The membranes purchased from Whatman had pore diameters of 100 nm as a nucleopore track-etched filter. Figure 8.7 shows, as an example, SEM pictures of a 10 μm pore-size membrane from Millipore and the 100 nm pore-size membrane from Whatman.

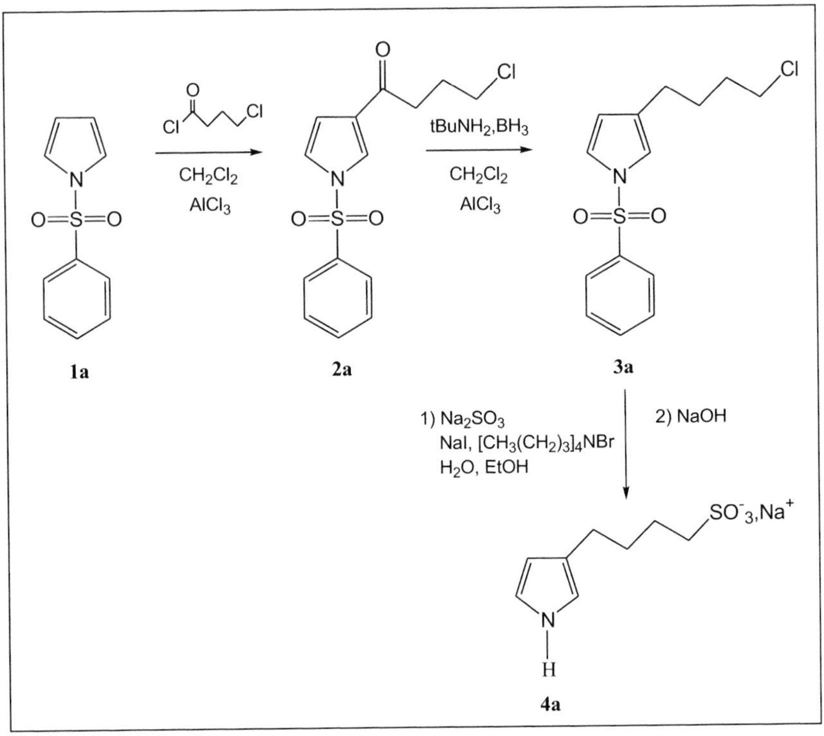

Figure 8.5. Schematic representation of the synthesis of the charged pyrrole monomer.

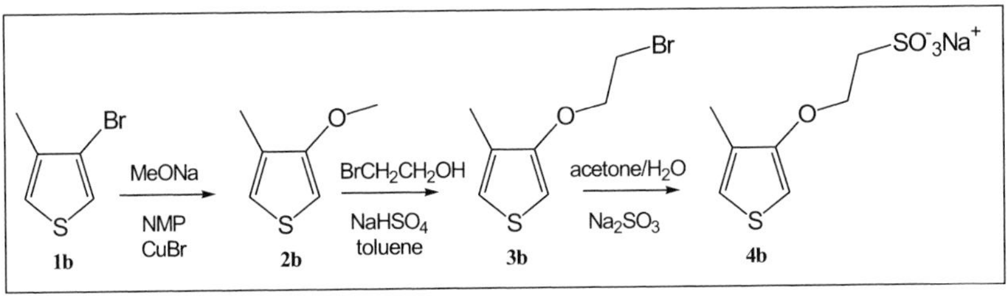

Figure 8.6. Schematic representation of the synthesis of the charged thiophene monomer.

Both the Millipore and Whatman membranes are made in polycarbonate with a thickness of 10 μm. The pore density is quite low, between 6 and 9% and represents a limiting factor for their application in stacked chips which have to contact the maximum number of organic wires to transport signals from one chip to the other with a high bandwidth. The commercial membranes are used in order to study their potential application as insulating support structures; however, it is clear that the pore density of the membrane has to be increased if the real

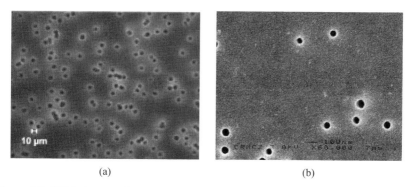

(a)   (b)

Figure 8.7. SEM pictures of commercial filtration membranes in polycarbonate with pore size: (*a*) 10 μm (from Millipore) and (*b*) 100 nm (from Whatman).

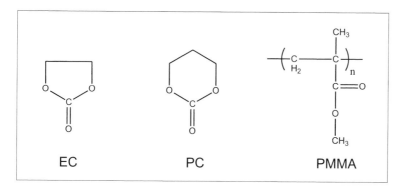

Figure 8.8. Chemical compounds of the gel solution.

application of such a polymer wire filled membrane is to be achieved. Tests and results are shown in the following sections.

8.3.2 Gel

As a second insulating layer with a higher elasticity and no limit in 'pore density', the use of a gel was investigated. A gel solution was prepared with the compounds chemically represented in figure 8.8.

The gel solution was prepared in the laboratory with the following composition (in weight ratio):

- 26.3% poly(methyl methacrylate) (PMMA) (molecular weight (MW) = 120 000)
- 36.8% propylene carbonate (PC) (MW = 102.09)
- 36.8% ethylene carbonate (EC) (MW = 88.06)

At room temperature and in an argon atmosphere, EC and PC were first dissolved in 20 ml anhydrous tetrahydrofuran (THF) before the polymer (PMMA) was added. A good dissolution and dispersion of all compounds in solution was obtained by stirring for several hours. The PMMA gel solution thus obtained is a translucent solution free from any particles.

Further investigations on the gel preparation itself have shown that deposited at room temperature on a glass substrate, the solvent evaporates, leaving a gel formed on the surface of the substrate after several hours. The gel conserves its sticky and pasty aspect on substrates heated to 50–60 °C for 1 hr. By incorporating nanosized Al_2O_3 particles (size: 5.8 nm) in the gel solution with a concentration of 6%, the gel deposited on a glass substrate keeps its gel aspect, with an eye-visible dispersion of Al_2O_3 particles all over the gel, until heated to 70 °C over 1–2 hr.

8.4   Chemical polymerization

Chemical polymerization (CP), a polymer growth process based on the membrane used, was only explored with the pyrrole (Py) monomer.

During the CP process, a membrane was used as a dividing wall in a two-compartment cell with one compartment containing an aqueous pyrrole solution and the second one an oxidant reagent ($FeCl_3$) solution. Purified water, obtained from a milli-Q (Millipore) water purification system, was used as the solvent for the polypyrrole (PPy) synthesis. Ferric chloride $FeCl_3$ (Janssen Chimica, 99%) was used without any purification as an oxidant. By diffusion of the monomer and the reagent through the membrane, schematically shown in figure 8.9, a polymer was formed inside the membrane. The CP was performed at room temperature. The polymerization time was varied between 3–11 min. After the polymerization, the membranes were removed from the cell and rinsed for several minutes with purified water. A number of tests were performed in the laboratory in order to obtain the best configuration for the two-compartment cell and to achieve homogeneous polymerized membranes with the maximum number of pores filled with polypyrrole.

8.5   Electrochemical polymerization

Electrochemical polymerization (ECP) was used as a technique to form conducting polymer wires in membranes as self-supporting layers and pyrrole as monomer. The ECP setup is represented in figure 8.10. A metallic layer (Au) was evaporated onto one side of the membrane with a thickness greater than the pore size in order to close the pores and to serve as an electrode in the ECP process. The ECP was performed at room temperature in a one-compartment cell with a Pt counter electrode and an Ag/AgCl reference electrode. The membrane was mounted on the cell so that the Au side was in contact with the flat anode formed

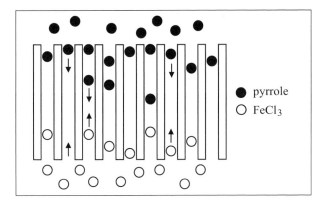

Figure 8.9. Schematic representation of the diffusion process of pyrrole and FeCl$_3$ molecules in the pores of the membrane.

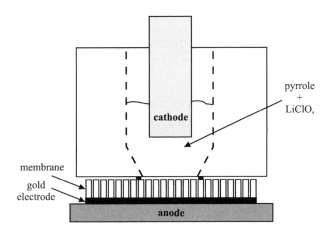

Figure 8.10. Schematic representation of the electrochemical polymerization setup with a flat anode.

by a Si wafer. The solution contained 0.1 M pyrrole with 0.1 M LiClO$_4$ was used as an electrolyte. The polymer was grown potentiostatically at 0.8 V. To create a homogenous growth over the whole membrane, a short pulse of a higher voltage (2.5 V) was applied during the first 10 s. This pulse guaranteed that the polymer started to grow over the whole surface simultaneously. Without this additional process step, the polymer grew only on those places where the membrane was in close contact with the anode.

8.6   Directional electropolymerization

Directional electropolymerization (DEP) was used as a technique to form patterns of conducting polymer wires in membranes or gels as self-supporting layers.

8.6.1   First generation of the DEP process (DEP-1)

The aim of this new technique was to control the polymer growth and, thus, to avoid crosstalk connections between the molecular wires. The main feature of the DEP technique was the appropriate choice of anode design. Its structure should be able to control the location and direction of growth of the conducting fibrils as represented in figure 8.11(*b*). The anode (figure 8.11(*a*)) was formed by a patterned Si wafer with top metal islands (Au/Cr, 300 nm thick) isolated by surrounding SiO_2. The pads had surface areas ranging from 300 μm $\times$ 300 μm down to 20 μm $\times$ 20 μm and a height of 5 μm with respect to the SiO_2 surface. A metallic layer (Al) was deposited on the back side of the anode to ensure a good contact with a metallic plate contact of the polymerization setup. Due to the metallic pads in Au/Cr, it was not necessary, in the case of the DEP process, to evaporate a gold layer onto the back side of the membrane. As in the ECP process, a solution containing 0.1 M pyrrole with 0.1 M $LiClO_4$ was used as an electrolyte. The polymer was grown potentiostatically at 0.8 V.

In the case of the gel as a self-supporting layer, the gel solution was injected in the polymerization setup and the DEP took place only at the Au/Cr pads (figure 8.11(*c*)). In this case, DEP was realized giving micropatterning of the gel as described in the results section. Furthermore, a two-step process based on electropolymerization of the polymer pads in water followed by incorporation of the gel via post-injection was used. The post-injection of the gel was obtained by spin coating in order to stabilize the system after drying the polymer pattern under nitrogen.

8.6.2   Second generation of the DEP process (DEP-2): gel-layer-assisted DEP

In the DEP-2 process, a new design for the DEP electrode was used (figure 8.12) that allows not only sub-10 μm resolution to be reached but also to perform volume and surface pattern deposition of PPy on polycarbonate membranes [13]. During the fabrication of these DEP electrodes, the same masks were used for bottom and upper chip production. Therefore, the DEP-2 process performed on the porous membranes produces a pattern of conducting polymer wires which is a direct copy of the electrode pattern on the chips and can be used as a transfer layer in the 3D chip stack after alignment with the chips.

The DEP-2 process is schematically shown in figure 8.13. Compared to DEP-1 electrode generation, the level of the conducting Au-electrodes is lower (650 nm) than the isolating SiO_2 layer. This should lead to two advantages: better

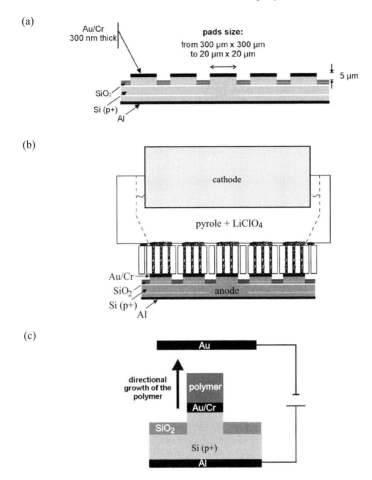

Figure 8.11. (*a*) Schematic representation of the textured anode used in the DEP process (DEP-1), (*b*) schematic representation of the DEP in a membrane by DEP-1 and (*c*) schematic representation of the DEP on one contact pad with a gel layer as self-supporting layer.

contact between the membrane and the DEP electrode to increase the contrast of the polymerization process on the membrane and the production of polymer wires which stick out (1000 nm approximately) above the membrane surface facing the electrode. Thus, the contact with the polymer wires via an external electrode on a chip should be improved as well.

Figure 8.14 shows the SEM images of the patterned anodes with square and circular Au pads embedded in the SiO_2 layer. While square pads have a side

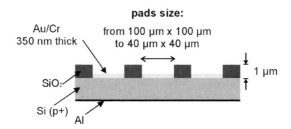

Figure 8.12. Schematic representation of the textured electrode in the DEP-2 process.

length between 30 and 100 μm, the circular pads in figure 8.14(*b*) with 8 μm diameters are fabricated to obtain sub-10 μm resolution in the DEP-2 technique.

In order to use such small structures on the electrode, the contact between the DEP electrode and the membrane had to be improved. We did several experiments using different solvents as the adhesion layers between the membrane and the electrode but, in the DEP process, no improvement in the patterning process could be observed, leading to a loss in contrast in the small structure. However, by using an intercalating electropolymerization medium between the anode and the membrane to be patterned in the form of an insulating gel layer (around 300 nm thick), an almost perfect translation of the pattern on the electrode onto the membrane could be obtained. The gel layer was deposited by spin coating, starting from a highly diluted solution. Then the membrane was mounted on the covered electrode where the gel worked like a glue to bring the membrane into very close contact with the electrode. Why the gel glues the membrane so efficiently to the anode is not yet understood but the phenomenon can be used to produce a perfect mechanical contact over a large surface area (several centimetres). Then a one-compartment cell is placed on top of the membrane prior to the addition of a pyrrole aqueous solution. When a negative voltage is applied between the anode and a large Pt electrode, covering the whole anode area above the membrane, polymer formation starts at the gold pads due to diffusion of the monomers through the pores of the membrane and the gel to the anode. After filling the space between the gold pad and the SiO_2 surface, the polymer grows across the gel. Since both the monomer and electrolyte are soluble in the gel medium, they can cross the gel barrier by a simple diffusion process. Inside the membrane, the polymer growth occurs along the main axes of the pores towards the outer surface of the membrane. The quantity of polymer grown during the DEP-2 process is controlled by the number of charges flown through the patterned electrode. Thus, at constant voltage, the variation in the polymerization time determines whether the polymer micropatterns are either grown across the whole substrate ($t1$) leading to a volume patterning of the PC membrane or deposited only onto the surface of the membrane ($t2$) with, however, limited penetration of the polymer wires inside the pores, which is expected to increase the mechanical adhesion of the deposit to the substrate. Typical polymerization times t for volume

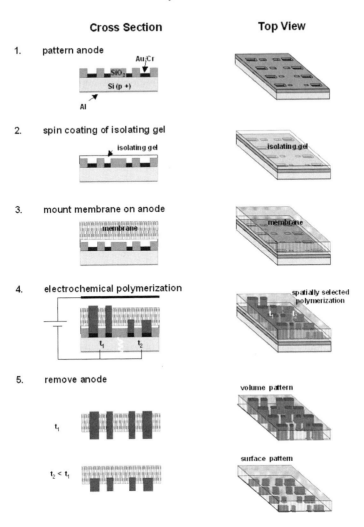

Figure 8.13. Schematic representation of the DEP-2 process on polycarbonate membranes.

and surface patterning were 4 and 1 min, respectively, at a constant voltage of 1.4 V against an Ag/AgCl reference electrode.

8.7   Conductivity and crosstalk measurements

In our experiments, the conductivity and crosstalk ratio were determined as follows: the polymerized layers were pressed between the upper and bottom

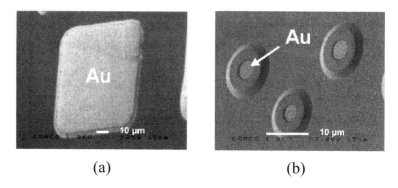

(a)   (b)

Figure 8.14. SEM images of square- (*a*) and circular- (*b*) shaped gold electrodes of a patterned anode.

chip and, by using I/V (figure 8.15) measurements, the resistance R of the membrane was obtained. The Si lower chip of the test rig was patterned with metal pads (300 nm Cu or Au/Cr layer), with a surface area of 300 μm $\times$ 300 μm or 100 μm $\times$ 100 μm and a height of 5 μm, in order to exercise pressure locally on the self-supporting layer. The flat Si surface was used as the upper chip. In the case of ECP-processed membranes, the Au-film on the back side of the membrane was removed prior to the crosstalk measurements. Such standard chips were used in order to compare conductivity values between the different polymerized layers. For the 3D chip-stacking technology (see section 8.15), the upper and lower chips had a different and adapted design.

The conductivity σ and σ_{CT} are calculated from

$$R = \frac{l}{\sigma A} \qquad (8.1)$$

where l is the thickness of the layer (when calculating σ) or the distance between two pads of the lower chip (when calculating σ_{CT}), A is the surface of the contact pads. σ_{CT} was measured by using an insulating upper chip in the setup. In this configuration, the current transport between two contact pads on the lower chip can go only via interconnected conducting fibrils inside the self-supporting layer or via polymer depositions on the surface of the layer and, thus, the crosstalk could be measured.

The crosstalk ratio is defined as

$$CT = \frac{\sigma}{\sigma_{CT}} \qquad (8.2)$$

The development of a process to produce a polymer filled layer with a high CT value combined with a high conductivity through the membrane (σ) was the objective of our study.

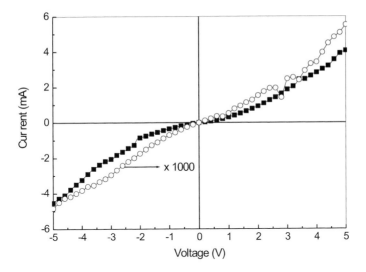

Figure 8.15. Typical I/V curves for the conductivity (filled squares) and for the crosstalk (open circles) measurements. The current shown for the I/V curve of the crosstalk is multiplied by a factor of 10^3.

8.8   Polymerization of commercial membranes by CP and ECP

Figure 8.16(*a*) shows the 10 μm-pore-size membrane before the polymerization, while figure 8.16(*b*) shows one polymerized pore of the same membrane after a 3 min polymerization process. In the large 10 μm pores, the polymer formed fibril networks which resembled polymer growth in a solution and gave a relatively disordered aspect to the wires, with a small amount of polymer on the membrane surfaces. It is worth mentioning that there was a large amount of polymer inside the pores compared to a thin film on the surface, which is obviously very favourable for a good crosstalk ratio combined with high conductivity σ. However, the size of the pores was too large to be used as conducting channels with contact pads of the same size deposited randomly on the filled membrane. In the case of the 1.2 μm pores (figure 8.16(*c*)) a solid one-phase polymer was obtained inside the pores after a 11 min polymerization time, whereas for the 100 nm pores and a 8 min polymerization time, the polymer formed tubules which had varying diameters (figure 8.16(*d*)). As a consequence, the diffusion of the monomer and the reagent through pores smaller than 10 μm was changed by the polymer which already existed in the pores and could easily create an interconnected polymer layer on the membrane surfaces for longer polymerization times. We established that, in principle, by using CP, it was possible to produce conducting channels inside the pores of the membrane even if

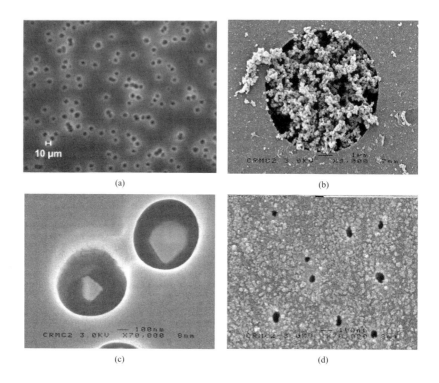

Figure 8.16. SEM pictures of the membrane surfaces with pore size: (*a*) 10 μm membrane before the CP process, (*b*) 10 μm membrane after a 3 min CP process, (*c*) 1.2 μm membrane after an 11 min CP process, (*d*) 0.1 μm membrane after a 8 min CP process.

the CP growth of the polymer through a membrane depends strongly on the pore size. However, the non-uniform growth of the polymer through the membrane creates certain performance limits concerning the crosstalk induced by polymer deposition on the surface if the polymerization time is not adequate [10]. Table 8.1 summarizes the values of conductivity obtained for each pore size in the case of maximum σ values (A–C).

The origin of the σ_{CT} conductivity in a polymerized membrane may be due to interconnected pores inside the membrane or to polymers deposited on the surface of the membrane which were created during the polymerization process. The first problem could be solved by the use of track etched membranes (Whatman membranes), which have a negligible density of interconnected pores. As it was difficult to control the growth of the polymer by the CP process, we used an additional after-treatment of the polymerized membranes. Ultrasonic cleaning in an ethanol solution completely removed the polymer from the surface,

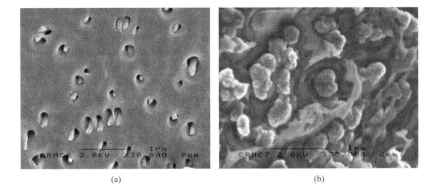

(a)   (b)

Figure 8.17. SEM pictures of the 0.1 μm-pore-size membrane (from Whatman) surfaces filled with polypyrrole by a 6 min CP process: (*a*) after an ultrasonic cleaning of the membrane and (*b*) after a polishing following with a secondary 1 min CP process on both sides of the membrane.

leaving conducting fibrils inside the pores sticking out above the surface (see figure 8.17(*a*)) without any sign of alteration on the membrane surface. The electrical characteristics of the cleaned membrane are given in table 8.1, (D1). A crosstalk value of 4×10^3 was obtained but with a drastic decrease in the conductivity σ compared to C in table 8.1. This value reveals the poor contact between the cleaned membrane and the chip contact pads. A second technique was to polish the chemically polymerized membrane on both sides with a 1 μm diamond paste and to realize a secondary CP process of just 1 minute on each side. The resulting surface state is shown in figure 8.17(*b*). A large number of separated polymers grown like mushrooms were found on the surface, which should be favourable in achieving a good contact to the polymer. However the conductivity measurements (table 8.1, D2) on these membranes showed no improvement, compared to the untreated membrane (table 8.1, C). The polymers formed during the second CP process do not form a good electrical contact to those produced first in the membrane and so only crosstalk connections are created.

During the ECP process, the amount of the polymer grown inside the membrane is controlled by the polymerization time at a certain voltage (1.2 V). In order to avoid crosstalk between the polymer wires, the polymerization has to be stopped prior to the formation of crosstalk interconnections on the surface. Figure 8.18 shows two surfaces of ECP-processed membranes with different polymerization times. Using an ECP process with a polymerization time of 2–3 min, the growth was found to be non-uniform over a large area (figure 8.18(*a*)) where we could find polymer wires sticking out of the membrane surface (area 4 on figure 8.18(*a*)) while other pores were either still unfilled (area 1–3) or

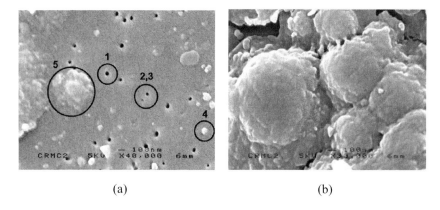

(a)   (b)

Figure 8.18. SEM pictures of the 0.1 μm-pore-size membrane (from Whatman) surfaces filled with polypyrrole by an ECP process with different process time: (*a*) $T_{poly} = 2$ min and (*b*) $T_{poly} = 4$ min.

polymer wires had already formed interconnections (area 5 on figure 8.18(*a*)) over the surface. Figure 8.18(*b*) shows details of the interconnected wires for a process time of 4 min. The conductivity measurements (table 8.1, E) reveal that a compromise must be made between conductivity and crosstalk ratio. The best result produced by ECP was polymer wires embedded in a membrane with a conductivity of 14 S m^{-1} and a crosstalk ratio of 46 where a secondary ECP process was performed. However, real control of the polymer formation could allow a much better performance for such a polymer matrix. Therefore, a modified ECP process called DEP was developed and is discussed in the following section.

8.9 Micropatterning of commercial filtration membranes by DEP-1

In the DEP process, polymer formation was limited to the areas corresponding to the conducting metallic pads on the anode. By placing a porous membrane in very close contact with the structured anode during a DEP process, the structure of the metallic pads was imprinted onto the membrane by polymer wires grown locally through it. Using this technique, micropatterning of the membrane with polymer wires was performed depending on the metallic pattern of the anode. In figure 8.19(*a*), a membrane containing a polymer pattern of 300 μm $\times$ 300 μm per pad is shown. The polymer pads were separated by highly insulating areas (i.e. the pure polycarbonate membrane) and finally there was no crosstalk between the polymer microwires (see table 8.1, F). In addition the polymer pads could be grown in such a way that they stuck several micrometres out of the membrane without any crosstalk which reduced the pressure needed to achieve

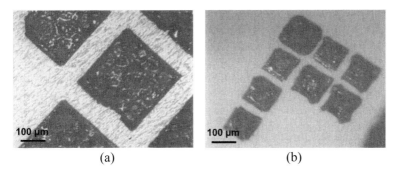

(a)   (b)

Figure 8.19. Optical images of polymerized membranes (Whatman, 100 nm pore size) obtained by a DEP process (DEP-1) with pads of size: (*a*) 300 μm × 300 μm and (*b*) 100 μm × 100 μm.

a good contact with the polymer wires in the test rig. Each polymer wire in an area of 300 μm × 300 μm is formed by numerous nano-wires ($\geq$1000). For smaller structures of 100 μm × 100 μm separated by a distance of 30 μm (see figure 8.19(*b*)), an enlargement of the micropattern imprinted on the membrane was observed. This might be due to the fact that the distance between the metal pads of 30 μm and the height of a pad of 5 μm were in the same order which led to a high leakage current under the membrane and, thereby, to polymer formation beside the pads. The conductivities σ of the patterned membranes were 50 and 30 S m^{-1} for pad sizes of 300 μm and 100 μm respectively, which are improved values compared to a standard ECP process, with a guarantee of no crosstalk between the polymer microwires.

8.10   Conductivity values of polymerized commercial filtration membranes

Table 8.1 summarizes the conductivity values (σ and σ_{CT}) and the crosstalk ratio (σ/σ_{CT}) of chemical polymerization, electrochemical polymerization and directional electrochemical polymerization processed membranes [12]. The corresponding figures are noted for a better understanding.

8.11   Polymer-based wires in a gel layer

8.11.1   Micro-patterning of a gel by DEP-1

As a second insulating layer with a higher elasticity and no limitations on pore density, a gel was used as a self-supporting layer for the polymer wires instead of a rigid polycarbonate membrane. Polypyrrole formation in the gel was studied by cyclic voltammetry. The cyclic voltammogram in figure 8.20 shows clearly

Table 8.1. Conductivity values (σ and σ_{CT}) and the crosstalk ratio (σ/σ_{CT}) of chemical polymerization (A–D), electrochemical polymerization (E) and directional electrochemical polymerization (F,G) processed membranes. d is the pore diameter and T_{CP} is the polymerization time.

No.	σ (Sm^{-1})	σ_{CT} (Sm^{-1})	σ/σ_{CT}	Remarks	Figure
A	50	0.56	89	Millipore $d = 10\ \mu$m, $T_{CP} = 3$ min	8.16(*b*)
B	2.6	0.31	9	Millipore $d = 1.2\ \mu$m, $T_{CP} = 3$ min	8.16(*c*)
C	100	1.8	55	Whatman $d = 0.1\ \mu$m, $T_{CP} = 3$ min	8.16(*d*)
D1	0.8	0.0002	4000	Whatman $d = 0.1\ \mu$m, $T_{CP} = 6$ min + ultrasonic cleaning	8.17(*a*)
D2	0.1	0.06	1.6	Whatman $d = 0.1\ \mu$m, $T_{CP} = 6$ min + secondary CP of $T_{CP} = 1$ min	8.17(*b*)
E1	50	NA	NA	with Au film, Whatman $d = 0.1\ \mu$m	
E2	0.01	0.001	10	as E1 but Au film removed	8.18(*a*)
E3	14	0.3	46	as E2 + secondary ECP	
F	50	—	—	Whatman $d = 0.1\ \mu$m, pad size 300 μm × 300 μm	8.19(*a*)
G	30	—	—	As F, but pad size is 100 μm × 100 μm	8.19(*b*)

a reversible oxidation and reduction wave at 1.2 V and 0.5 V, respectively, indicating polymer formation in the PMMA gel.

The polymer was only formed on the Au pads and followed exactly the shape of the Au pad up to the thickness of the gel layer (around 2 μm) as demonstrated in figure 8.21.

In order to determine the electrical properties of the polymer grown in the gel, I/V measurements were done with Au contact pads mounted on the top of the polymer wires in the gel. In the case of a patterned gel layer, figure 8.22 shows an I/V curve that proves a diode behaviour for the polymer wire–Au contact and a non-ohmic characteristic was obtained as was the case for the polymer wires produced in the membranes. The different behaviour is explained by the fact that the salt ions, which have the task of doping the polymer, can still move when the polymer is in contact with the gel. Thus, depending on the polarization, the dopants can leave the polymer, partially creating a zone of an insulating, wide band-gap semiconductor.

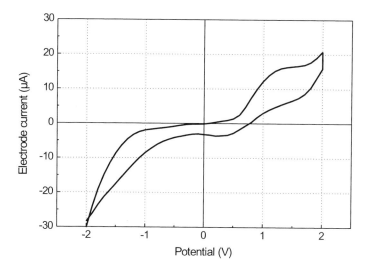

Figure 8.20. Cyclic voltammetry (second cycle) of the polypyrrole formation (10^{-1} M) in a solution of LiClO$_4$ (10^{-1} M)/H$_2$O, $v = 100$ mV s^{-1}, ref: Ag/AgCl.

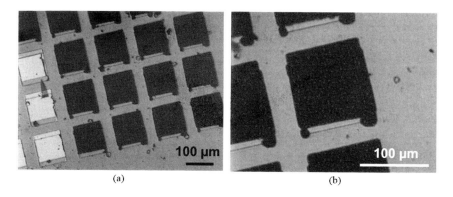

Figure 8.21. Optical images of a polymerized gel obtained by a DEP process with surface electrodes of 100 μm × 100 μm: (*a*) view of a large surface and (*b*) view of one pad.

8.11.2 Directional polymerization in water followed by post-injection of a gel

In order to use a gel as an isolating support layer but to prevent any problems with monomer and salt ion diffusion in the gel after the polymer wire has formed, a new two-step process was used. First, a DEP process was performed on a structured electrode using only an aqueous solution containing the reagents (monomer and salt). Then the electrode possessing the polymer pattern was rinsed in deionized

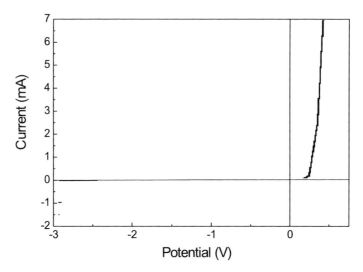

Figure 8.22. I/V curve obtained with an Au contact mechanically mounted on the polypyrrole wires formed in gel.

water and dried under nitrogen flow. In the second step, the gel was deposited on top of the polymerized electrode by spin coating as an isolating support medium, with the aim of stabilizing mechanically the polymer pattern on the electrode. The polymerization in water gives the possibility of a high-resolution process, which can be seen in figure 8.23(a), where each corner of the polymerized contact pads is reproduced in detail by the polymer. Figure 8.23(b) shows the non-uniform growth over one single pad with a maximum height of 30 μm in the middle of the polymerized pad. The mushroom shape should be very advantageous for electrical contact to metal pads for chip stacking.

In order to estimate the minimum size and, thus, the maximum contrast of a molecular wire electrochemically deposited on a metallic pad, the polymer growth was studied as a function of polymerization time and size of the Au pads. First of all, with increasing polymerization time and, thus, thickness of the polymer wires, a loss of contrast in the DEP process was observed. Above a thickness of 40–50 μm, an important enlargement of the polymer over the Au-pad could be observed. Investigation of the influence of pad size on polymer growth also revealed that a certain critical size exists, where the growth of a very small thickness leads to an enlargement of the polymer. In the optical images in figure 8.24, this behaviour can be clearly observed. The Au pads on the electrode have sizes from 50 μm $\times$ 50 μm to 20 μm $\times$ 20 μm. For pads of the same size (figure 8.24(a)). A uniform polymer formation is found with the polymer following the shape of the Au pads. Figures 8.24(b) and 8.25 show polymer wires produced on electrodes with varying pad sizes and two different polymerization

Figure 8.23. Optical images of polymer wires obtained by a DEP process in water and post-injection of a gel with surface electrodes of 100 μm × 100 μm: (*a*) view of a large surface and (*b*) view of one pad.

Figure 8.24. Optical images of a polymerized gel obtained by a DEP process in water and post-injection of a gel with surface electrodes of (*a*) 50 μm × 50 μm and (*b*) from 50 μm × 50 μm to 20 μm × 20 μm.

times. While in figure 8.24(*b*), the polymer wires follow still the pad shape of every pad, a strong enlargement of the polymer wires and, thus, a loss of DEP resolution is observed for the small pads when the polymerization time is increased (figure 8.25). The enlargement depends clearly on pad size, leading to a reduction in the maximum height of the polymer with decreasing pad size. As a consequence, the polymer wires grown on a 20 μm × 20 μm metallic pad have a maximum height of only 1 μm.

8.11.3 Area selective directional polymerization

The feasibility of local polymerization on a single selected gold pad in a gel layer as well as in an aqueous solution with post-injection of the gel was investigated.

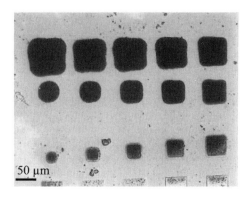

Figure 8.25. Optical image of a polymerized gel obtained by a DEP process in water and post-injection of a gel with surface electrodes from 50 μm $\times$ 50 μm to 20 μm $\times$ 20 μm and a longer polymerization time.

Whereas previous polymer-based microwires were fabricated simultaneously, the possibility of selective polymerization over the electrode area is important for the deposition of different polymer types on the supporting medium as would be necessary for future sensor applications for example. A structured electrode that possesses separately-addressable metal pads at selected areas was used, as shown in figure 8.26. The dark pads on the electrode present two polymer wires that were produced successively by a DEP process in a gel with different polymerization times. Each polymerized pad was connected to an external circuit where a voltage was applied. The results reveal the feasibility of selective fabrication of polymer wires locally on a chip. By successive polymerization of different polymers on a structured chip, use of this technique for applications such as biosensors could be imagined.

8.12   DEP process based on charged monomers

The polymerization of the functionalized monomers of pyrrole and thiophene (see section 8.2.1) were performed in different solutions (H_2O, H_2O + Acetonitrile mixture). In a polar solution, the functionalized monomers are charged negatively because of the SO_3^-. In figure 8.27, the potentiodynamic electropolymerization in aqueous solution of polypyrrole and polythiophene based on charged monomers are shown.

We observed a very low adhesion of both polymers on the working electrode. Only a small part of the polymer stayed at the electrode while the rest dropped off and dissolved in the solution. One possible reason for this behaviour could be the high solubility of the polymer in an aqueous solution because of the charges substitution on the monomers. But even a mixture of water with other, less polar, solvents could not increase the adhesion. Another reason could be

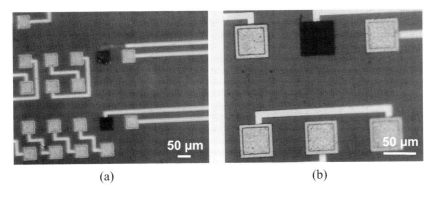

Figure 8.26. Optical images of (*a*) two successive polymerized pads and (*b*) one polymerized pad.   The polymerized gel obtained by a DEP process in water and post-injection of a gel with selected surface electrodes of 50 μm × 50 μm.

related to the fact that the SO_3^- group on the monomers tends to create a repulsion between the polymer chains and leads to a loss in the adhesion of the polymer chains to each other, leaving only a thin polymer layer on the electrode. This interpretation is supported by the fact that the deposition of a monolayer of thio-pyrrole covalently bonded to an Au electrode followed by electropolymerization could not improve the polymer deposition. The repulsion could also be a reason for the difficulties we found in growing these polymers inside the pores of the membranes.   Only the membranes with a 1.2 μm diameter could be used to deposit the polymer inside the pores. When the diameter of the pores was smaller, then there was no inside polymer deposition. In conclusion, the functionalized monomers could not improve the control of the polymer growth and, in addition, new problems were found, probably related to the overcharge of the polymer chains. However, the time-dependent conductivity measurements, which were discussed in section 8.14, led to interesting results concerning the dopants bonded covalently to the polymer.

8.13   Micropatterning of commercial filtration membranes and gels by DEP-2

8.13.1   Volume and surface patterning of polycarbonate membranes

The DEP-2 process is based on the combined use of preformed anodes that possess pattern-transfer elements with a negative of the desired topography and a gel layer as an intercalating electropolymerization medium between the anode and the membrane to be patterned. It allows volume and surface patterning of porous polycarbonate membranes with high precision [13]. In order to increase the adhesion between the membrane and the textured anode, a very thin layer of an

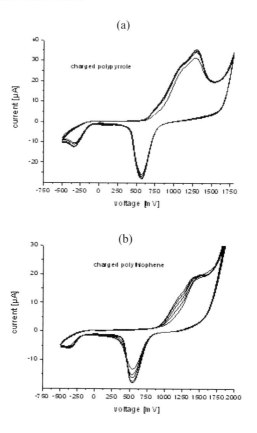

Figure 8.27. Potentiodynamic electropolymerization of the charged polypyrrole (*a*) and the charged polythiophene (*b*) in aqueous solution.

insulating gel (around 300 nm thick) was introduced. The gel layer was deposited by spin coating on the electrode, then the membrane was mounted directly on the gel film which acted as a glue bringing the membrane (Millipore, 1 μm pore size) into very close contact with the electrode. Afterwards, a standard DEP process was performed which produced polymer wires starting from the Au pad, crossing the gel and continuing their growth through the membrane pores onto the surface of the membrane. The isolating gel avoided any leakage current between the membrane and the Au pads on the electrode. Only when the Au pads are located on the electrode, does the polymer cross the gel and form a wire through the pores.

We first present some examples of PPy volume micropatterning of PC membranes that are produced as interconnection layers for 3D chip stacking. Figures 8.28(*a*) and (*b*) show SEM images of a structured anode with statistically distributed circular gold pads (8 μm in diameter) and the corresponding polymer micropattern of the membrane. In figure 8.28(*b*), the PPy microstructures on

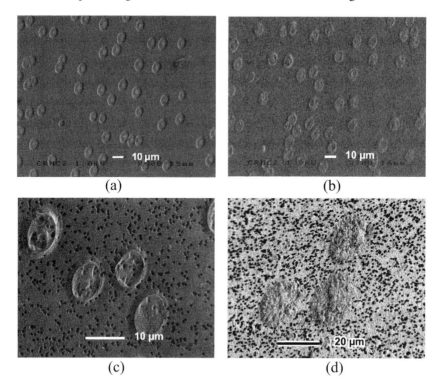

Figure 8.28. SEM pictures of PPy volume micropatterning of a polycarbonate membrane obtained by the DEP-2 process: (*a*) the patterned anode used in the DEP process; (*b*) and (*c*) PPy micropattern on the front side of the membrane; (*d*) polymer micropattern on the back side corresponding to PPy wires grown across the membrane.

the surface of the membrane that was turned towards the anode, i.e. front side of the membrane, show circular polymer pads 8 μm in diameter and a height of approximately 1 μm, representing a perfect replica of the gold pad pattern of the anode. At higher resolution (figure 8.28(*c*)) it can clearly be seen that the pores of the membrane, located only a few hundreds of nm beside the polymerized areas, are still open and, thus, unfilled, which reveals the precision of the DEP technique. The PPy micropattern obtained on the opposite side of the membrane, i.e. back side, is shown in figure 8.28(*d*). The circular microstructures correspond to polypyrrole wires grown across the membrane. They clearly stick out of the substrate surface, which is very important for achieving good mechanical contact with the polymer wires by external metal electrodes in a 3D chip stacking. Contrary to the front-side pattern, an enlargement of the polymer structures compared to the metal pads of the anode is observed, probably due to

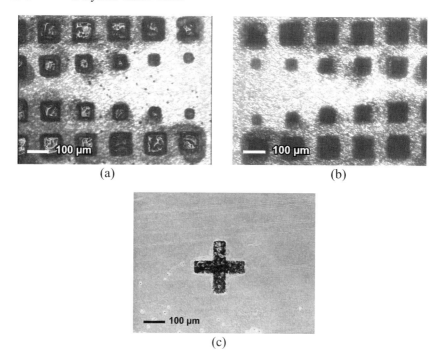

(a)   (b)

(c)

Figure 8.29. Optical images of PPy volume micropatterning of a polycarbonate membrane obtained by the DEP-2 process: (*a*) polymer microstructures on the front side of the membrane and (*b*) on the back side corresponding to PPy wires grown across the membrane; (*c*) cross-shaped polymer pattern, used as an alignment mark.

interconnected pores inside the membrane and the preferential 2D growth of the polymer when reaching the free surface [17].

In figure 8.29, another example of a PPy volume micropatterning consisting of square pads of different sizes is shown. It can be clearly seen that, for large-area patterning a direct replica of the anode structure can be produced on the front side (figure 8.29(*a*)) and through the volume to the back side of the membrane (figure 8.29(*b*)). Furthermore, a cross-shaped mark shown in figure 8.29(*c*) was polymerized through the membrane. This cross shape corresponds to the cross-shaped alignment marks which exist both on the lower and upper chip. Therefore, the alignment marks in the form of conducting polymers allow chip stacking, with the polymer wires in the transfer layer aligned relative to the electrode on the lower and upper chips, respectively.

The electrical characterization of the volume pattern results in conductivity values of 50 S m^{-1} for the pattern as in figures 8.29(*a*) and (*b*), and 1 S m^{-1} for the statistical pattern in figures 8.28(*c*) and 8.28(*d*). While the lower conductivity value of the second pattern is related to the low density of polymer wires over the

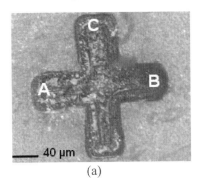
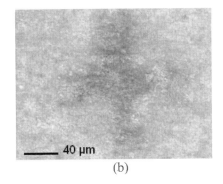

(a)   (b)

Figure 8.30. Optical images of a PPy cross deposited as surface micropatterning on a polycarbonate membrane obtained by the DEP-2 process: (*a*) front side and (*b*) back side of the membrane.

membrane surface, an improved DEP electrodes design should lead to a strong increase in the electric performance of such patterned membranes.

Regarding the 3D chip stacking problem, the polymer patterns shown in figures 8.29(*a*) and (*b*) have to be aligned with the metal contacts of the test chips. The circular polymer patterns (figure 8.28), due to the small size of the PPy structures in comparison with the large metal pads as well as their statistical distribution, could be measured without any alignment, offering an important advantage for technical applications in 3D chip stacking.

In order to perform surface patterning of PC membranes, we produced shorter polymerization times during the DEP-2 process. The polymer cross that is shown in figure 8.30 is only formed on the front side of the PC membrane with a high resolution, but no polymer formation, only the shadow of the pattern produced on the front side due to the transparency of the membrane, is observed on the back side (figure 8.30(*b*)) revealing a pure PPy surface patterning. This result demonstrates the general applicability of the DEP-2 process to perform both volume and surface patterning of conducting polymers with high resolution on flexible substrates such as PC membranes. The feasibility of both surface and volume polymer patterning on the same membrane could be demonstrated by the use of a more complex electrode (see section 8.11.3) that possesses separately-addressable metal pads in combination with different polymerization times for electropolymerization at selected areas.

Electrical characterization of the surface pattern results in a very high conductivity of 2000 S m^{-1} along the surface and no conductivity across the membrane revealing the high performance of the DEP-2 technique for surface and volume patterning of nanoporous membranes.

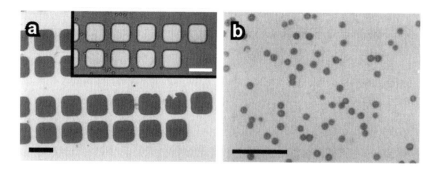

Figure 8.31. Optical images of a PPy volume micropatterned gel layer obtained by the DEP process: (*a*) square-shaped polymer pads and (*b*) circularly shaped polymer pads. Insert of (*a*): anode used for the DEP patterning of the gel layer. Scale bar: 100 μm.

8.13.2   Micro-patterning of gel layers

A self-supporting layer, based on an insulating gel layer, was micropatterned using the DEP-2 process with conducting PPy wires. The thickness of the insulating gel layer, which was deposited directly onto the structured anode of the DEP-2 process via spin coating, was 2–5 μm. During the DEP-2 process, polymer wires were grown across the gel layer. Results from the PPy volume patterning of such gel layers using the DEP-2 process are shown in figure 8.31. For both types of anode, i.e. a structured anode with statistically distributed circular gold pads and an electrode consisting of square-shaped pads, we demonstrate the versatility and high resolution of the DEP process on PMMA-based gel layers. In order to use such a polymer wires containing gel layer as interconnection layer in a 3D chip stack, the patterning of the gel, due to its strong adhesive property, has to be performed directly on the lower chip. Moreover, in a three-chip stack, a combination of the two types of interconnecting molecular layers, PC membranes and gel layers with different physical and mechanical properties, could offer great advantages for 3D chip stacking and all-plastic electronics.

8.14   Time dependence of the conductivity

In order to evaluate the possible use of polymer wires for interconnects, it is of particular interest to study the time evolution of the conductivity. We measured the time dependence of the conductivity of the polymer wires at different voltages (1 V and 10 V), where the polymerized membranes were placed between two Cu-electrodes as shown in figure 8.32.

The dc conductivity at 1 V of polypyrrole wires which were formed with the DEP-2 process in a 1.2 μm porous membrane shows some fluctuations and an average decrease in conductivity of 20% in 5 hr as can be seen in

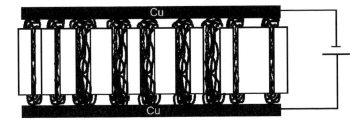

Figure 8.32. Schematic setup for measuring the time dependence of the polymer conductivity of the polymer wires.

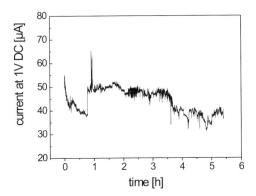

Figure 8.33. Time dependence of the conductivity of the polypyrrole polymer wires at the voltage of 1 V.

figure 8.33. Thus, at a voltage of 1 V, the conductivity measurement revealed a time stable conductivity for the polymer wires of 50 S m^{-1}. However, when the voltage is increased to 10 V, the polymer wires show a completely different behaviour (figure 8.34). A significant reduction in the conductivity of two orders of magnitude with time can be observed. Furthermore, when the polarity is changed at the same contact (figure 8.34(*b*)), the conductivity continues to drop by a further two orders of magnitude. Both curves demonstrate an exponential time dependence of the conductivity, see the inset in figure 8.34(*a*). The results indicate a reduction in the conductivity induced by the diffusion of the negatively charged dopants (ClO_4^- ions) in the electric field. The migration of the dopants creates undoped and, thus, insulating polymer regions located at the negative electrode, and, therefore, reduces the conductivity of the wires by several orders of magnitude. However, the results obtained at lower voltages of 1 V indicate that there is a certain binding energy of the dopants to the polymer, keeping the

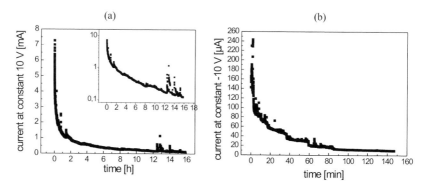

Figure 8.34. (*a*) Time dependence of the conductivity of the polypyrrole polymer wires at a voltage of 10 V; (*b*) the same contact after the change of polarity.

polymer conductive at low voltages, and only when a high voltage is applied, do the dopants start to migrate.

The migration of the dopants in the polymer matrix is possible because these ions are not attached directly to the polymer chains and can move along the chains. One possible way to prevent the migration is to attach the dopants covalently to the polymer matrix. The functionalized monomers of pyrrole and thiophene (see section 8.2.1) have an SO_3^- group substituted with the monomer via an alkyl chain. This charged group acts like a dopant in the polymer and, therefore, should lead to a conductivity almost independent of the electric field applied to the polymer. The time dependence of the conductivity measured on polymer wires based on a functionalized polypyrrole reveals strong fluctuations as well as a constant average conductivity as can be seen in figure 8.35. The fluctuation in the conductivity can be explained by the large number of charges inside the polymer, specifically one charge per monomer unit, which reacts with compression and repulsion in parallel under the electric field and leads, therefore, to a cyclic response to the conductivity through the polymer wires. Furthermore, repulsion between the charged monomers, which disturbs the close packing of the polymer chains and, thus, the effective charge transport inside the polymer, can be used as an explanation for the low conductivity observed in these functionalized polymers in general. However, the constant average of the conductivity indicates that the attachment of dopants to the polymer matrix presents a promising technique for preventing dopant migration, although the conductivity of these polymers is too low at the moment. In order to increase their conductivity, the fabrication of a copolymer of alternating functionalized and normal monomers could improve the organization of the polymer chains and, thus, the transport through the polymer.

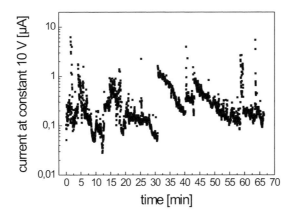

Figure 8.35. Time dependence of the conductivity of the functionalized polypyrrole wires at a voltage of 10 V.

8.15 3D chip stack

For the functional modelling of complex structures with high-density 3D connections (such as the visual cortex), a simple structure was proposed, with through-chip connections and molecular wires between layers. On the road to a 3D chip stack, we investigated the feasibility of the electrical transmission of a signal from a lower chip to an upper chip by the molecular conducting wires in a two-chip and three-chip stack represented on figures 8.36(a) and (b), respectively [14]. In figure 8.36(a), the input electrical signal (I_{input}) starts from the lower chip and a continuous conducting path is made from the lower chip, through a molecular wire connection to the upper chip, then back from the upper chip to the lower chip where the output signal (I_{output}) is collected. In the case of the three-chip stack, a middle chip equipped with the through-wafer plugs connecting pads on the upper and lower surfaces ensured a continuous path between the two intermediate conducting molecular layers. In such vertical alignments, diagonal paths due to the connectivity between molecular wires on the same intermediate layer, will not achieve high conductivity values.

The 3D chip-stacking technology, presented here and described in more detail in chapter 2, involves structured chips as represented in figure 8.37. The lower chip shown on figure 8.37(a) has a silicon or glass substrate. The profile structure of one patterned Au/Cr pad on the Si substrate is schematically represented in figure 8.38(a). On top of the SiO_2 layer, a layer of metal is deposited and patterned to form a number of metal bonding pads (not shown on figure 8.37(a)—connected to the electronic part of the chip carrier), each of

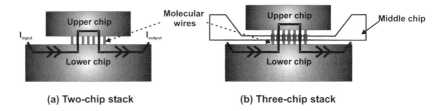

(a) Two-chip stack   (b) Three-chip stack

Figure 8.36. Schematic representations of (*a*) two-chip and (*b*) three-chip stacks with a continuous conducting path between the different layers.

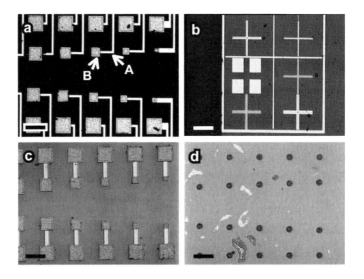

Figure 8.37. Surfaces of chips used in two- and three-chip stacks: (*a*) square-shaped pads in the lower chip, (*b*) cross-shaped pads in the lower chip, (*c*) square-shaped interconnected pair pads of the upper chip and (*d*) middle chip. Scale bar 100 μm.

which is connected via a metal track (A) to a smaller pad (B) that is used for attachment of the molecular material. The tracks and bond pads are fabricated in the same metal layer, a Al:1% Si layer (0.6–0.8 μm thick). The metal structures are coated with an electrically insulating layer (silicon oxide, 0.8 μm thick) containing windows to allow connections to the pads (bonding and attachment pads) while preventing electrical connection to the tracks.

The 'molecular-wire' attachment pads are of a different material (Au/Cr) deposited over and in electrical contact with the bond pad/track metal layer. The size of the attachment pads will define the size of the conducting molecular wires used for the electrical signal transfer. The attachment pads (B) have a square size ranging from 100 μm × 100 μm to 20 μm × 20 μm. Cross-shaped alignment

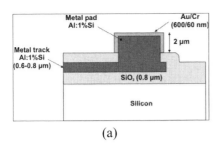
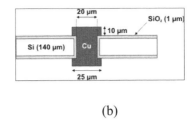

(a)   (b)

Figure 8.38. Schematic representations of the cross section: (*a*) a Au/Cr pad on lower and upper chips on a silicon substrate and (*b*) a Cu-plug in the partly thinned middle chip.

marks (figure 8.37(*b*)) are placed in each corner of the lower chip. Such marks are deposited in order to facilitate the alignment process of chips using a microscope. The lower and upper chips are fabricated on the same starting material (silicon or glass wafers) with the same preparation process and successively separated in two different layers as a last fabrication step. The upper chip (figure 8.37(*c*)) has electrode pads with the same size and design as those on the lower chip. Pairs of pads are connected together to provide electrical continuity on the upper chip before going back to the lower chip through the interconnection medium. The upper chips have alignment marks etched on their back surfaces to correspond with matching marks defined in metal on the lower chips.

In order to form a three-chip stack, a partly thinned middle chip having through-wafer copper plugs connecting pads on the upper and lower surfaces is used (figure 8.37(*d*)). Copper plugs as small as 20 μm $\times$ 20 μm and as deep as 140 μm are obtained—for details see chapter 11. Figure 8.38(*b*) gives a schematic representation of the Cu plug. The test rig (figure 8.39, described in chapter 2) is a multiple-axis translation/rotation system that allows the upper chip to be moved and positioned over the lower chip by the manipulating arm equipped with micromanipulators. An optical microscope is used for checking alignment (accuracy ±2 μm). The lower chip is mounted in a 40-pin chip carrier on a printed circuit board connected to the electronic part of the test rig. This allowed signals to be sent on and received from the chip. In the two-chip stack, the upper chip is directly attached to the manipulating arm while, in the three-chip stack, a vacuum-head-equipped arm is used. In this case, the middle chip is first positioned and deposited on the lower chip equipped with the molecular wires and, in the last step, the upper chip vacuum attached to the arm is aligned to the middle chip covered on its top surface by the second molecular wires based layer. In both cases (two- and three-chip stacks), the manipulating arm allows a controlled pressure to be exerted on all the interconnected layers.

The vertical interconnections between the chips are based on conducting polymer-based wires such as PPy volume micropatterned in transparent flexible

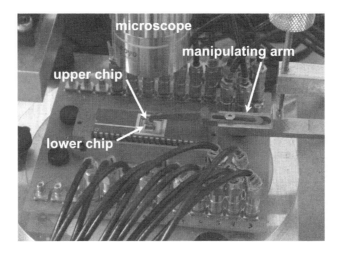

Figure 8.39. The test rig for chip positioning and conductivity measurements.

layers (see section 8.13). The conducting molecular wires are exact copies of the metal attachment pads (square- and cross-shaped pads) in order to be accurately aligned with the chips using a microscope. To measure the 3D conductivity, we put the polymer-patterned layers in contact with the metal pads of the chips, combining the test rig with a Hewlett-Packard 4140B pico-amperemeter-dc voltage source. By using the cross-shaped alignment marks placed in each corner of the chips and reproduced in the intermediate polymer-based layers, an alignment process can be realized step by step in the two- and three-chip stack. Figure 8.40 gives a schematic fragmented 3D representation of the three-chip stack with the detailed design of each layer (lower, middle and upper chips as well as intermediate 'molecular interconnect' layers). The full bottom-up vertical lines represent the electrical signal path in the three-chip stack starting from an input electrical signal in the external connection noted 1 and going out from the external connection noted 2.

The ideal case is to have high-density, self-assembled molecular wires, which connect only those pads which are in vertical alignment (measured by the conductivity σ) but with no 'diagonal' connectivity (measured by the crosstalk conductivity σ_{CT}). As already described in section 8.13, the DEP-2 process offers a high quality of growth control which could be proved by measuring both the 3D conductance between an external connection pair, 1 and 2 as shown in the example in figure 8.40, and the crosstalk conductance between the external connection 1 and a connection from 3 to 16. Table 8.2 summarizes the conductivity values (σ) and the crosstalk ratio (CT) in the two- and three-chip stacks. For each reported value, the nature of the intermediate patterned layers and the surface

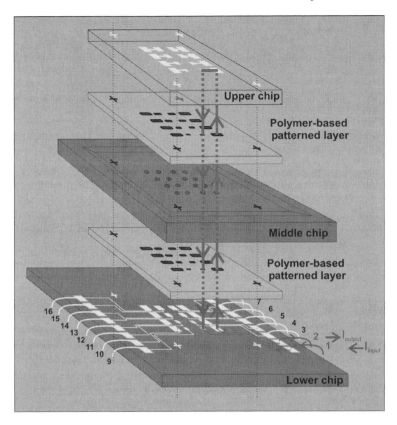

Figure 8.40. Schematic fragmented 3D representation of the three-chip stack using polymer volume micropatterned PC membranes.

design of the upper chip used in the 3D-stack are noted. The lower chip used in each stack involving membranes is as represented in figure 8.37(*a*). Initial conductivity measurements, as described previously in section 8.10, with a flat surface metal upper chip on a polymer-patterned membrane by DEP-2 is reported in table 8.2 as a reference for comparison with the conductivity obtained on such layers in which the upper chip is equipped with pads. Preliminary conductivity measurements on a polymer-patterned gel layer were realized on a gel layer as in figure 8.31, obtained directly on a DEP-2 anode on which a flat-surface upper chip had been deposited. Results for such a two-chip stack involving a gel layer give a conductivity of 20 S m^{-1} and a crosstalk ratio of 10^5, which are comparable with those of patterned membranes obtained in the same conditions.

The conductivity and crosstalk measurements reported here in effective 3D chip stacks using pads-equipped upper chips are obtained on interconnection layers consisting of polymer volume micropatterning of PC membranes. In a

Table 8.2. Conductivity value (σ) and the crosstalk ratio (CT $= \sigma/\sigma_{CT}$) in the two- and three-chip stacks. For each reported value, the nature of the intermediate patterned layers and the surface design of the metal upper chip used in the 3D-stack are noted. The lower chip used in each stack process involving membranes is as represented in figure 8.37(*a*). With a gel as the intermediate layer in the two-chip stack see the text for specific conditions.

3D chip stack	Intermediate patterned layer	Metal chip surface of the upper chip	σ (S m^{-1})	CT
Two-chip	(a) membrane	flat	80	—
Two-chip	(b) membrane	pads equipped	50	$\gg 10^9$
Two-chip	(c) gel	flat	20	$\gg 10^5$
Three-chip	(d) membranes	pads equipped	10^{-4}	$\gg 10^4$

two-chip stack, the patterned membrane is aligned in a sandwich configuration between the upper and the lower chips. The electrical characterization results in conductivity values of 50 S m^{-1} for patterns such as the one in figure 8.29(*a*) which is in the same range as the 80 S m^{-1} reference conductivity and both values are in good agreement with conductivity values measured on polypyrrole wires obtained by other pattern techniques [18]. In the case of the circular polymer patterns (figure 8.28(*b*)), due to the small size of the PPy structures compared to the large metal pads as well as their statistical distribution, their conductivity values could be measured without any alignment. A drop in the conductivity to 1 S m^{-1} (not reported in table 8.2) is observed which we assign to the low wire density of the pattern. However, an increase in the density should lead to a substantial improvement in the conductivity. It has to be emphasized that no crosstalk exists for either polymer patterns (CT $\gg 10^9$) giving a direct proof of the principle of our new 3D-chip stacking technology involving molecular interconnections. Furthermore, we realized a three-chip stack such as that represented in figure 8.40. A conductivity value of 10^{-4} S m^{-1} and a crosstalk ratio of CT $\gg 10^4$ for the polymer-patterned pads were obtained. The poor conductivity of the whole polymer wire system inside the 3D stack compared to that of the single polymer wire layer can be attributed to the very simple alignment technique used in this work. Misalignment of the different layers (three chips and two polymer-based wire layers) with respect to each other as well as poor mechanical contacts between the patterned membranes and the chips are possible reasons for conductivity losses. It has been noted that, at this stage of the stack process, the three-chip stack was not optimized like the two-chip one where we could demonstrate the feasibility of 3D chip stacking using polymer wire based interconnection layers with a highly directional conductivity needed for such electrical transfer.

8.16   Conclusion

Some important steps on the road to self-supporting polymer-based wiring layers of high quality, used in 3D-chip stacking, have been experimentally achieved. Self-assembled polymer wires have been produced by different polymerization techniques like chemical polymerization (CP), electrochemical polymerization (ECP) and a new directional electropolymerization technique (DEP). Standard conducting polymers such as polypyrrole as well as functionalized charged polymers based on pyrrole and thiophene monomers were used for the fabrication of the polymer wires. The study of polymer growth through the membranes, as an insulating self-supporting layer, gave important results towards a deeper understanding of the possibilities and limits for the fabrication of conducting channels. We could demonstrate good values for the conductivity and crosstalk ratio of polymer wires produced in nanoporous membranes and investigated strategies to improve the performance towards an optimized self-supporting polymer-based wiring layer. Although conductivities up to 14 S m^{-1} were found for ECP-processed membranes, it was difficult to avoid a low crosstalk ratio of 46. The conductivity measurements revealed the need for a new technique in order to operate with real growth control, thus obtaining a high conductivity value without additional crosstalk at the same time. Therefore, a new polymerization technique was developed which was based on the combined use of preformed anodes that possess pattern-transfer elements and a gel layer as intercalating electropolymerization medium between the anode and the membrane to be patterned. With this so-called DEP, we could demonstrate a high-resolution patterning of a porous membrane with conducting wires with a diameter less than 10 μm and without any crosstalk between the wires. Thus, a very powerful technique for fabricating transfer layers based on polymer wires inside porous membranes was developed that has a hundred fold higher wire density than actual chip stacking techniques based on solder balls. The conductivity measurement obtained on the polymer wires produced by the DEP process revealed a high conductivity of the polymer wires up to 50 S m^{-1}. The time-dependence measurements of the conductivity at different voltages demonstrate that, at low voltages, the conductivity of the polymer is stable but when higher voltages are applied, a migration of the dopants of the polymer at the electrode interface occurs leading to important losses in conductivity. The use of functionalized monomers with a substituted dopant presents a promising approach towards stabilizing of the conductivity. We have successfully designed and fabricated a two-chip stack with a perfect vertical electrical transfer between layers without any appearance of crosstalk. The measured values of the conductivity for the polymer-based wires were as high as 50 S m^{-1}. Furthermore, we could demonstrate the feasibility of a three-chip stack by using two intermediate polymer-based patterned layers and a special middle chip with through-chip metal connections.

Acknowledgments

We are especially grateful to C Rebout (LMMB, Marseille France) for his helpful technical assistance in CP and ECP measurements and to C Pain (LMMB, Marseille France) for the synthesis of charged monomers (pyrrole and thiophene). We would like to thank Professor F Fages (LMMB, Marseille, France) for fruitful discussions on charged monomers and the DEP-2 process. We are also grateful to Ing. S Nitsche from the CRMCN of Faculté des Sciences de Luminy in Marseille, France, for SEM pictures.

References

[1]   Zotti G 1992 *Synthetic Met.* **51** 373
[2]   Holdcroft S 2001 *Adv. Mater.* **13** 1753
[3]   Angeli A 1916 *Gazz. Chim. Ital.* **46** 279
[4]   Friend R H 1993 *Conductive Polymers II—From Science to Applications (Rapra Review Reports)* **6**
[5]   Duchet J, Legras R and Demoustier-Champagne S 1998 *Synthetic Met.* **98** 113
[6]   Ando E, Onodera S, Lino M and Ito O 2001 *Carbon* **39** 101
[7]   Fichou D (ed) 1999 *Handbook of Oligo- and Polythiophenes* (Weinheim: Wiley)
[8]   Martin C R 1994 *Science* **266** 1961
[9]   Martin C R 1991 *Adv. Mater.* **3** 457
[10] Ackermann J, Crawley D, Forshaw M, Nikolic K and Videlot C 2003 *Surf. Sci.* **532–535** 1182
[11] Ackermann J, Videlot C, Nguyen T N, Wang L, Sarro P M, Crawley D, Nikolic K and Forshaw M 2003 *Appl. Surf. Sci.* **212–213** 411
[12] Crawley D, Nikolic K, Forshaw M, Ackermann J, Videlot C, Nguyen T N, Wang L and Sarro P M 2002 *J. Micromech. Microeng.* **13** 655
[13] Ackermann J, Videlot C, Nguyen T N, Wang L, Sarro P M and Fages F 2004 *Adv. Mater.* accepted for publication
[14] Videlot C, Ackermann J, Fages F, Nguyen T N, Wang L, Sarro P M, Crawley D, Nikolic K and Forshaw M 2004 *J. Micromech. Microeng.* submitted
[15] Havinga E E, ten Hoeve W, Meijer E W and Wynberg H 1989 *Chem. Mater.* **1** 650
[16] Chayer M, Faïd K M and Leclerc M 1997 *Chem. Mater.* **9** 2902
[17] Shan J, Yuan C and Zhang H 1997 *Thin Solid Films* **301** 23
[18] Zhou F, Chen M, Liu W, Liu J, Liu Z and Mu Z 2003 *Adv. Mater.* **15** 1367

Chapter 9

Discotic liquid crystals

A McNeill, R J Bushby, S D Evans, Q Liu and B Movaghar
Centre for Self-Organizing Molecular Systems (SOMS),
Department of Chemistry, and Department of Physics and
Astronomy, University of Leeds

This chapter summarizes the work performed by the SOMS centre at the University of Leeds to demonstrate the feasibility of very high-density, three-dimensional (3D) molecular 'wires' between electrical contacts on separate, closely spaced, semiconductor chips or layers. The Leeds group have investigated the use of molecular wires based on discotic liquid crystals [1–13].

9.1   Introduction

The properties of metals and semiconductors are well known and their use in device applications is well documented. However, with increasingly more complicated electronic systems being devised, materials with equally complicated and novel properties are being sought. One particular area that has recently been exploited is that of organic semiconductors. Organic semiconductors offer comparable levels of performance but with the increased functionalization and an ability to be used on the molecular level to allow the production of true nanoscale electronic devices. One particular system that falls into this category is that of discotic liquid crystals.

Conventional materials usually have two condensed states of matter: the crystalline solid phase which has the lowest temperature and, hence, the greatest amount of order and, in contrast, the isotropic phase which contains no order and consists of the molecules in a constant random motion. The liquid crystal phase or mesophase is a completely separate condensed phase that lies between the two and has molecular orientational order but only has positional order (if at all) in one or two dimensions. This orientational order within a liquid crystal leads to a high degree of anisotropy for many of the properties including electrical

Figure 9.1. Representation of a discotic liquid crystal column. The central light grey area shows the rigid core that exhibits strong π–π interactions therefore allowing effective charge conduction along the centre of the column. The dark grey area is the insulating alkyl chains that electrically insulate the central column from it's neighbours.

conduction. It is this anisotropic behaviour of the conduction that makes these materials attractive in novel electronic devices.

Discotic liquid crystals are liquid crystals whose molecules have one molecular axis shorter than the other two and, hence, are disc-shaped. This flat disc-shaped molecule results in extensive molecular π orbitals above and below the plane of the disc. Favourable interactions between the π orbitals of neighbouring molecules leads to an attraction between the molecules and they then prefer to align flat against one another. This interaction is the mechanism that gives rise to the mesophase. Multiple π–π interactions between many molecules give rise to the disc-like molecules self-assembling into a column of molecules and, hence, the discotic columnar phase Col_h as shown in figure 9.1.

The practical importance of these systems can be seen when the structure of these columns is examined in closer detail. The individual discotic molecules consist of a rigid core containing the overlapping π orbitals and a flexible periphery consisting of alkyl chains. Due to the overlapping nature of the central π orbitals and their columnar assembly, a highly conductive one-dimensional pathway is created along the centre of the stack. Not only is there conduction along the stacks but also each individual column is insulated from neighbouring stacks by the flexible alkyl side chains. This mesomorphic configuration of the molecules leads to the bulk sample having highly anisotropic conductive behaviour, with favourable conduction along the core of the columns. This anisotropy has been reported to be as high as 10^3.

The discotic liquid crystal column therefore consists of a central column of conducting aromatic cores surrounded by insulating alkyl chains. This means that charge can be transported in one dimension along the conducting cores but it is significantly shielded from its surroundings. This is a 1D molecular wire. By their very nature, molecular wires are delicate objects, prone to defects and difficult to manipulate. Herein lies the interest in discotic liquid crystal wires: not only do they self-align between electrodes, they also self-heal if broken. The specific objective of the research is to demonstrate the feasibility of very high-density, molecular 'wires' between electrical contacts on separate, closely spaced, semiconductor chips or layers.

9.2   Conduction in discotic liquid crystals

At first glance, these self-aligned columns possess an enviable array of properties that would make them ideal for use as molecular wires. Nevertheless, problems with these systems do exist, reducing their effectiveness. First, the mobility of charge carriers within the system is intrinsically low (7.1×10^{-4} cm^2 V^{-1} s^{-1} for hole mobility in HAT6 in the columnar phase at $110\,^{\circ}$C cf Si $\mu = 10^2$ cm^2 V^{-1} s^{-1}). Second, the intrinsic charge carrier concentrations are very low and charge injection is required. These limitations reduce the effectiveness of liquid crystals for use in electronic applications. Methods of improving the electrical conduction in these systems have been attempted in a number of ways. First, their low oxidation potential allows comparatively easy formation of radical cations upon the introduction of electron acceptors such as AlCl$_3$ and NOBF$_4$ into the mixture. It has been shown that the presence of these radical cations enhances the charge carrier conductivity along the columns in these systems. Another method for achieving higher conductivities has been achieved by increasing the size of the core of the liquid crystal column. The mobility achieved by this method is as high as 0.46 cm^2 V^{-1} s^{-1} for holes in the Col$_{h3}$ phase of peri-hexabenzocoronenes at $192\,^{\circ}$C. An alternative method for increasing the conduction in these systems is to increase the molecular order within the columnar stack. Liquid crystal systems can be chosen so as to maximize the π–π interactions and, in turn, increase the columnar order. For example 2,3,6,7,10,11-hexakis(hexylthio)triphenylene has a reported mobility of 1×10^{-2} cm^2 V^{-1} s^{-1}. Increasing the columnar order can also be achieved by mixing two or more complementary liquid crystals as shown in the 'CPI' mixture HAT6-PTP9. These systems optimize favourable interactions between alternating molecular species to increase the columnar order and, hence, the charge carrier mobility.

By sandwiching the discotic liquid crystal between two ITO (indium tin oxide) electrodes spaced at a distance of a few micrometres and annealing just below the I/Col transition temperature, the columns can be made to self-assemble electrode-to-electrode. I/V curves can then be measured. This typically gives rise to an I/V curve as shown in figure 9.2.

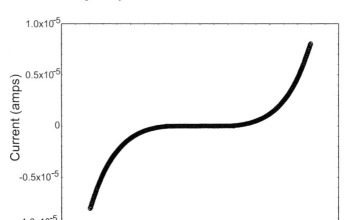

Figure 9.2. Typical I/V curve for a thin homeotropic discotic liquid crystal layer sandwiched between ITO electrodes.

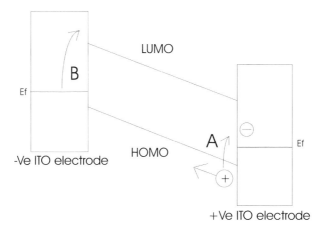

Figure 9.3. Band diagram showing energy levels of discotic liquid crystal with migration of holes.

The graph in figure 9.2 can be explained by splitting it into two regions, the ohmic region (below 0.05 MV cm^{-1}), and the charge injection region (greater than a critical field which, for HAT6, is 0.05 MV cm^{-1}). The ohmic region is highly resistive due to the potential barriers existing between the electrodes and the discotic liquid crystal. Figure 9.3 shows the potential barriers for the case with ITO electrodes and an applied field.

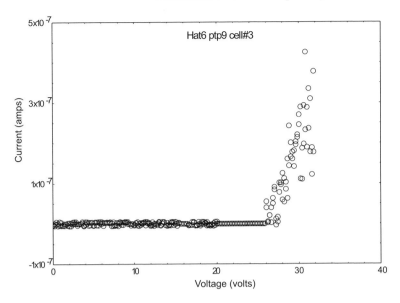

Figure 9.4. I/V curve for the ITO/HAT6-PTP9/ITO system at 175° spaced at 6 μm. Onset of the high conduction can clearly be seen due to the 'field anneal'.

If the field is sufficiently high it is possible to create an electron–hole pair and to promote the electron across gap A (the hole then accelerates towards the negative electrode). The ITO is, therefore, a hole injector. Although route A may be the preferred route for conduction, at room temperature the electrons will not have enough energy to overcome the potential barrier at A. Therefore, a potential has to be applied to reduce the barrier at A and to increase the conduction. If we take the case where the Fermi level of the ITO electrode has reached the level of the HOMO band, then significant charge injection of holes can occur.

9.2.1   Liquid-crystal-enhanced conduction—the field anneal

The behaviour described earlier is typical for most liquid crystal systems and for the HAT6 PTP9 system at low temperatures. However, one of the most striking results discovered at Leeds is the effect known as the field anneal. If the system is heated into the isotropic phase (175 °C) and an increasing field applied, the liquid crystal undergoes a transition. Figure 9.4 shows the onset of this transition.

The flat portion of this graph is consistent with the conduction as shown in figure 9.2 and has a current flow of about 10^{-8} A. However, at some critical field E_c, the conduction increases by five to six orders of magnitude and remains even after the field and elevated temperature are removed. It is only with a combination of high temperature and high field at the same time that the enhancement of the conduction occurs.

Figure 9.5(*a*) shows the I/V curve for the same sample as shown in figure 9.4 taken directly after this high field annealing. For equivalent fields, the current has increased from microamps to milliamps and the I/V curve shows a much weaker dependence on space charge injection. These systems also exhibit rectifying behaviour. If the high field annealing procedure is now repeated a second time, with a reversed polarity, the symmetric I/V plot shown in figure 9.5(*b*) can be obtained. This behaviour shows that the enhancement is probably due to some modification of the interface. Similar enhancements of the conductivity have been observed in HAT6, 1-fluoro-HAT6 and 1-nitro-HAT6. Measurements have been taken using Al and Au electrodes and they also show the enhancement behaviour. It should be noted that on cooling the HAT6–PTP9 mixtures take several months to crystallize but, upon crystallizing, the high conduction is lost and the sample requires 'field annealing' again to restore the high conduction.

9.2.2   Explanation of the field anneal phenomenon

A clear explanation of the field anneal process is vital when wishing to optimize the conduction. There are several possible causes. First, it is possible that the field anneal has created large amounts of ionic species within the sample that give rise to the enhanced conduction. Second, it may be that some improvement in alignment of the discotic columns is significantly improving charge conduction in the sample. Third, it is possible that domains within the sample could have been thermally annealed removing defects and therefore trapping sites for the charge carriers. There is also the possibility of the formation of ITO fibrils connecting one electrode to the other. Finally, it may be that some change at the interface is significantly improving the limiting behaviour of the potential barrier. This interfacial change could take one of a number of forms. There could be electrochemistry occurring at the interface: that is coating the electrode with a layer of charged polymeric or monomeric species, thereby reducing the barrier to charge injection from the electrode. The enhancement of the conduction could be due to a more simple process such as the discotic having much better wetting of the electrode. Finally, there is the possibility of the impurity ions forming a layer next to the electrode, again reducing the barrier to charge injection. This type of process has been observed in luminescent organic materials.

9.2.3   The domain model

At first sight, the domain model would appear to be the most reasonable since, to achieve higher conduction via domain removal, one would need to apply high fields and high temperatures just as used in the field anneal. This treatment would drive the system through a structural transition and could induce a sudden change in conduction. However, the domain model does not fit all of the experimental features. If we were to remove domains thermally, then there ought to be some

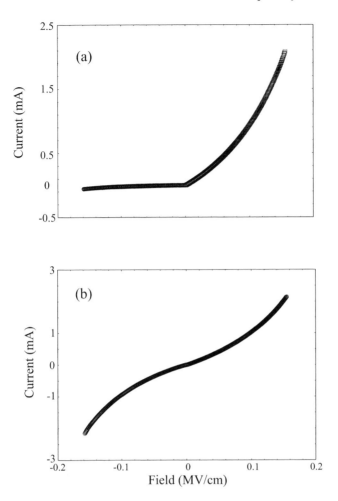

Figure 9.5. (*a*) Graph shows the field anneal but for only one field direction giving rise to the rectification behaviour. (*b*) Graph shows removal of rectification when field anneal is repeated in opposite field direction.

quasi-reversible behaviour and it should be possible subsequently to de-align the sample. In practice, this is not found. It is also very unlikely that domain removal would be able to generate the magnitude of the enhancements that have been observed.

9.2.4   The molecular alignment model

If it is assumed that the increased conduction is due to better alignment of the discotic columns, then possible explanations of this phenomenon due to the

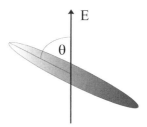

Figure 9.6. Discotic molecule aligned at angle θ to applied electric field.

presence of a field (as used in the field anneal) are not entirely obvious. First, we have to consider the behaviour of an individual discotic molecule aligned at some angle θ to an applied electric field (figure 9.6).

The molecule is anisotropic with respect to the polarizability depending on whichever molecular axis is measured. This means that, if the molecule is in a field E, then it will have different values for the polarization arising from its anisotropy. These polarizations are: $P_\parallel = \alpha_\parallel E \cos\theta$ and $P_\perp = \alpha_\perp E \sin\theta$. From this, the energy arising from the polarization can be obtained:

$$\text{Energy} = W = -\mathbf{P} \cdot \mathbf{E} = -\alpha_\parallel E^2 \cos^2\theta - \alpha_\perp E^2 \sin^2\theta.$$

The polarizabilities along the short and long molecular axes (for molecule core only not side chain) are: $\alpha_\parallel = 135.3 \times 10^{-40}\,\text{F}\,\text{m}^2$ and $\alpha_\perp = 172.5 \times 10^{-40}\,\text{F}\,\text{m}^2$. These values give an energy of interaction of the order of 10^{-7} eV, cf $kT \approx 0.03$ eV. From this it can be seen that rotation of individual molecules in a field is highly unlikely at room temperature; however, the numbers approach kT when the CPI discotic liquid crystal columns are examined.

In the binary mixtures composed of HAT6 and PTP9, greater intermolecular forces between molecules mean that columns are more ordered and have a much greater coherence length. This increased coherence length has implications for the polarization energy because, in a stack of n molecules, the polarization energy is the sum of individual energies in the stack: $\delta W = n\alpha E^2$. Therefore, in a stack of approximately 10 000 molecules the value for the polarization energy approaches the order of kT.

Another consequence of molecules stacked close to each other is that, if a dipole is induced in one molecule, then the field induced can affect the molecule next to it. This effect is known as the Lorentz correction and it modifies the previous equation as follows:

$$\delta W = \frac{n\alpha E^2}{1 - 4/3\pi \alpha n}.$$

This correction also increases the value for the interaction energy and makes alignment in a field almost energetically favourable.

The final consideration of the polarization energy is that of band formation. In a column of discotic molecules, the molecules sit closely together, which can result in the overlap of π orbitals and the corresponding sharing of electrons. In turn, this leads to the column developing its own band structure where electrons are shared along the length of the stack. By taking into account the development of a band structure along the length of the discotic column, the polarizability is calculated by using the following equation:

$$\alpha = \frac{2e^2 \sum_i \langle \psi_0 | x | \psi_i \rangle}{(E_0 - E_{ex}) - B}$$

where B is the modification due to banding and is equal to half the electronic bandwidth. In general, the polarizability is inversely proportional to the size of the band gap and the band gap is a function of the size of the physical gap between molecules.

When these corrections have been taken into account, then it may be possible that discotic columns will align in an electric field in such a way as to optimize the conduction. However, this explanation is very unlikely as it is reliant upon the fact that the average column coherence length is $\sim 10\,000$ and this is the extreme upper bound of the expected coherence length of the columns in a CPI mixture of HAT6 and PTP9 but is far too high for HAT6 on its own. This explanation is rendered further more unlikely by the fact that the field anneal is performed at elevated temperatures and, indeed, seems to require the higher temperature. Under these conditions, thermal fluctuations would dominate the ordering of the liquid crystal and destroy any ordering as soon as the field was removed.

9.2.5 Creation of ionic species

The second possibility for explaining the enhanced conduction in a field annealed sample is the production of large amounts of new ionic species. By applying a high voltage, it is possible that the molecules within the columns have been chemically modified. The resulting ionic species produced could well be responsible for the high conduction. One method of probing the origin of the charge carriers is to apply a dc field and monitor the current as a function of time. If the high conduction is due to the creation of ionic species, then the current will drop as a function of time as ions migrate to the electrodes and a double layer forms.

Figure 9.7 shows the current flow as a function of time for a sample that has undergone a field anneal. It can be seen clearly that the current drops exponentially as a function of the time. This indicates that some ions are present and that they are migrating to the electrode. However, the current decreases but still stabilizes to a high value indicating that the enhanced conduction is not solely due to the creation of ionic species.

This argument can be further discounted by looking at the dc electrical conduction measurements immediately after the first field anneal. Typically,

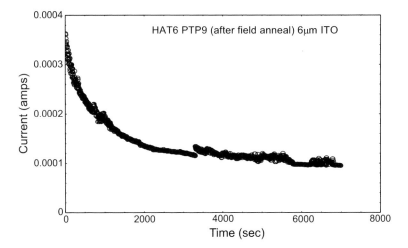

Figure 9.7. Time dependence of dc conductivity for HAT6 TTPA in a 6 μm ITO sandwich cell.

after one field anneal, the sample as a whole acts as an efficient rectifier, this rectification would not be observed in a sample that had had large numbers of ions produced in it. Furthermore, the production of ions would also be a permanent effect and results have shown that the field anneal disappears after a number of months and the sample returns to its original state.

9.2.6 Fibril growth

There have been a number of reports in the literature from people using sandwich cells with ITO electrodes. At elevated temperatures and fields, it has been reported that fibrils of ITO can extend from the surface of the electrode and form a conducting 'bridge' from one electrode to the other. This would lead to results consistent with *some* of the results observed in the field anneal. However, there are a number of important differences that rule out this possibility. First the rectification behaviour observed in the field-annealed samples is inconsistent with the hypothesis of fibril growth. Second, fibril growth from the ITO would give rise to a permanent increase in the conduction of the discotic liquid crystal and it has been shown that the field anneal is only a temporary effect.

9.2.7 Modification of the electrode/discotic interface

In figure 9.2, the flat highly resistive portion of the graph was attributed to the potential barrier at the interface, that has to be overcome before charge can be injected into the system. However, in figure 9.4, this flat portion of the curve is no longer present. This would therefore indicate that the potential barrier to charge injection at the electrode/discotic interface no longer exists. There are a number of possible explanations that might account for this. First, that the field anneal has changed the discotic liquid crystal molecules in some way (perhaps in creating some ionic species) and that this new material has a work function that is closer to that of the electrode and, therefore, charge can be injected into the system. This would not, however, be compatible with the rectification results nor the fact that the effect disappears after a number of months. Second, it is possible that the liquid crystal is wetting the electrode better. This would generate an improved interface and, in consequence, charge injection efficiency would be increased. The possibility of enhanced electrode wetting is not compatible with the dielectric results shown later in this chapter and it is unlikely that such a dramatic effect would be seen simply from electrode wetting.

A third possibility is that of polymerization on the surface of the electrode. Similar systems have shown that polymerization of liquid crystal molecules can occur on electrode surfaces when large enough fields are applied. This extra layer on the surface might have a work function somewhere between that of the electrode and the liquid crystal, which could aid the injection of charge from the electrode. This explanation seems unlikely as it would have a permanent effect on the conduction of the system and, in practice, it has been shown that the field anneal is only a temporary effect.

There is one final interfacial theory that could explain the field anneal effect and that is the formation of a layer of ions at the electrode/discotic interface due to the presence of intrinsic ionic impurities. Although the DLC used in this project are of high purity, there is the possibility that these trace impurities could migrate across the cell and accumulate. The mobility of these ions would be enhanced by the elevated temperature that would lower the viscosity of the discotic. The presence of even small numbers of ionic impurities concentrated in the interfacial region would have significant implications for the band structure of the discotic local crystal locally in the interfacial region which would, in turn, affect the charge injection properties.

9.2.8   Dielectric loss via migration of ions in a sandwich cell

The migration of ions across the cell and their subsequent build-up in the interfacial region has a large impact on the dielectric properties of the cell as a whole. Naemura *et al* [14] have investigated this type of dielectric loss. They looked at the attributes of mobile ions in a liquid crystal sandwich cell by analysing the complex dielectric constants in the low-frequency region. They looked at sandwich cells with ITO electrodes spaced at a distance of 9.2 μm containing the nematic calimitic liquid crystal 5CB. Measurements were taken in the frequency regime below 1 kHz. They then simulated the dielectric response of the liquid crystal cell by assuming a model that focuses on the migration of ions through the cell due to the presence of an electric field and the subsequent build-up of ions in the interfacial region next to the electrode. This dielectric model has

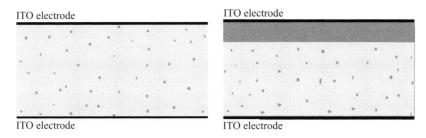

Figure 9.8. Schematic diagrams showing the presence of an ionic layer adjacent to the electrode area of a liquid crystal cell.

been used as a template for the investigation of the field anneal. The dielectric constant can be split up into real and imaginary components: $\varepsilon_i^* = \varepsilon_i' - i\varepsilon_i''$. Each component is treated individually as shown:

$$\varepsilon_i' = -\left(\frac{nq^2 D}{\varepsilon_0 \omega k T A}\right)\left|\frac{1 + 2\exp(A)\sin(A) - \exp(2A)}{1 + 2\exp(A)\cos(A) + \exp(2A)}\right|$$

$$\varepsilon_i'' = \left(\frac{nq^2 D}{\varepsilon_0 \omega k T}\right)\left|1 + \frac{1 - 2\exp(A)\sin(A) - \exp(2A)}{A\left(1 + 2\exp(A)\cos(A) + \exp(2A)\right)}\right| \quad (9.1)$$

with $A = d\sqrt{\omega/(2D)}$. In these equations, n is the ionic concentration, q is the charge on an electron, D is the diffusion constant of the ions, ε_0 is the permittivity of free space, ω is the frequency, k is the Boltzmann constant, T is the temperature and d is the thickness of the cell or layer.

9.2.9   Application of Naemura dielectric model to field anneal data

If an electric field is applied, the drift of ions in the columnar phase gives rise to a current of the order of pico-ampères. However, if the sample is heated to 175 °C, a transition to the isotropic liquid phase occurs and the viscosity is significantly reduced, therefore enabling the ions to migrate through the liquid more easily. If the sample is heated with a field for relatively long periods of time, then even though we have small numbers of ions present within the sample, they accumulate in the interfacial region forming a layer of ions next to the electrode. This behaviour is described schematically in figure 9.8. Dielectric measurements have been made for samples that have been both field annealed and non-field annealed. Measurements were taken in the frequency region of 100–0.001 Hz and the applied ac voltage was 1 V (rms).

Both curves in figure 9.9 are for the same sample. However the first (black curve) is for a sample that has not been treated in any way and has dc conduction similar to that as shown in figure 9.2 with symmetric I/V curves exhibiting charge injection behaviour. The second (grey curve) is for the same sample;

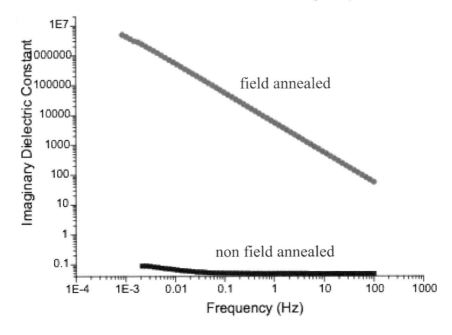

Figure 9.9. Imaginary part of the dielectric constant as a function of frequency. The black curve is for a sample that has not been treated in any way whilst the grey curve is for the same sample that has been field annealed in one field direction only.

however, in this case it has been treated with a single field anneal in one field direction. This single field anneal induces high conduction in the sample for one field direction only (the field direction in which the field anneal was performed) and the sample acts as a rectifier as shown in figure 9.5(*a*). This means that, in the negative field direction (the opposite direction of the field anneal), both samples show exactly the same dc current voltage characteristics but have nearly eight orders of magnitude difference in their imaginary dielectric constant values in the milliHertz region of the curve. The only possible reason for such a disparity in the dielectric constants for this sample when they have identical negative dc current characteristics is the presence of 'large' amounts of ions in the interfacial region next to electrode.

The equations developed by Naemura *et al* were then fitted to the data in figure 9.9. It was found that the model fitted well to the sample before it had been field annealed and fitted very poorly after the field annealed had been performed. The main problem with the fitting of the field-annealed sample was the magnitude of the response compared with the fitted data, although this can be easily understood when considering the structure of the field-annealed cell. In a cell that has been field annealed, there is a layer of ions held adjacent to the electrodes that has a set of physical properties completely different to the bulk material. When the

sample is cooled following the field anneal, the viscosity is significantly increased and the ions are 'held' in the interfacial region. Therefore, the bulk and interfacial regions should be considered independently with individual contributions to the dielectric properties. This means that the Naemura equations need to be expanded to include contributions for two independent regions giving rise to the following new equations for the dielectric constant:

$$\varepsilon_i' = -\left(\frac{n_1 q^2 D_1}{\varepsilon_0 \omega k T A}\right)\left|\frac{1 + 2e^A \sin(A) - e^{2A}}{1 + 2e^A \cos(A) + e^{2A}}\right|$$
$$- \left(\frac{n_2 q^2 D_2}{\varepsilon_0 \omega k T B}\right)\left|\frac{1 + 2e^B \sin(B) - e^{2B}}{1 + 2e^B \cos(B) + e^{2B}}\right|$$

$$\varepsilon_i'' = \left(\frac{n_1 q^2 D_1}{\varepsilon_0 \omega k T}\right)\left(1 + \frac{1 - 2e^A \sin(A) - e^{2A}}{A\left[1 + 2e^A \cos(A) + e^{2A}\right]}\right)$$
$$+ \left(\frac{n_2 q^2 D_2}{\varepsilon_0 \omega k T}\right)\left(1 + \frac{1 - 2e^B \sin(B) - e^{2B}}{B\left[1 + 2e^B \cos(B) + e^{2B}\right]}\right)$$

where $A = d_1 \sqrt{\omega/(2D_1)}$ and $B = d_2 \sqrt{\omega/(2D_2)}$.

These equations show the real and imaginary parts of the dielectric constant for a cell that consists of a liquid crystal in the bulk with a small number of ions and a separate layer of liquid crystal with a much higher concentration of ions at the interface. D_1 is the diffusion constant of ions in the bulk whilst D_2 is the diffusion constant of ions in the interfacial region, n_1 and n_2 are the concentration of the ions in their respective layers, d_1 is the thickness of the cell and d_2 is the thickness of the interfacial area. The values for the concentration of ions in the bulk and the thickness of the cells were already known. The value for the diffusion constant of the bulk was also calculated by the fitting of the one layer Naemura equation to the original sample that had not been field annealed. The two-layer equations were then fitted to a sample that had been field annealed. The values for D_2, n_2 and d_2 were adjusted so as to minimize the error between the fitted values in the model and the experimental results. The low-frequency dielectric spectroscopy measurements were repeated at intervals of two weeks and the results are shown in figure 9.10. Included in this diagram are theoretical fits to the data using the Naemura model.

In figure 9.10, the measurements are taken at intervals of two weeks, starting with the outer data points and moving through to the inner data points. After the initial fitting to the first curve, the only parameter that needed to be changed in going from one data set to the next over a period of months was the number of ions in the interfacial layer. This indicates that the only change within the sample over a series of months is the number of ions in the interfacial region as they drift away due to natural repulsion of the ions. This is in keeping with the observed behaviour of the field anneal where the high conduction of the liquid

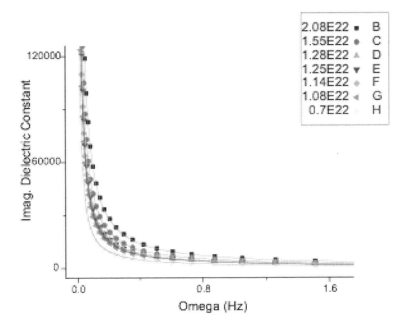

Figure 9.10. Graph showing the imaginary part of the dielectric constant as a function of frequency for a single-field-annealed sample that has been measured at intervals of two weeks. The full curves on the graph are theoretical fits to the data using the Naemura model.

crystal remains for 6–9 months but then returns to its original state. The number of ions in the interfacial region as a function of time is shown in figure 9.11.

As well as the number of ions in the interfacial layer, the thickness of the layer can be obtained. From the fitting the thickness has been found to be 4 nm. The diffusion constant of the ions in the interfacial region has also been calculated via the same method and has been shown to be 5×10^{-13} m^2 s^{-1}. The calculated number of ions in a 4 nm thick layer next to the electrodes is equivalent to one molecule in every thousand being an ion.

9.2.10   Effect of interfacial layer on conductivity of liquid crystal cell

Typical conduction in liquid crystal sandwich cells is dominated by the potential barrier at the electrode interface and subsequent injection of charge through this barrier. The presence of relatively large numbers of ions concentrated in the interfacial region reduces the potential barrier at the interface by concentrating the field applied across the cell at the interfaces rather than being dropped evenly across the cell. This phenomenon is similar to the build-up of interfacial charge and subsequent increase in photoluminescence that other groups have reported.

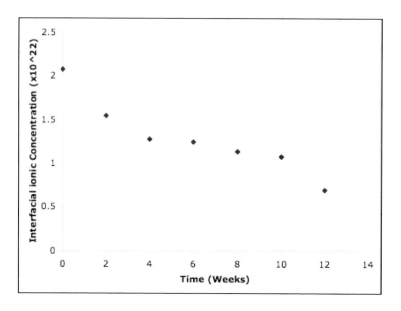

Figure 9.11. Variation in the number of ions in the interfacial region of the cell as a function of time.

9.3   Cortex stacked chips

9.3.1   Construction

The CORTEX chips were constructed using the chip aligner supplied by University College London (UCL), described in chapter 3. The first chips were made of two active layers, the bottom layer being attached to the electronic part of the test rig. This allowed signals to be sent on and received off the chip. The second part of the rig was the manipulating arm. The top chip was attached to this arm and micromanipulators allowed careful movement of the arm so that the top and bottom chips could be aligned exactly above each other. Some 5 μm thick PTFE sheets were cut to size and used as the spacing between the chips: this method has proved extremely accurate when it comes to reproducing the thickness of the cell.

9.3.2   Electrical conduction measurements of stacked chips

Three separate sets of chips have been made. Those with electrode sizes of 300 μm $\times$ 300 μm, 100 μm $\times$ 100 μm and those with 20 μm $\times$ 20 μm electrode pads. Initial measurements with the 300 μm pads showed conduction with currents in the range of picoamps before the field-anneal samples. Conduction after a field anneal through the molecular wires is shown in figure 9.12 for a

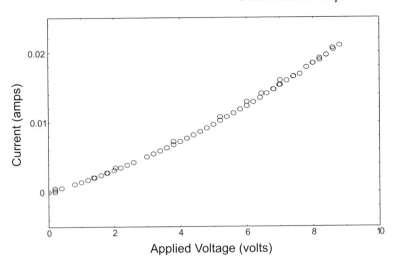

Figure 9.12. Graph showing the I/V curve for a CORTEX chip with 300 μm × 300 μm electrode pads.

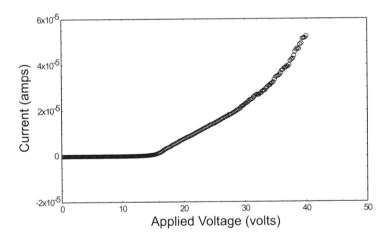

Figure 9.13. Graph showing the I/V curve for a CORTEX chip with 100 μm × 100 μm electrode pads.

CORTEX chip with 300 μm × 300 μm electrode pads. The I/V curve for the 100 μm × 100 μm electrode pads is shown in figure 9.13. Measurements have also been made on a third type of CORTEX chip with 20 μm × 20 μm sized electrodes (figure 9.14). Shown on the same graph is the conduction for the cross-talk between two non-connected electrodes on separate layers. It is so small as to be beyond the limits of the instrument at 10^{-12} A.

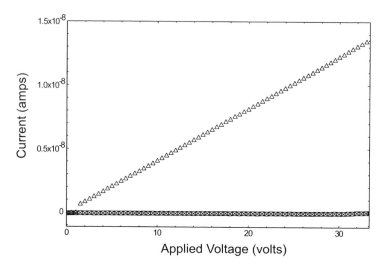

Figure 9.14. I/V graph for a Cortex chip with 20 μm $\times$ 20 μm electrodes spaced at 5 μm taken at 90 °C. The flat line along the bottom shows the leakage current between two non-connected electrodes on separate chip layers and is less than 10^{-12} A.

Table 9.1.

Electrode size (μm)	Current (A)	Current density (A m^{-2})
300 $\times$ 300	0.02	220 000
100 $\times$ 100	2×10^{-5}	2000
20 $\times$ 20	2×10^{-8}	25

One interesting point to note about these graphs is in the current density. As the electrode size is decreased, one would expect that the current per unit area would remain the same. However, we found a reduction in the current density. One possible explanation could be disorder of the liquid crystal around the edge of the electrodes. If we assume that the area near to the electrode edge is disordered and that this disorder penetrates into the liquid crystal by around 6–9 μm then by recalculating the electrode area for only the remaining ordered area, one obtains a modified electrode area in roughly the same ratio as the current densities. The same reduction in 'useful' electrode area could be explained by the misalignment of one layer with another. If all the chips of each electrode size were misaligned with respect to each other by 5–10 μm then the resulting overlapping electrode area would account for the variation in current densities.

Figure 9.15. Two pictures showing different designs of mask for use in photolithography. These masks are used to pattern the surface of the 3D chips with an SU8 photoresist, therefore allowing easier stronger construction of the chips.

9.3.3   Alignment solutions—the snap and click chip

Alignment of the chips was found to be a problem for chips of less than $100\ \mu\text{m} \times 100\ \mu\text{m}$ and, as a consequence, a new alignment method was developed. This used photolithography to pattern a structure onto the surface of the chips to allow the chips to fit together in much the same way as children's building blocks do.

The process involves spinning on the surface of the chip a photoresist, SU8, as a layer that is $5\ \mu\text{m}$ thick. A patterned mask is then placed on top of the chip and the whole thing is illuminated with UV light. Where the mask is unpatterned, the UV light is able to pass through and polymerize the SU8 forming a solid structure. Where the SU8 remains un-polymerized, it is removed by use of a solvent leaving the polymerized structures standing on the surface of the chip. By designing complementary structures on the bottom and top chips, the two can be made to snap together like building blocks. This idea can be applied to the top and bottom of many chips and provides an easy and cheap way of aligning many chips into a 3D stack. Figure 9.15 shows two separate designs for SU8 structures on the surface of the chip and how the top and bottom parts of the chip fit together.

Figure 9.16 shows a working example of these structures but this time fabricated on glass. This design gives the two surfaces an alignment accuracy of $\pm 20\ \mu\text{m}$. Better accuracy can be achieved down to less than $\pm 1\ \mu\text{m}$ if better mask alignment facilities are used. The first picture shows the chips after they have been clicked together, whilst the second picture shows the effect of applying a force onto the two surfaces and trying to shear them apart.

9.4   Thermal cycling

Thermal cycling represents the on–off environmental condition for most electronic components and, therefore, is a key factor that defines reliability. Thermal-cycling measurements have been taken for the CORTEX chips to

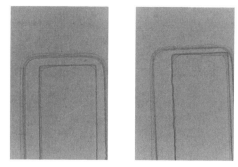

Figure 9.16. Pictures taken of the SU8 structures on the surface of two glass slides. The SU8 was spun onto the surface of the glass at a thickness 5 μm. It was then patterned into the shapes shown. The inner part of the structure is on the lower glass slide and fits inside the outer structure that is on the top piece of glass.

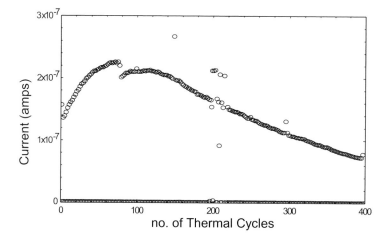

Figure 9.17. Graph showing the thermal cycling of HAT6 between 30 and 100 °C heated at a rate of 10 °C min^{-1}. The two curves show the conduction as measured with 10 V applied for high and low temperature.

establish how reliable the chip is and, therefore, its expected lifetime. In this particular case, it is to assess not only the chip as a whole but also the role of the liquid crystal molecular wires in defining the reliability.

Typical conditions for such an experiment are to cycle between −40 and 135 °C ramping at about 5 °C per minute and then holding at these temperatures for about 5 min. This ramping is usually repeated several thousand times. Figure 9.17 shows the thermal cycling behaviour for the liquid crystal HAT6 in a 2 μm thick ITO cell. In this experiment, the liquid crystal is heated between 30 and 100 °C with the temperature being ramped at a rate of 10 °C min^{-1}. When the lower temperature is reached, a voltage of 10 V is applied and the current measured; the sample is then heated and the measurement repeated.

References

[1] Kreouzis T, Scott K, Donovan K J, Boden N, Bushby R J, Lozman O R and Liu Q 2000 Enhanced electronic transport properties in complementary binary discotic liquid crystal systems *Chem. Phys.* **262** 489–97

[2] Kreouzis T, Donovan K J, Boden N, Bushby R J, Lozman O R and Liu Q 2000 Temperature-independent hole mobility in discotic liquid crystals *J. Chem. Phys.* **114** 1797–802

[3] Bushby R J, Evans S D, Lozman O R, McNeill A and Movaghar B 2001 Enhanced charge conduction in discotic liquid crystals *J. Mater. Chem.* **11** 1982–4

[4] Boden N, Bushby R J, Cooke G, Lozman O R and Lu Z B 2001 CPI: A Recipe for Improving Applicable Properties of Discotic Liquid Crystals *J. Am. Chem. Soc.* **123** 7915–16

[5] Lozman O R, Bushby R J and Vinter J G 2001 Complementary polytopic interactions (CPI) as revealed by molecular modelling using the XED force field *J. Chem. Soc. Perkin Trans.* **2** 1446–52

[6] Boden N, Bushby R J, Donovan K, Liu Q Y, Lu Z B, Kreouzis T and Wood A 2001 2,3,7,8,12,13-Hexakis[2-(2-methoxyethoxy)ethozy]-tricycloquinazoline: a discogens which allows enhanced levels of n-doping *Liquid Crystals* **28** 1739–48

[7] Boden N, Bushby R J and Movaghar B 2001 Electronic properties of discotic liquid crystals and applications *Encyclopedia of Materials* ed S Chandrasekhar (Weinheim: VCH)

[8] Wegewijs B R, Siebbeles L D A, Boden N, Bushby R J, Movaghar B, Lozman O R, Liu Q, Pecchia A and Mason L A 2002 Charge-carrier mobilities in binary mixture discotic triphenylene derivatives as a function of temperature *Phys. Rev. B* **65** 245112

[9] Bushby R J and Lozman O R 2002 Discotic Liquid Crystals 25 years on *Curr. Opin. Colloid Interfacial Sci.* **7** 343–54

[10] Boden N, Bushby R J and Lozman O R 2003 Designing better columnar mesophases *Molec. Cryst. Liquid Cryst.* **400** 105–13

[11] Boden N, Bushby R J and Lozman O R A comparison of CPI and charge-transfer two-component phases *Molec. Cryst. Liquid Cryst.* accepted for publication

[12] Boden N, Bushby R J, Donovan K, Kreouzis T, Lozman O R, Lu Z B, McNeill A and Movaghar B Enhanced conduction in the discotic mesophase *Molec. Cryst. Liquid Cryst.* accepted for publication

[13] Bushby R J, Liu Q, Lozman O R, Lu Z B and McLaren S R New functionalised derivatives of hexaalkoxytriphenylene *Molec. Cryst. Liquid Cryst.* accepted for publication

[14] Sawada A, Sato H, Manabe A and Naemura S 2000 *Japan. J. Appl. Phys.* **39** 3496

Chapter 10

Scaffolding of discotic liquid crystals

*M Murugesan, R J Carswell, D White, P A G Cormack and
B D Moore*
University of Strathclyde, Glasgow

10.1 Introduction

This chapter is a summary of the results in the field of construction of peptide-based scaffolding for the columnar assembly of discotic liquid crystals. Discotic liquid crystals (DLCs) are disc-shaped molecules containing rigid aromatic cores and flexible alkyl side-chains. Over certain temperature ranges they spontaneously form columnar mesophases, with the nature of the core and the alkyl side-chains controlling formation of the mesophase and the columnar ordering. Significantly for the 3D interconnects, numerous DLCs have been shown to exhibit electrical conduction along the lengths of the columnar stacks [1, 2]. However, for practical applications, it would be useful to increase the operating temperature ranges, have improved control over the surface alignment of discotics and enhance the efficiency of charge injection into the discotic stacks.

The aim of our work is to develop molecular 'scaffolding' that can be used with discotic molecules to enhance their molecular ordering. This was a novel approach in the DLC field but we reasoned that if it were successful, it would have major advantages. This is because it would then be possible to enhance and stabilize the alignment of the discotic molecules without having to resort to lengthy synthetic procedures. It might also lead to improved charge injection. We chose peptides as target scaffold materials because they can form well-defined secondary structures such as α-helix and β-sheets and can be synthesized in a modular fashion. The results have shown that the basic concept of scaffolding DLCs can be used as a way to modify the liquid crystal (LC) and conductance properties. In the case of diamines (chosen as model compounds for peptides), it was shown that the DLC properties were dramatically improved in terms of stability and conductance. Measurements in the test rig confirmed this behaviour.

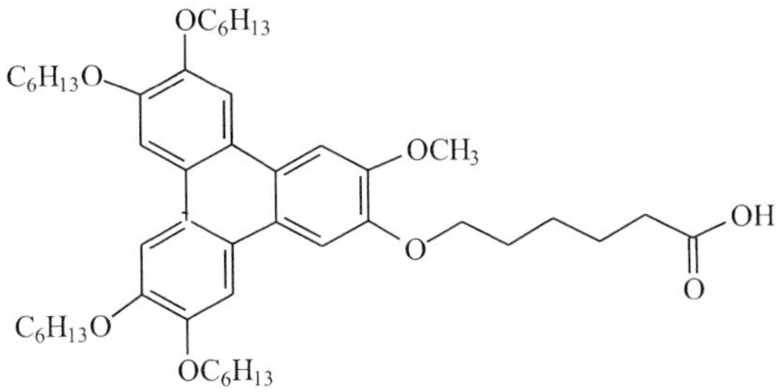

Figure 10.1. Structure of the discotic molecule **I**.

The results for the many α-helix and β-sheet scaffold peptides synthesized were equivocal, with phase separation being a significant competing process to formation of the mesophase. Samples mixed by shearing in the melt showed promising DLC behaviour but could not be prepared in cells used for conduction measurements. Finally, thiol-terminated helical peptide-discotic conjugates were prepared and used to form self-assembled monolayers at gold electrodes. Changes in charge injection behaviour were observed.

10.2 Synthesis and characterization of peptide-reinforced discotic liquid crystals

The long-term aim of our work is to establish whether peptides of defined secondary structure can be used to enhance and stabilize the alignment of DLCs within submicrometre dimensioned cavities. If this can be achieved, it should be possible to use them to generate stable conductive pathways suitable for chip-to-chip connections. The first goal was to synthesize molecular structures suitable for such an application.

10.2.1 Synthesis of mono-carboxy substituted hexa-alkyloxy triphenylene derivatives

Asymmetrically substituted triphenylene-based DLCs bearing carboxylic acid groups in one of their side-chains were synthesized by modification of a literature procedure [3] and characterized using IR, MS, NMR and elemental microanalysis. The carboxy substituent was chosen because it provides a versatile site for interfacing with peptides containing amino acids such as lysine (Lys), arginine (Arg), glutamic acid (Glu) or aspartic acid (Asp), via ion-pairing or hydrogen bonding. The structure of such a discotic molecule is given in figure 10.1.

The mesophase behaviour of the discotic molecule **I**: (2-[(6-hexanoic acid) oxy]-3-methoxy-6,7,10,11-tetrahexyloxy triphenylene) was determined using differential scanning calorimetry (DSC) and hot-stage polarizing optical microscopy (HS-POM). It was established that **I** exhibited liquid crystalline

Figure 10.2. Optical micrograph of the mesophase formed by the discotic molecule *I* at 75 °C upon cooling (magnification ×400). For the full-colour picture see the website.

behaviour, showing a monotropic discotic phase upon cooling. The optical micrograph of the discotic molecule is shown in figure 10.2. Interestingly, the methyl ester of *I* did not show any liquid crystalline behaviour, perhaps suggesting that dimerization via the carboxylic acid groups is instrumental in favouring liquid crystal formation. Mesophase behaviour of *I* = K 120 I 76 LC 57 K

10.2.2 Conjugates of discotic molecules with various diamines and monoamines

Rather than moving directly on to peptide–discotic hybrid materials, we thought it prudent to test the effect of complexing the carboxy-substituted discotic *I*, with some simple alkyl mono- and di-amines.

The mesophase behaviour of the discotic molecules in the presence of various monoamines and diamines were investigated using DSC and HS-POM. For the mono-amines, a 1:1 molar ratio of the amine to discotic was used, whereas for the diamines a 1:2 ratio of the diamine to discotic was used.

Conjugates of monoamines and *I* showed no sign of LC behaviour and underwent reversible K–I transitions. The melting temperature was found to vary with the amine chain length, however, indicating that a complex between the carboxyl group and the amine had been formed. In contrast, the conjugates prepared between 1:2 mixtures of various diamines with the discotic showed enantiotropic LC behaviour with extended mesophase ranges. The mesophase

Table 10.1. Liquid crystal (LC) behaviour of *I* in the presence of various diamines and monoamines.

Conjugate	Amine	Discotic	LC behaviour	
			Heating	Cooling
—	—	*I*	K 120 I	I 76 LC 57 K
1	1,2-diaminoethane	*I*	K 66 LC 129 I	I 125 LC G
2	1,4-diaminobutane	*I*	K 62 LC 143 I	I 136 LC G
3	1,6-diaminohexane	*I*	K 62 LC 120 I	I 116 LC G
4	1,9-diaminononane	*I*	K 72 LC 120 I	I 112 LC G
5	1,12-diaminododecane	*I*	K 72 LC 104 I	I 100 LC G
6	1-aminohexane	*I*	K 84 I	
7	1-aminododecane	*I*	K 120 I	

behaviour of *I* in the presence of various diamines and monoamines is given in table 10.1.

DSC and x-ray diffraction measurements showed that the diamine–discotic conjugates formed glassy phases at room temperatures and crystallized over a period of days. Thus, the addition of only a very low weight fraction of the diamine completely transforms the LC behaviour of *I*. This bodes well for our proposed strategy of non-covalent binding of peptides to the discotics. Of particular interest are the vast differences in properties observed between adding monoamines as opposed to diamines to the discotics. The requirement for two binding sites indicates that inter- or intra-stack interactions between discotic molecules are beneficial. This observation was used in the subsequent design of the peptides.

10.2.3   X-ray diffraction studies

The x-ray powder diffraction patterns for the diamine–discotic conjugates showed diffuse reflections in the LC phase, which were consistent with a columnar rectangular arrangement of the molecules with the lattice parameters shown in table 10.2. There is no obvious trend in the lattice parameters on going from 1,6-diaminohexane to 1,12-diaminododecane. This suggests that the structure of the mesophase is able to accommodate the difference in the length of the alkyl chain between the two amino groups.

The x-ray diffractogram of Conjugate 1 in the small angle region shows diffuse reflections in the LC regions corresponding to a columnar hexagonal arrangement of the molecules with an intercolumnar distance of 18.0 Å and an intracolumnar distance of 3.4 Å. A diffuse band at 4.8 Å corresponds to the diffraction from the fluid alkyl chains.

Figure 10.3. Optical micrograph of the mesophase formed by Conjugate 1 at 55 °C upon cooling (magnification ×650). For the full-colour picture, see the website.

Table 10.2. Lattice cell parameters obtained from x-ray measurements for the conjugates prepared from a 1:2 ratio of various diamines to discotic *I*.

Diamine	$T(°C)$	a (Å)	b (Å)	c (Å)
1,2-diaminoethane	25	47.9	17.5	3.5
	80	48.7	18.0	3.4
1,6-diaminohexane	25	41.7	23.1	3.5
	80	57.3	21.1	3.6
1,12-diaminododecane	25	48.0	19.3	4.4[a]
	60	46.0	17.6	4.4

[a] Looks like mixtures of two phases, hexagonal and orthorhombic.

10.2.4 Conjugates of discotic molecules with β-sheet peptides

Based on the results from diamine–discotic conjugates, we have synthesized various β-sheet forming peptides based on arginine (Arg, R) and valine (Val, V). In β-sheet structures, peptide chains are fully extended and are stabilized by intermolecular hydrogen bonding between strands. The amino acid arginine was chosen because the guanidine side chain can form very strong electrostatic/H-bond interactions with carboxylic acid groups as found on discotic *I*. The amino acid valine was chosen because it is a strong promoter of β-sheets. It can be seen that all four peptides 1–4, are capable of binding simultaneously to two discotic molecules since each peptide contains two arginine residues.

1. Ac–Arg–Arg–NH$_2$: **RR**
2. Ac–Arg–Val–Arg–NH$_2$: **RVR**
3. Ac–Arg–Val–Val–Arg–NH$_2$: **RVVR**
4. Ac–Arg–Val–Arg–Val–NH$_2$: **RVRV**

Peptides 1–4 were synthesized using a Novasyn Crystal Solid-Phase Peptide Synthesiser via conventional peptide coupling protocols. The purity of the peptides was analysed using high performance liquid chromotography (HPLC) and they were characterized by mass spectrometry.

The 1:2 ratios of peptide to discotic were prepared by direct weighing. In order to ensure thorough mixing of the discotic with the peptide, the two substances were dissolved in a 1:3 mixture of trifluoroethanol: dichloromethane. The solution was then evaporated to dryness in a vacuum centrifuge and the process repeated thrice. Finally, the residue obtained was used for polarizing optical microscopy and differential scanning calorimetry studies.

Conjugate	Peptide	Discotic	Peptide:Discotic (mol:mol)
8	**RR**	I	1:2
9	**RVR**	I	1:2
10	**RVVR**	I	1:2
11	**RVRV**	I	1:2

The β-sheet structures of the peptides in the conjugates were confirmed by IR spectroscopy. For example, the IR spectrum of Conjugate 9 was measured as a KBr dispersion. It showed peaks at 1518 and 1633 cm^{-1}, which is characteristic of the β-sheet structure of peptide 2 in the conjugate.

Discotic **I** on its own shows clear, sharp melting behaviour at 120 °C which is characteristic of the melting of the crystalline to isotropic phase. The DSC thermogram of the Conjugates 8–11 were recorded and, unlike the amine conjugates, showed no sharp exotherms or endotherms. However, broad, weak melting exotherms were observed which, by comparison to HS-POM data, corresponded to the LC to isotropic transitions. The absence of any K–LC or K–I transition would seem to indicate that these conjugates form a glass on cooling. This seems to be confirmed by the HS-POM results.

10.3   Conductance measurements

The main aims here were to demonstrate alignment of scaffolded discotics between electrodes and to measure their conductance behaviour. We have shown that good alignment can be obtained using gold electrodes similar to those on the test chips (manufactured at TU Delft). Conductivity measurements have been carried out with a range of discotic–diamine conjugates and relatively high values obtained even with highly purified samples.

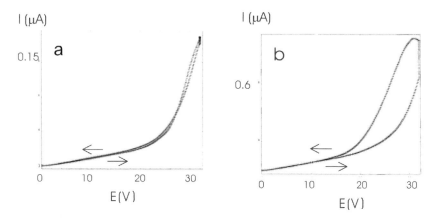

Figure 10.4. *I–E* curve for Conjugate 1 using ITO electrodes spaced at 6 μm (*a*) $T = 120\,°C$ and (*b*) $T = 110\,°C$.

10.3.1 Conductivity studies using ITO coated cells

At the start, to gain some insights into the aligning and conductivity behaviour of the conjugates prepared between peptides and the discotic, initial experiments were first carried out on diamine–discotic conjugates in ITO-coated glass cells of 6 μm thickness and surface area 10 mm × 5 mm (EHC Ltd, Japan). The conductance measurements were carried out using a Thunlby Thandar PLP electrometer and a Keithley 617 programmable electrometer in the SOMS Centre, Leeds, UK. Samples were aligned by being heated into their isotropic phase followed by slow cooling into their mesophase at a rate of $0.1\,°C\ min^{-1}$. Polarizing optical microscopy showed the samples to be homeotropically aligned.

The LC behaviour of the discotic molecule alone is K 120 I 76 LC 57 K. To fill the cell with neat LC, the discotic molecule *I* was placed as a solid at the edge of the cell and taken into the isotropic state (120 °C). Unexpectedly, the cell did not fill efficiently by capillary action and, indeed, the discotic did not fully wet the whole surface area of the electrode. Nevertheless, it was cooled ($0.1\,°C\ min^{-1}$) to 75 °C (i.e. into the mesophase region) but, unfortunately, no homeotropic alignment was induced. Conductivity measurements were, therefore, not carried out on the neat LC.

A small amount of Conjugate 1 was placed on one edge of the cell and heated to the isotropic state (129 °C). The conjugate in its isotropic state did not flow inside the cell smoothly and, consequently, the whole electrode surface was not fully wetted. Encouragingly, however, on cooling slowly ($0.1\,°C\ min^{-1}$) to 120 °C, columnar alignment of the conjugate was observed by polarizing optical microscopy. Conductivity measurements were carried out using the partially filled cell.

Figure 10.4(*a*) shows the conductivity behaviour of Conjugate 1 at 120 °C using ITO-coated electrodes. The conductivity was found to be approximately one order of magnitude higher than symmetrically substituted discotics such as HAT6, which was very encouraging. The *I–E* curve in the high field region showed a crossover on cycling the voltage. Conductivity measurements were also carried out at different temperatures in the mesophase region. The *I–E* curve measured at 110 °C is shown in figure 10.4(*b*). The *I–E* curves measured at temperatures below 120 °C generally show significant hysterisis.

The presence of hysterisis, the step-like behaviour during the field anneal and the lowering of the conductivity following the field anneal, could be due to the presence of small amounts of ionic impurities in the conjugates. This may also explain the enhanced conductivity relative to the symmetrical precursor. Alternatively, the results could be due to greater susceptibility of the conjugates to electrochemical reactions in the potential window explored (32 V) leading to decomposition of one or more of the components at the ITO/LC interface and the formation of ionic species during measurements at higher potentials.

Attempts were made to fill ITO-coated electrodes using Conjugates 3, 5, 9 and 10. However, the conjugates were found not to fill the cell sufficiently well to enable useful measurements to be made.

10.3.2 Alignment and conductivity measurements of the discotic molecule in gold-coated cells

The surface-wetting problem faced when using ITO electrodes was eliminated by using gold-coated home-made cells. The thin gold-coated electrodes were washed thrice with methanol and with chloroform prior to their use. A 6 μm thickness spacer was placed between them and the filling procedure carried out similarly to the ITO-coated electrodes.

The discotic molecule *I* was filled in the thin gold-coated cell and heated into its isotropic state. It was then cooled from the isotropic state to 74 °C at a rate of 0.1 °C min^{-1}, as it had been shown that the discotic molecule formed columnar structure in the range 76–57 °C upon cooling. Unfortunately, crystallization of the discotic molecule was observed under the cross-polarizers, even upon very slow cooling from its isotropic state. It is, therefore, difficult to align the discotic molecule by this method.

Attempts were made to align the discotic molecule *I* by keeping it in its mesophase range for a long period of time. To assess the stability of the mesophase in terms of time, the discotic molecule was taken into its isotropic state in gold cells and then cooled very rapidly to its mesophase at a temperature of 74 °C. The discotic mesophase was formed, as was clear from its texture under the polarizing optical microscope, but was only stable for about 30 min. Thus, aligning such a discotic molecule with a metastable mesophase is not really feasible.

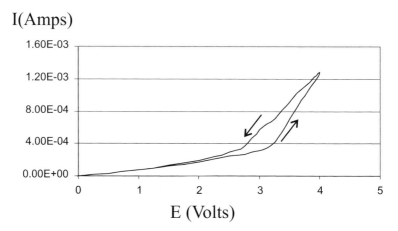

Figure 10.5. *I–E* curve for Conjugate 1 in the aligned columnar state (at 100 °C) using gold-coated electrodes spaced at 6 μm.

Conductivity studies were carried out, nevertheless, in the isotropic state, mesophase region and crystalline state on non-aligned samples in the gold cell. The measurements were carried out between 0 and 10 V to eliminate the possibility of electrode reactions at high voltages. The discotic molecule shows conductance in the isotropic and mesophase state but no conductance in the crystalline state. Moreover, while moving from isotropic to mesophase region, a decrease in conductance is observed.

10.3.3   Alignment and conductivity measurements of the conjugates

Conjugate 1, prepared from a 1:2 ratio of 1,2-diaminoethane to discotic, was filled in a gold-coated cell and heated to its isotropic state (129 °C). The conjugate was then cooled to 100 °C with a very slow cooling (rate 0.1 °C min⁻¹) in order to promote alignment into columnar stacks. The absence of birefringence under the cross-polarizers indicated the homeotropic alignment of the conjugate. Figure 10.5 shows the conductivity of Conjugate 1 at 100 °C in the aligned columnar state.   There is a decrease in conductance as expected at lower temperatures for an ionic mechanism but there also appears to be a threshold voltage around 2.5 V, above which the conduction increases much more rapidly, which is puzzling.

The same conjugate was then field annealed in the gold-coated cell at 100 °C and 3 V (figure 10.6). During annealing, a decrease in the current was observed with respect to time. The *I–E* curve for the conjugate after field annealing is shown figure 10.7.

The field anneal appears to enhance the conductance of the aligned discotic in the mesophase by a factor of around three. This is again puzzling since, for an

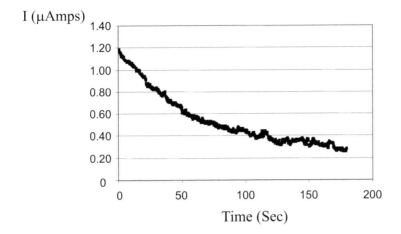

Figure 10.6. Field annealing of Conjugate 1 at 3 V ($T = 100\,°C$).

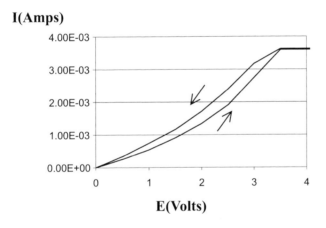

Figure 10.7. *I–E* curve for Conjugate 1 after field annealing at 3 V using gold-coated electrodes spaced at 6 μm ($T = 100\,°C$).

ionic conductance mechanism, it would be expected that the field-anneal process would lower the concentration of charge carriers and lead to lower conductivity.

Conductivity measurements were also carried out for Conjugate 5 prepared from a 1:2 ratio of 1,12-diaminododecane to discotic molecule **I**. It was filled in a gold-coated cell and taken to its isotropic state (110 °C) and then aligned in its columnar state at 100 °C by slow cooling (0.1 °C min^{-1}).

Initially the measurements were made in the range 0–10 V but, unexpectedly, on the reverse scan, a negative current was found upon returning to 0 V. To

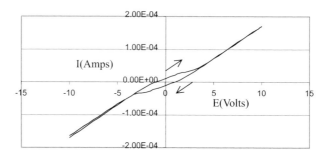

Figure 10.8.   *I–E* curve for the conjugate prepared from a 1:2 ratio of 1,12-diaminododecane and discotic molecule in the aligned columnar phase using gold-coated electrodes spaced at 6 μm ($T = 100\,^{\circ}$C).

understand the data, further measurements were carried out in the range -10 V to 10 V. This showed a clear hysterisis loop around 0 V (figure 10.8). The reason for the hysterisis is unknown but it is interesting to note that the loop closes at a similar potential as the change in gradient observed for Conjugate 1.

Conjugate 10 was filled in the gold-coated cell and slowly cooled from the isotropic state to align it in its columnar state. Unexpectedly, crystallization of the conjugate was observed which prevented us from carrying out any further measurements at that time.

10.4   Synthesis of newer types of conjugates for conductance measurements

Based on the previous results, a second generation of hybrid materials were designed to give molecular ordering and enhanced conduction. Our previous results showed that coupling discotic molecules with various β-sheet forming peptides enhanced the degree of columnar ordering. We have synthesized further β-sheet peptides and α-helical peptides containing both arginine and lysine side-chains. Also we have synthesized α-helical peptides having surface binding properties and demonstrated changes in conduction following formation of monolayers at gold surfaces.

It was expected that α-helical peptides might exhibit different scaffolding properties when compared with the β-sheet peptides because of their rod-like geometry. In addition, helical peptides were expected to be monomeric and produce less viscous materials. In β-sheet systems, there are three types of organizing forces. One is the intermolecular hydrogen bonding between peptide monomers for the formation of β-sheets, another is the π–π interaction between discotic molecules for the formation of columnar arrangements and the third is the H-bond/salt-bridge between the discotic and the peptide. It is expected that the overall assembly depends on the balance between these. In order to investigate whether changes to the H-bond/salt bridge strength were important, a series of peptides with lysine side-chains rather than arginine side-chains were made. These were expected to have the advantage of easier synthesis of both the peptide and conjugate.

The following α-helical peptide based on arginine, α-aminoisobutyric acid and alanine was synthesized and characterized by mass spectrometry.

$$\text{5. Ac–RAZRAZRAZRA–NH}_2$$

Attempts were made to introduce formyl and thiol groups at the amino terminus but the usage of very high percentages of trifluoroacetic acid to remove the 'Pbf' protecting group on the guanidine side chain of arginine was found to cleave them off again.

Conjugate 12 was prepared by mixing the α-helical peptide 5 with the discotic molecule in a 1:4 ratio after removing the TFA counter-ion in the guanidine side-chain by treating with sodium methoxide. The HS-POM shows melting at 180 °C and LC properties at 175 °C. The results are encouraging when compared with the very small LC window of the discotic molecule:

K 180 I 175 LC 50 G/K

HS-POM shows the mesophase formation is not uniform in all places, with some isotropic regions observed in the LC phase. This may be due to incomplete removal of the TFA counter-ion in the guanidine side chain.

The results indicated that conjugates based on α-helical peptides could be expected to have higher melting temperature and extended mesophase region. Based on the promising behaviour of conjugates of discotic with the α-helical peptide and of those with the diamines, it was decided to combine the two properties and make some lysine-based peptides for further experiments. A potential advantage over the previous system was that removal of the TFA counter-ion from the amino side-chain of lysine is simpler compared with the arginine side-chain.

The following β-sheet-forming peptides 6, 7 and 8 were prepared and characterized by mass spectrometry. The removal of the TFA counter-ion was effected by treating with 2-chlorotrityl chloride and piperidinomethyl polystyrene resins that were scavengers for the TFA ions. The absence of TFA was ascertained by the absence of peaks at 115 and 163 ppm in ^{13}C NMR spectra.

$$\text{6. Ac–KVVK–NH}_2$$
$$\text{7. Ac–KVVVK–NH}_2$$
$$\text{8. Ac–KVVKVVK–NH}_2$$

The peptides 6 and 7 were mixed with the discotic molecule in a 1:2 ratio and peptide 8 in a 1:3 ratio. The β-sheet structure of the peptide and the complex formation between the peptide and discotic molecule was confirmed by IR spectroscopy.

Conjugate	Peptide	Discotic	Ratio
13	Ac–KVVK–NH$_2$	I	1:2
14	Ac–KVVVK–NH$_2$	I	1:2
15	Ac–KVVKVVK–NH$_2$	I	1:3

The mesophase behaviour of conjugates 13, 14 and 15 was investigated using DSC and HS-POM. Unfortunately, DSC showed no characteristic LC phase transitions on heating up to 200 °C with fusion of the discotic at the normal K–I transition temperature, the only peak of note. HS-POM confirmed this result. Thus, the change from arginine to lysine appears to lead to only a very weak interaction between the discotic and peptide which is insufficient to change the mesophase behaviour. In fact, POM suggests that, on melting, the discotic phase actually phase separates from the peptide.

Peptide 9, which is a lysine-based analogue of helical peptide 5, was also synthesized and characterized.

9. Ac–NH–KAZ–KAZ–KAZ–KA–CONH$_2$

The TFA counter-ion was removed using the same procedure as used for β-sheet systems. Conjugate 16 was prepared by mixing peptide 9 with discotic molecule in a 1:4 ratio. The mesophase behaviour of the Conjugate 16 was determined using DSC. It indicates monotropic LC behaviour with only a slightly extended range.

K 115 I 77 LC 45 K

Peptide 10, with a thiol group at the carboxylic terminus of peptide 9, was also prepared and characterized.

10. Ac–NH–KAZ–KAZ–KAZ–KA–CONH–CH$_2$–CH$_2$–SH

Conjugate 17 was formed by mixing peptide 10 with the discotic molecule in a 1:4 ratio. The HS-POM studies for the gold-covered electrodes gave the following results

K 114 I 74 LC 54 K.

It was anticipated that if a monolayer of conjugate 17 was used, this might alter the stability of pure discotic used to fill the 6 μm gap between the two slides. As can be seen from the results, the mesophase temperature range is comparable to the parent discotic. However, it was noted that if the temperature was held within the mesophase range the discotic crystallized much more slowly suggesting the surface modification has reduced the number of potential nucleation sites. In addition, it was noted that filling the cells was much faster indicating the

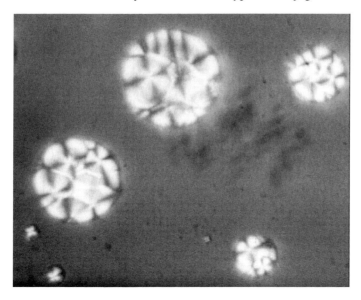

Figure 10.9. Optical micrograph of the mesophase formed by Conjugate 10 at 63 °C while cooling (magnification ×650). For the full-colour picture, see the website.

peptide monolayer helps the wetting process. This could have important practical applications.

In view of the disappointedly poorer behaviour of the lysine-based peptides relative to the arginine derivatives studied initially, it was decided to revisit the arginine β-sheet systems and look for better ways to align them. To this end, the effect of shearing in the mesophase region was studied to test whether it would help induce the columnar arrangement. Conjugate 10 prepared from peptide 3 with the discotic molecule in a 1:2 ratio was taken to the isotropic state, sheared well and left overnight for annealing. This conjugate was found to form a mesophase at 63 °C which was stable to room temperature overnight and while heating to 110 °C. It should be noted, however, that the shearing process appeared to generate a much more viscous material that tended to separate into discrete droplets on the microscope slide. The enhanced LC stability may, therefore, be a kinetic effect arising from the lack of nucleation sites in these microdroplets. All attempts to obtain large areas of aligned peptide–discotic conjugates were unsuccessful. Interestingly, it was noted that, where it was possible to obtain a large wetted area of conjugate on cooling it would form a metastable LC phase, then crystallize and finally undergo a further reorganization which could potentially be attributed to phase separation of the peptide from the discotic.

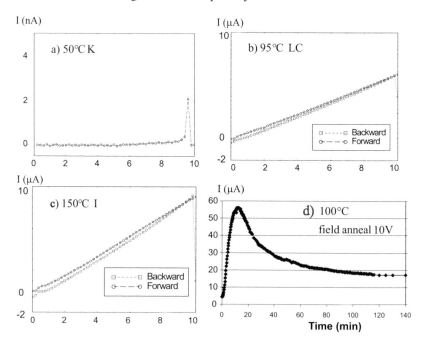

Figure 10.10. (*a*)–(*c*) I/V curves measured for Conjugate 1 in crystalline, LC and isotropic states. (*d*) Field anneal of Conjugate 1.

10.5   Conductance of scaffolded DLC systems in the test rig

Having gone through a number of development iterations to obtain scaffolded DLC systems with improved mesophase stability, the final milestone was to test the conductance of the best candidates within the CORTEX test rig.

It was clear from the POM studies that the best systems for these measurements would be the diamine–DLC conjugates since these could be straightforwardly filled into a conductance cell by capillary action and quickly aligned. The diamine–discotic Conjugate 1 was chosen as representative. In order to be able to interpret data from the test rig, it was important first to characterize fully the conductance-temperature behaviour of a highly purified sample of Conjugate 1. Experiments were, therefore, carried out in a number of the standard test cells. Figures 10.10(*a*)–(*c*) show the results obtained at temperatures corresponding to the crystalline, LC and isotropic phases.

In the LC phase, reasonable ohmic conductance was observed with minimal hysteresis on scanning between 0–10 V. On moving to the isotropic phase, conductance was retained and even increased whilst, in the crystalline phase, the material was insulating. This is indicative of an ionic rather than electronic conduction mechanism. Field anneals were also carried out as shown in

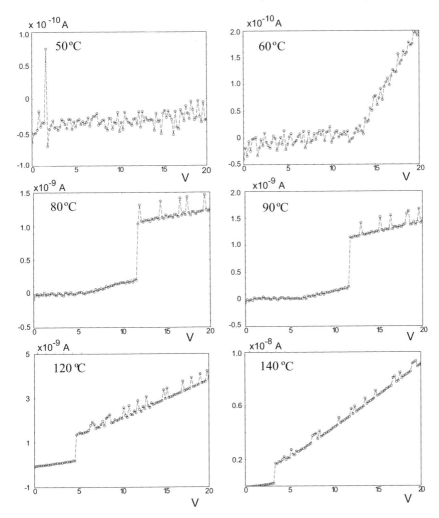

Figure 10.11. I/V curves of Conjugate 10 measured in the test rig using a 300 μm electrode chip.

figure 10.10(d). The anneal at 10 V showed an initial ten-fold rise in current followed by a gradual decay to a level four times greater than the initial current. This behaviour is similar but less pronounced than that observed with other DLC systems. The high retention of charge carriers following the field anneal is interesting and suggests ions are continuously being generated via an electrochemical mechanism. The diamine scaffolded DLC sample was then tested in the CORTEX test rig using the 300 μm contact-pad chip assembly. The I/V curves measured at a series of different temperatures are shown in figure 10.11.

Figure 10.12. POM photo for a new structure (RVVR+DIS)+PTP-9 (magnification ×400). For the full-colour picture, see the website.

(The current step seen in each trace is an artifact arising from an automated scale change.) It can clearly be seen from figure 10.11 that, in the crystalline phase, the material is insulating while, above this temperature, it shows ohmic behaviour up to 20 V. As seen in the test cells, the conductivity increases with temperature but interestingly not as steeply. There is a four-fold increase on moving from 80 to 140 °C at which temperature the measured current was ∼10 nA at 20 V. Of significant note, however, was the very high reproducibility of the measurements. Near identical traces were obtained on cooling or heating the sample over a number of cycles. This indicates that even these low performance discotics can provide a consistent electronic contact between chips that persists over a wide temperature range.

From these studies, it is clear that pure conjugates of the current generation of scaffolded DLCs offer clear advantages over conventional non-scaffolded materials, e.g. extended mesophase ranges and enhanced alignment properties. It should be noted, however, that blends of the pure conjugates with conventional materials are also possible. For example, addition of ∼20%w/w Conjugate 10 to HAT6 was found to expand the mesophase range of HAT6 by 10 °C at the low-temperature end.

References

[1] Chandrasek S 1998 *Handbook of Liquid Crystals* vol 2B, ed D Demus *et al* (Weinheim: Wiley–VCH)

[2] Boden N and Movaghar B 1998 *Handbook of Liquid Crystals* vol 2B, ed D Demus *et al* (Weinheim: Wiley–VCH)

[3] Stewart D, McHattie G S and Imrie C T 1998 *J. Mater. Chem.* **8** 47

Chapter 11

Through-chip connections

P M Sarro, L Wang and T N Nguyen
Delft University of Technology, The Netherlands

11.1 Introduction

The primary goal of our work was to investigate the possibility of fabricating thin silicon circuit substrates with closely-spaced through-chip connections. The ultimate target is to achieve an interconnect grid with a 25 μm pitch which will allow one million connections for die area of 25 mm $\times$ 25 mm.

Initial activities have concentrated on a theoretical study and on the identification of the potentially suitable fabrication techniques [1–4]. Based on some initial results, the task of preparing a middle chip with through-wafer connectors has been pursued.

With respect to the proposed goal, research on through-wafer metal plugs formation has been performed [5–8]. Next to the combination of wet and dry etching to form the through-wafer vias, a combination of mechanical lapping and dry etching is investigated as well. In addition, an alternative method aiming at the realization of high-density, high-aspect-ratio, closely spaced interconnects based on macroporous silicon formation is explored.

11.1.1 Substrate thinning

11.1.1.1 Substrate bending

Although silicon substrates can be thinned down to a fraction of a micrometre, this is only possible if they are bonded to a supporting, insulating substrate (as in silicon-on-insulator technologies). It is important to know how thin, free-standing, silicon chips can be made and how flat they will be. Applying the beam-bending theory, a simple analysis has been performed to estimate the mechanical bending of self-supported silicon substrates. Calculations of the deflection of a silicon beam under its own weight as a function of its length and thickness show that, for the targeted die dimensions of 25 mm × 25 mm, a substrate thickness of 50 μm is sufficient when bending up to 30 μm is acceptable. The situation will complicate when multilayer substrates have to be considered. The various layers with different thermal expansion coefficients and/or any internal stress within them can cause excessive bending after the substrate is thinned to its target thickness and need, therefore, be taken into account.

11.1.1.2   Thinning methods

For substrate thinning, several methods can be used. The most suitable ones with an indication of the major advantages/disadvantages are mentioned here.

- Wet etching: simplicity, acceptable removal rate, uniformity problems.
- Grinding: very high removal rate, damage to the substrate.
- Chemical–mechanical polishing (CMP): planarization capability, complicated processing.
- Atmospheric plasma etching: simple processing, acceptable removal rate, expensive equipment.

For demanding applications, usually a combination of the first three techniques is applied. First, the bulk volume is removed by grinding. The mechanical damage at the substrate surface is subsequently removed by wet etching. If high planarity is required, the processing sequence is finished by CMP. For less demanding applications (partial thinning, uniformity is not critical), wet etching is the optimum solution.

11.1.2   High-density interconnect

The realization of a high-density through-wafer interconnect will require at least the following three steps:

(i)   the formation of high-aspect-ratio vias (through-chip holes),
(ii) deposition/growth of an insulating layer and
(iii) deposition and patterning of a conductive layer.

Two different approaches should be considered:

(a) forming the interconnect before the substrate is thinned to its targeted thickness. Here, the interconnect is fabricated in the form of isolated trenches filled by a conductive layer. Afterwards, the initially thick wafer is thinned from the wafer back-side to its targeted thickness making access to the conductive vias. This option can have a significant advantage due to potentially easier handling of initially thick wafers.
(b) Forming the interconnect after the substrate is thinned to its targeted thickness. In this case, special handling of the thinned wafers during interconnect formation will be required and/or only substrates with a certain minimum thickness could be processed.

11.1.3   Through-chip hole-forming

A number of techniques that can be applied for through-wafer hole-forming in silicon substrates are readily available. Some of these techniques together with their major characteristics are mentioned here:

- anisotropic wet etching—simple equipment, high etching rate, but low aspect ratio ($\sim$1),
- high-density plasma etching—expensive equipment, acceptable etching rate, very high aspect ratio (up to 400),
- photoanodic etching in aqueous HF—complicated setup, high aspect ratio (200), limitations in substrate orientation and doping and
- sand blasting—expensive equipment, high etching rate, low aspect ratio, substrate damage, low resolution.

The small pitch of $\sim$25 μm and the intended application will require the use of one of the technologies that provide high-aspect-ratio etching capabilities. From the previously mentioned techniques, high-density plasma etching is probably the most suitable. The Alcatel 601 high-density plasma etching system installed at the Delft Institute of Microelectronics and Submicrontechnology (DIMES) has been used for the through-chip hole experiments.

11.1.4   Insulating layer

After the high-aspect-ratio trenches or holes are formed, an insulation layer has to be formed to avoid electrical connection through the semiconductive silicon substrate. If the thermal budget is not an issue, thermal oxidation will be the most suitable technique. Another option is to use low pressure chemical vapour deposition (LPCVD) or plasma enhanced chemical vapour deposition (PECVD) of dielectric layers which require lower deposition temperatures.

11.1.5   Conductive interconnect

For conductive layer forming, any low-resistivity material should be considered (e.g. metals, highly doped semiconductors). Filling small diameter high-aspect ratio vias is rather demanding and requires processes with excellent step coverage. LPCVD of *in situ* doped polySi and chemical deposition of metals from aqueous solutions were investigated. For less demanding geometries, sputtered layers were used.

11.2   Test chips

11.2.1   The middle chip

We investigated issues related to chip stacks, in particular testing multiple molecular wire connections. Our work was focused, in particular, on the

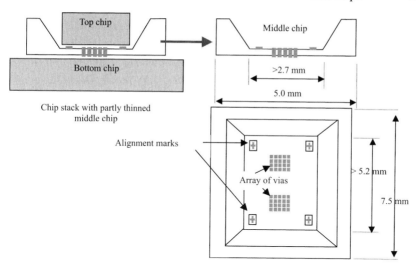

Figure 11.1.   Schematic drawing of the chip stack and layout for the middle-chip configuration.

intermediate (or middle) chip (see figure 11.1). As the long-term idea is to stack thin ($< 100 \ \mu$m thick) chips, a configuration as depicted in figure 11.1 has been used, as the locally thinned chip closely resembles the envisioned chip thickness but, in this initial exploratory stage, it is somewhat easier to handle.

The first middle chip design (CX1069) is shown in figure 11.2. Three types of cells were considered to investigate the potential and limits of the technique used. Modified versions of this design have subsequently been realized to overcome some of the problems encountered during chip fabrication, as explained in section 11.3.1. The final version of the middle chip (CX1132/CX1137) contains essentially only cell 0 and cell 1 structures with some expanded spacing and modified alignment patterns. Further, highly irregular structures, initially placed to perform some additional testing of the process, were removed.

There are two possible approaches to realize the described structures. In the first one, the silicon is locally etched from the back side until the desired membrane thickness is reached. Then the dry reactive ion etching (DRIE) of silicon from the front side is performed until the membrane is etched through. The second approach is first to DRIE the vias from the front side until the desired depth is reached and then etch the bulk silicon from the back side in an anisotropic wet etchant like KOH or TMAH. Both approaches have been investigated. Details on the advantages and disadvantages of both methods, as well as on alternative approaches, are reported in the next section (section 11.3.1). The second approach has been selected to realize the middle chips. The process flow used to realize the middle chip is schematically presented in figure 11.3.

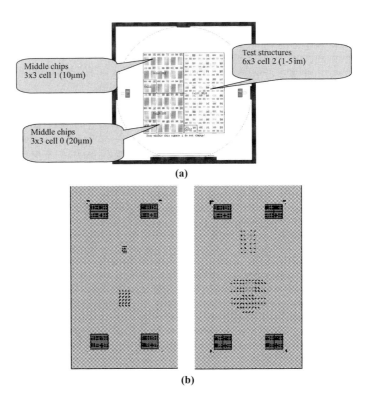

Figure 11.2. (*a*) Layout of the middle chip: cell 0 used in combination with top and bottom chip; cell 1 and 2 to investigate further downscaling of the process. (*b*) A close-up of a cell.

Fabrication of the middle chips has been completed only recently, due to several technological problems encountered with the new design. The reduced size and spacings of some structures in combination with the presence of large structures (such as the alignment marks) resulted in Cu overgrowth in the larger opening regions while attempting to fill the small size structures. This overgrowth should be avoided as it can cause shorts between adjacent structures. To solve this problem, the middle chip has been redesigned, in order to keep only the arrays of vias and remove additional test structures. The spacing between adjacent vias was enlarged to avoid shorts related to overgrowth. Alignment marks are now covered by resist during plating. Additional problems were encountered with the patterning of the electrodes on one side of the wafer. Electrodeposition of the resist is used for this step. A new system has been installed in DIMES and some initial fine-tuning of the process, especially the development step, was needed to obtain uniform, pinhole-free resist layers. The use of both evaporated and sputter

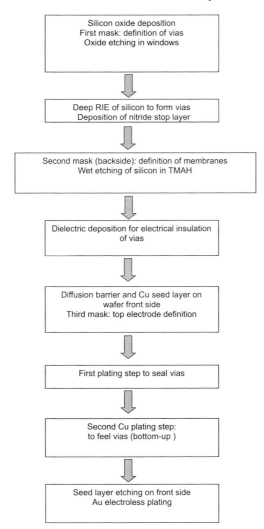

Figure 11.3. Schematic overview of the middle-chip process flow.

seed layers has been investigated as well controlling undesired resist plating inside the vias.

11.3 Electronic layer fabrication

Within the CORTEX project, research has been performed to demonstrate Cu filling into high-aspect-ratio vias (for aspect ratio up to 7). From the three approaches investigated, one-side plating and bottom-up filling has been selected

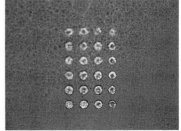

Figure 11.4. Optical images of a middle-chip cell: (left) back-side view of filled arrays of vias; (right) front-side view of one such array after Au plating.

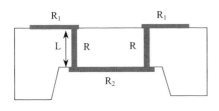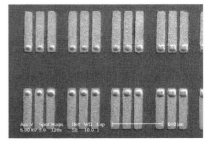

Figure 11.5. Electrical characterization of the Cu plugs: (left) schematic drawing of a test structure for resistance measurements; (right) SEM image of a series of the realized test structures (front-side view).

as the most suitable one for further investigation [8]. In order to use this approach for realizing of the middle chip, the filling of smaller vias as well as the influence of pattern variation has been studied. For larger structures and regular patterns, the process can be rather well controlled and reproducible results obtained. However, when the size of the vias is reduced and the design contains irregular patterns, several problems are encountered.

11.3.1   Through-wafer copper-plug formation by electroplating

A first design, CX1019, was used to study the bottom-up Cu-filling process. Furthermore, this design contains test structures to characterize the Cu plugs electrically.   A schematic drawing of a resistance/impedance measurement structure is shown in figure 11.5, together with an SEM image of the realized test structures. This structure includes two vias with diameters of 40 μm and lengths L (resistance $= R$), two contact pads (R_1) and the bottom resistance (R_2). This structure is realized on different wafers with different membrane thickness (or different L). The total resistance R_{total}, given by

$$R_{total} = 2R + 2R_1 + R_2.$$

The via resistance R can be extracted for different values of the via thickness. As a result, an average value of the copper resistivity inside the electroplated vias can be derived. The average copper resistivity inside the vias is 2.5×10^{-5} Ω cm. Compared with the resistivity of doped polysilicon (greater than 10^{-2} Ω cm), which is normally used for via filling, these Cu-filled vias have an extremely low resistivity. This is very attractive particularly for applications that require very high-density interconnections and, therefore, low resistivity is crucial.

11.3.2 Approaches for middle-chip fabrication

As mentioned in the previous section, when using a combination of wet and dry etching to thin the silicon wafers locally and open the through-wafer vias, two approaches can be followed as schematically depicted in figure 11.6. In the first approach, the deep dry etching of silicon is done after the wafer has been locally thinned down to 100 μm or less by wet etching. An Alcatel DRIE system equipped with He cooling that allows a substrate temperature of $-120\,°C$ is used as in this way vias less than 10 μm in size and 100 μm deep can be etched with a high anisotropy. However, a complete etch-trough of the silicon results in a He (used to cool the substrate during etching) leak through the substrate, thus stopping the etch process. To avoid this problem, a carrier wafer must be used or an additional stopping layer on the wafer back-side is required. However, the presence of an additional wafer or a thermally insulating layer between the cooled chuck and the wafer alters the wafer cooling and, consequently, the characteristics of the etch process. For very high-aspect-ratio structures, in particular, the cooling is very critical as this highly affects the anisotropy. As illustrated in figure 11.7, the presence of the stop layer alters the anisotropy of the process causing an enlargement of the via at the bottom. This effect is somewhat dependent on the over-etch time and size of the vias. Further research is needed to optimize the DRIE process and minimize this effect.

The second approach (figure 11.6) is to start with the etching of the vias with DRIE. The etch step is stopped after the desired depth is reached (see figure 11.8). The etch depth is also dependent on the opening size, thus, for openings of the same size, roughly the same depth is achieved. Optimization of the etching process is required for layouts containing structures with large variations in size. Then a dielectric coating is applied to create an etch stop for the wet etching of silicon from the back side. This layer is subsequently stripped to open the vias completely to proceed to the Cu filling step. Two additional steps (dielectric coating and subsequent removal) are needed is this case.

At present, the second approach has been selected for the realization of middle chips containing arrays of vias placed in a regular pattern or for special test structures as shown in figure 11.9.

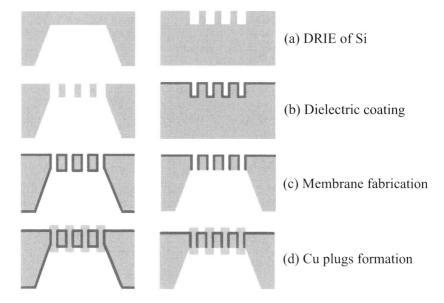

(a) DRIE of Si

(b) Dielectric coating

(c) Membrane fabrication

(d) Cu plugs formation

Figure 11.6. Two approaches for realizing through-wafer interconnects: (left) the vias are etched after local thinning of the substrate and (right) the vias are etched before local thinning of substrate.

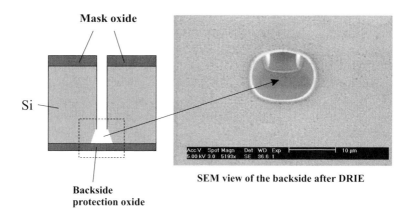

Mask oxide

Si

Backside
protection oxide

SEM view of the backside after DRIE

Figure 11.7. The effect of an oxide etch stop layer on the etch profile of the vias.

An alternative is the use of a lapping and polishing process to thin the wafer instead of the anisotropic wet etching. In this case, the variation in etch depth due to different opening size is less important since the wafer thinning step can be adjusted to the shallower structure. However, wafers with deeply etched cavities

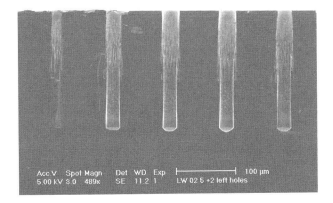

Figure 11.8. SEM image of the cross section of vias etched by DRIE in bulk silicon.

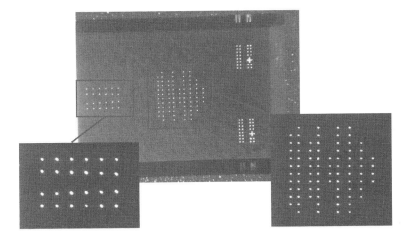

Figure 11.9. Arrays of through-wafer interconnects (Au-coated Cu plugs) test structures: back-side view of the membrane and close-ups of the two arrays (front-side view).

(vias and trenches), i.e. micromachined wafers, must be thinned mechanically. The results of this investigation are presented in the next section.

11.4   Thinning micromachined wafers

The techniques described so far and used to realize this first series of middle chips employ local wafer thinning. A method to thin the entire wafer in a controlled and reliable way is an interesting alternative. Lapping and polishing, a frequently used technique for wafer thinning, can offer this alternative. However, as in our case,

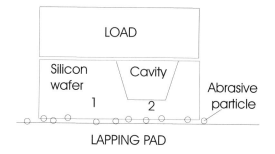

Figure 11.10. A schematic view of the lapping configuration for a wafer with a cavity structure on the front side.

through wafer interconnects are needed and the wafers present a large number of deeply etched structures, creating a new situation for the mechanical lapping process that needs to be investigated.

11.4.1 Cavity behaviour during mechanical thinning

In mechanical thinning, the lapping rate strongly depends on the effective pressure between the wafer bottom surface and the pad [9]. Therefore, the presence of cavities creates a different situation (see figure 11.10) and has to be considered. The lapping rate of the specific position depends on the relative speed of the wafer to the pad and on the effective pressure, i.e. the specific effective load L_{eff} over the effective contact area A_{eff}, and it is related to the pad and the slurry. During the lapping, when the load is applied to the whole wafer, the effective pressure at position 1 and 2 will be different when the membrane at position 2 has reached a certain thickness. This leads to the possible redistribution of the abrasive particles and may cause local accumulation of the particles under the membranes. As a result, (a) vertical strain near the edge of the membrane can happen and cause a sudden break of the membrane (especially if the lateral size is large) or (b) the lapping rate can be different on different positions on the wafer. This effect strongly depends on the geometry of the membranes as well as the load and the abrasive size of the slurry.

11.4.2 Thinning of wafers with deep trenches or vias

The situation previously described is encountered if wafers containing vias or trenches are mechanically thinned. When the trenches or vias are met, the configuration of the effective load as well as the effective area will change. This may cause an increase in the thinning rate at such regions for a period of time, thus resulting in a fluctuation of the thickness *versus* the area ratio occupied by holes or trenches. Another aspect is the edge damage problem. In fact, vias or trenches can act as trap centres for the abrasive particles. During the movement between

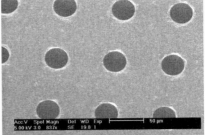

Figure 11.11. Trenches and holes after lapping and polishing. Abrasive particles size for lapping before CMP: 3 μm.

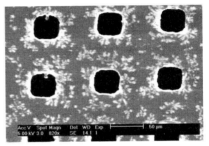

Figure 11.12. Trenches and holes after lapping and polishing. Abrasive particles size for lapping before CMP: 9 μm.

the pad and the wafer, some of the abrasive particles hit the edge and can cause damage until they enter the cavity that acts as a trapping area. Although such an edge attack cannot be avoided, the effective contact area strongly depends on the diameter of the abrasive particles. The larger the difference between L_{eff} and A_{eff} is the more serious the damage can be. Figures 11.11 and 11.12 show micrographs of samples lapped by using 3 and 9 μm abrasive particles, respectively and followed by a chemical mechanical polishing (CMP) for 30 min. As can be seen by using a larger particle size, more severe damage is produced. Furthermore, as the effective load strongly depends on the geometric structure of the holes or trenches, structure damage can be more pronounced for closely packed structures.

In case of the wafers coated with the oxide or nitride, as the polish or lapping rate of the oxide or nitride is much lower than that of silicon, some additional fence-like structure at the boundary is observed (figure 11.13). This fence is parallel to the plane of polish and can generate some particles in the post process. Ultrasonic treatment can remove such fences efficiently to avoid such a disadvantage.

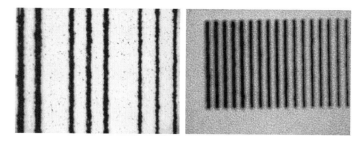

Figure 11.13. Wafer lapping and polish on the wafer coated by a nitride (left) and oxide (right).

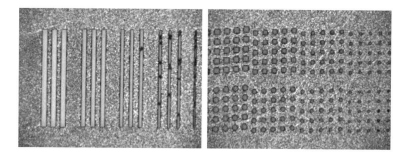

Figure 11.14. Optical micrographs of the wafer back side after lapping and chemical mechanical polishing.

Finally, when the size of the holes or trenches is smaller than that of the abrasive particles, the evaluation of the damage of the edge is more complex. It will depend much more on the spatial distribution of the structures and on the pad velocity since there is no trap for abrasive particles.

11.4.3   Particle trapping and cleaning steps

When the lapping reaches the holes or trenches, these may act as a trap centre for abrasive particles. Another source of particles comes from the silicon itself during lapping and small pieces of the broken structures. Moreover, the CMP process also generates particles that are very small (e.g. 0.3 μm) and are always charged. So the removal of such particles after lapping and polishing has to be addressed. We have tested various cleaning procedures. A commonly used process is ultrasonic agitation followed by hydrogen-peroxide-based (RCA) wet cleaning. In figure 11.14, optical images of holes and trenches after the final cleaning step are shown. In some cases, particles trapped in trenches or holes can be still observed. Some new CMP cleaning procedures such as Megasonic cleaning might further improve the removal of residual particles [10]. Further investigations to optimize the cleaning procedures are still in progress.

11.5   High-aspect-ratio through-wafer interconnections based on macroporous silicon

As device dimensions shrink, the connecting pad sizes are also reduced. This requires a large number of anisotropic through-wafer conductive connections with a very small pitch (≤ 5 μm). A suitable technology for realizing through-wafer holes with an opening size of only a few micrometres is needed. This is very difficult with DRIE as it is very sensitive to area distribution and microloading effects. In this section, we present a novel approach to realize very large arrays of closely spaced through-wafer interconnections using a process based on macroporous silicon formation, wafer thinning and metal electroplating. Macroporous silicon formation by electrochemical etching has been investigated for micromachined components and optical devices [11, 12]. For our investigation very deep pores (100–200 μm) only a few micrometres in size are needed. This means that the process for macroporous silicon formation needs to be tuned to achieve these very-high-aspect-ratio structures. Furthermore, the large number of through-wafer holes in a very small area demands a very good uniformity, i.e. very high anisotropy on a large area. Finally, the metal filling of the pores by Cu electroplating to form the through-wafer plugs is investigated.

11.5.1   Formation of high-aspect-ratio vias

The process flow used to prepare these samples is schematically depicted in figure 11.15. Low-stress LPCVD silicon nitride is deposited on the front side as the masking layer and an array of squares with a selected size and spacing is opened by dry etching. Sharp pits are then formed by etching the silicon in a 25wt% TMAOH solution at 80 °C for a few minutes. These pits represent the starting point for the macropores formation.

The macroporous structures are obtained by anodization in a solution of 5% HF mixed with ethanol. The sample holder has an open window for the back-side illumination provided by a 75W tungsten lamp. The back-side illumination is necessary to generate holes that take part to the reaction [13]. The counter electrode is a platinum grid. Values between 2.5 and 20 mA cm^{-2} for the current density and between 0.5 and 1.5 V for the voltage can be applied. The HF concentration and the current density are varied to find the optimal working conditions [14], i.e. sufficient etching uniformity and high anisotropy, to achieve pores depth down to 300 μm. Arrays of starting pits, 2 μm × 2 μm in size and with a spacing of 2 or 3 μm, are used for these experiments. The addition of ethanol to the electrolyte is essential. In fact, a sensible improvement in both uniformity and anisotropy of the pores is observed with respect to a solution using the same HF concentration but without the surfactant. An SEM image of a cross section of an array of pores is shown in figure 11.16. With a current density of about 10 mA cm^{-2} and a voltage of 1.2 V, 250 μm deep pores with good structural characteristics can be generated in about 3 h etching.

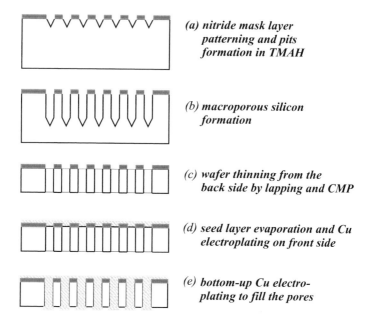

(a) *nitride mask layer patterning and pits formation in TMAH*

(b) *macroporous silicon formation*

(c) *wafer thinning from the back side by lapping and CMP*

(d) *seed layer evaporation and Cu electroplating on front side*

(e) *bottom-up Cu electro-plating to fill the pores*

Figure 11.15. Schematic drawing of the process flow for the array of through-wafer copper interconnects.

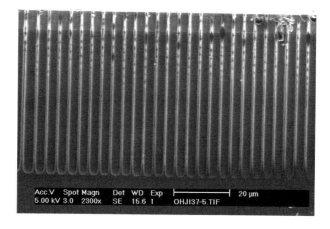

Figure 11.16. Cross section of an array of pores formed in a 5%HF/ethanol electrolyte.

11.5.2   Wafer thinning

After the formation of porous silicon, the sample is thinned down by lapping and polishing from the back side using a λ-Logitech LP50 machine. The type of slurry

Figure 11.17. Optical images of the (*a*) front and (*b*) back side of a high density array of 2 μm $\times$ 2 μm micropores after lapping and polishing.

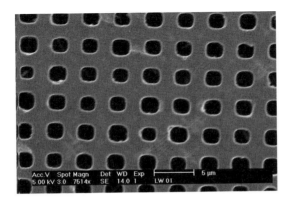

Figure 11.18. SEM image of through-wafer holes after lapping and CMP (back side view).

and particle size can be varied to obtain the proper lapping rate. By choosing the appropriate lapping rate and abrasive size, no remarkable damage to the pores structure is observed. A lapping rate of several micrometres per minute can be obtained with large-size-particle slurry but, for these samples, a 3 μm Al$_2$O$_3$ slurry is selected as this gives better results in terms of uniformity and surface damage. The lapping process is stopped just before the bottom of the pores is reached. Then a CMP step follows to remove most of the damage and slowly reach the bottom of the pores and open them. A 0.3 μm silica slurry (pH $\sim$ 10.2) with a polishing rate of 0.1–0.3 μm min^{-1} is used, followed by a specific cleaning process as discussed in the previous section. This procedure results in a damage-free surface as can be seen in figure 11.17 where the optical image of both front side and back side of a large array of pores after wafer thinning is shown. In figure 11.18, a close-up of the wafer back side after polishing is shown.

Figure 11.19. An SEM image of the sample's back side after the final electroplating step. The defined area for pore arrays can be clearly seen.

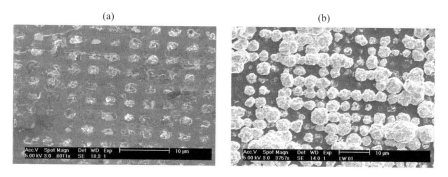

Figure 11.20. SEM images (close-up view) after the bottom-up plating: (*a*) the pores are (almost) completely filled and (*b*) over-plating resulting in lateral overgrowth.

11.5.3   Through-wafer interconnects

In order to fill the through-wafer holes with a conductive layer, a method similar to the one recently developed in our laboratory to form Cu plugs [15] is used. First, a very thin (20 nm) adhesion or diffusion barrier layer, either evaporated Cr or sputtered TiN, followed by a 60 nm Cu seed layer is applied at the wafer front side. A first Cu-electroplating step (5–6 μm) is performed to block the openings completely. At this point, the sample is turned upside down and a controlled bottom-up Cu-electroplating step is applied to fill the holes completely up to the other side. A current density of about 30 mA cm^{-2} resulting in a plating rate of about 40 μm hr^{-1} is generally employed. Figure 11.19 shows a large area of the die containing several high-density arrays of through-wafer holes completely filled after the bottom-up plating step. The selectivity of the process can be clearly seen as no Cu plating takes place outside the array regions.

(a) (b)

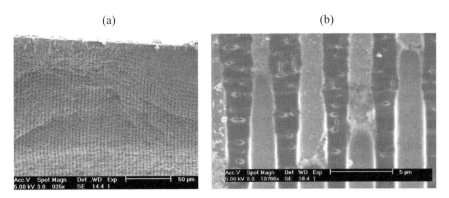

Figure 11.21. SEM images of a filled array: (*a*) cross section of the very deep narrow pores filled with copper and (*b*) close-up of a central region.

A close-up view of one such array where the Cu has reached the surface or is about to come up from the holes is shown in figure 11.20(*a*). If over-plating takes place, a lateral overgrowth is observed as can be seen in figure 11.20(*b*). The successful filling of the high aspect ratio vias is also confirmed by SEM inspection. The images of a sample cross section is shown in figure 11.21. Although it is quite difficult to cut the sample so to show the same via from surface to surface (due to the small size and small pitch of the vias and the relatively large amount of copper present in the sample), the images clearly indicate almost no voids and good uniform filling. Further investigation will include realization of electrical isolation as well as lithographic process for selective plating.

11.6 Conclusion

- A through-wafer Cu-plug interconnect technology has been pursued as an initial step towards the objectives envisioned in the CORTEX project. Based on the current findings, a process flow to realize through-wafer interconnects has been defined. Further investigation and optimization of the process is needed for downscaling of the via size.
- Investigation on wafer thinning by lapping and polishing on micromachined wafers has been carried out. Wafers containing a large number of vias and trenches have been successfully thinned down to 30 μm. The effect of surface damage and proper cleaning procedures is presented. Issues related to wafer handle and temporary bonding need further investigation.
- A potentially interesting technique suitable for very high density closely spaced interconnects based on macroporous silicon formation has been investigated. Large arrays of 2 μm $\times$ 2 μm with 2–3 μm spacing vias down to more than 200 μm deep have been realized. Mechanical thinning of the wafers containing such arrays of macropores has been successfully

performed. Moreover, the bottom-up Cu-plating technique has been applied to these wafers. Successful filling of these arrays has been demonstrated.

- Further characterization of the pattern dependence of the Cu-plating method needs to be done to increase the applicability of this process to other types of configurations.
- The developed through-wafer interconnects process based on the bottom-up Cu-plating method although requiring further investigation has already shown its potential for other applications, such as RF MEMS and MEMS packaging.

References

[1] Soh H T, Yue C P, McCarthy A, Ryu C, Lee T H, Wong S S and Quate C F 1999 Ultra-low resistance, through-wafer via (TWV) technology and its applications in three dimensional structures on silicon *Japan. J. Appl. Phys.* **38** 2393–6
[2] Xiaghua Li *et al* 2001 High density electrical feedthrough fabricated by deep reactive ion etching of Pyrex glass *Proc. MEMS 2001* pp 98–101
[3] Boellaard E *et al* 1999 Copper electroplating for integrated RF-devices *Proc. SAFE99 Workshop (Mierlo, The Netherlands, November)* pp 733–8
[4] Craciun G *et al* 2001 Aspect ratio and crystallographic orientation dependence in deep dry etching at cryogenic temperatures *Proc. 11th Int. Conf. Solid-State Transducers (Munchen, June 10–14)* vol 2, pp 612–15
[5] Nguyen N T *et al* 2001 Though-wafer copper electroplating for 3D interconnects *12th Micromechanics Europe Workshop (MME 2001) Proc. (Cork, Ireland)* pp 74–7
[6] Nguyen N T, Boellard E, Pham N P, Kutchukov V G, Craciun G and Sarro P M 2001 Through-wafer copper plug formation for 3-dimensional ICs *Proc. SAFE 2001 (Veldhoven, The Netherlands)* pp 141–4
[7] Wang L and Sarro P M 2001 Wafer thinning for highly dense 3D electronic structures *Proc. SAFE 2001 (Veldhoven, The Netherlands)* pp 214–19
[8] Nguyen N T, Boellaard E, Pham N P, Kutchoukov V G, Craciun G and Sarro P M 2002 Through-wafer copper electroplating for 3D interconnects *J. Micromech. Microeng.* **12** 395–9
[9] Preston F 1927 *J. Soc. Glass Technol.* **11** 214
[10] Francois Tardiff 2000 Post-CMP clean *Semiconductors and Semimetals Vol 63 Chemical Mechanical Polishing in Silicon Processing* ed S H Li and R O Miller (New York: Academic)
[11] Lehmann V 1993 The physics of macroporous silicon formation in low doped n-type porous silicon *J. Electrochem. Soc.* **140** 2836–43
[12] Ohji H 2002 Macroporous Si based micromachining *PhD Thesis* Delft, The Netherlands
[13] Kohl P A 1997 Photoelectrochemical etching of semiconductors *IBM Research*
[14] Nichelatti A *MSc Thesis* TU Delft
[15] Nguyen N T, Boellaard E, Pham N P, Craciun G, Kutchoukov V G and Sarro P M 2001 *Proc. MME (Cork, Ireland, 16–18 September)* ed M Hill and B Lane, pp 74–7

Chapter 12

Fault tolerance and ultimate physical limits of nanocomputation

A S Sadek, K Nikolić and M Forshaw
Department of Physics and Astronomy,
University College London

Computers are intrinsically probabilistic machines, constrained by the reliability of their algorithms and component physical parts. This was no more apparent than at the inception of digital computation over five decades ago. The ENIAC of the 1940s suffered a component (permanent) failure on average every two days, necessitating a shutdown and search and replace procedure for the burnt-out vacuum tubes. In addition, the occurrence of transient faults meant that its output had a reliability of only 90% in the best circumstances. These early problems with reliability prompted the first theoretical investigations into reliability of computation, notably by von Neumann [1]. The development of MOS integrated-circuit processors solved most of the problems of poor component reliability and, thus, computing for the mass market did not require the fault tolerant schemes for computation developed by von Neumann and others. In fact, until quite recently, device reliability in VLSI processors had a trend of *increasing* with increased scales of integration. Nevertheless, the high human, scientific and financial cost of space missions meant that ultra-reliable computation was paramount here, and thus the space program was the driving force for the development of computers engineered with highly fault tolerant architectures. The Saturn V guidance computer employed triple-modular redundancy, whilst the *Voyager* space probe computers, employing dynamic redundancy, are still operational more than 25 years later, despite cosmic and solar radiation, freezing temperatures and a harsh journey of billions of miles.

The remarkable period of unabated growth in computing performance through the progressive miniaturization of CMOS technology now seems to be close to an end. Around 2015, Moore's law is expected to breakdown due to increasing transient errors [2] and permanent faults and the enormous difficulty in scaling CMOS technology down to feature sizes approaching 10 nm. Even at present, with feature sizes around 90 nm and chips containing 10^7–10^8 devices, faults are becoming increasingly problematic, necessitating on-chip error correction routines, spare memory columns and greater burn-in times to screen out increased levels of latent defects.

The continued progression of computing power into the future calls for us to manipulate matter even further down at the nanoscale to develop *nanocomputers* consisting of more than 10^{12} logic resources implemented in three dimensional (3D) architectures. These may be crafted from novel devices such as molecular transistors, organic FETs, carbon nanotubes, single-electron transistors, spintronic devices or devices that have yet to be invented. Computing at this scale, however, will eventually bring us up against the limits to computation imposed by the laws of physics [3]. Faults that occur transiently during operation or permanently through manufacture will become extremely dominant [3–6] and the theoretical solutions to them have intimate connections with thermodynamics, quantum physics, communication theory and complexity theory. Some of the architectural solutions to these problems are, in fact, implemented by the mammalian brain, which performs remarkable feats of computation at a number of hierarchical levels from the macroscale right down to the nanometre level. Thus, at the nanoscale, noise, thermodynamics and quantum mechanics become tied up with the theory of computation itself and then we must ask: what are the ultimate limits to nanocomputation? How far can we push computation before we reach the barriers of speed, feature size, memory density and computing power whilst counteracting problems of energy consumption, power dissipation and faults? In this chapter we explore some of the solutions to noise and defects in nanocomputers and their limitations in the context of physics. There are some parallels to these solutions in the brain, with implications for nanocomputer architecture and its perhaps ultimate synthesis with neural systems [3].

12.1 Introduction

In analysing the effects of faults on nanocomputation, we need to consider separately the effects of device defects on computation and, more fundamentally, the effects of noise on *representation*, *communication* and *computation* in information systems. The basic approach to any kind of fault tolerant computation is to implement *redundancy* in either space (physical computing resources), time or both. Thus, the goal of theory is to elucidate what the maximum tolerable fault level is for a given redundancy via fundamental physical laws if possible and the engineering challenge is to design a system architecture that can achieve this limiting performance.

Permanent defects in the resources (i.e. logic gates, transistors etc) that make up a computer processor may arise as a result of imperfect manufacture or, during the course of operation, they may occur through failure of engineering or environmental insults. These kinds of faults are currently tackled mainly by making the manufacturing process as reliable as possible, so that after a routine of outright testing and burn-in time, a reasonable proportion of processors (5–90%) pass. This is appropriate for low defect probabilities such as $\sim 10^{-7}$ as is the case currently but, for a higher probability of defects, the optimum solution is to use *reconfiguration* or *self-repairing architectures* [5, 7–10]. A fundamental theory describing reconfiguration for combating permanent defects has been put forward, based on combinatoric optimization problems in mathematics [11] and further work may elucidate even deeper underlying physical principles.

The principles underlying information-processing systems may be phenomenologically broken down into those of communication, representation and computation [3]. Communication is the process of information transmission across space and/or time, the measure of information being the bit. Representation is the physical implementation by which a single bit, the minimum measure of information, is characterized in spacetime. From the most fundamental perspective, representation of bits may be through fields or particles. Computation, however, is the *manipulation* of information. What is popularly understood to be computation can, in fact, be separated into two different strands: one is the mathematical algorithms for processing information, whilst the other is the physical implementation of those algorithms. If we perceive of computation as the flow of information through a computer processor, then from our interpretation it is, in fact, the engineered flow of mass–energy through spacetime under the constraint of physical laws.

Transient errors that occur during computation, i.e. errors in resource operation that recover in a small, finite time, can be generally classified under the effects of noise. Noise is intimately connected to thermal fluctuations. As computing devices are scaled down to the nanometre level, noise effects in device operation are, hence, expected to rise exponentially because of these thermal fluctuations. The theory of noise-tolerant computation is then based in thermodynamics and statistical mechanics and, thus, it may form part of a greater physical theory of classical computation together with that of reversible and irreversible computation and information theory. Shannon and von Neumann suspected that the entropy in information theory was, in fact, related to thermodynamic entropy. Landauer's thermodynamical interpretation of irreversible computation as bit erasure leading to entropy increase and heat production [12] supported this view. Thus, noise, causing information erasure in a computer, may be what is missing in the development of a unified physical theory of classical computing. The controversy over the relation between information entropy and thermodynamic entropy is primarily due to the objection that information systems are generally never close to thermal equilibrium. The development of nanoscale information systems may, however, eventually fulfil this condition, which is why it will be important to understand the thermodynamical limits to computation in the presence of noise.

The fundamental theory of communication is Shannon and Szilard's *information theory* [13]. This theoretical treatment, based in thermodynamics, sets fundamental bounds on information transmission and storage in the presence of noise. The theory does not explicitly state the system by which informational noise tolerance may be achieved but numerous *error-correcting codes* have been developed that approach very closely the limits set by information theory [14,15].

However as has been described, no fundamental theory has been developed to describe *computation* in the presence of noise, although extensive work on this has been carried out [12, 16–29]. Whilst they address noise, the approaches of computer scientists have mostly considered the theoretical limits of simple Boolean circuits. Physicists on the other hand have emphasized thermodynamics and mostly overlooked the limits of computing tasks. The development of a generalized theory would set fundamental noise limits for reliable classical information processing, quantifying the tradeoffs in space, time and energy. Many examples of noise-tolerant computing architectures have been developed over the years [1, 3, 16, 30–39].

In this chapter, we investigate the efficacy of R-modular redundancy (module replicated R times), cascaded R-modular redundancy and parallel restitution (based on von Neumann's technique of multiplexing) for noise-tolerant nanocomputation. Reconfiguration, which may be able to deal with the high levels of manufacturing defects expected in nanoscale processors, is not discussed to any extent (but see [5, 7–10]).

The results of our work and others have shown that noise in computing systems is best tackled through the maintenance of parallel redundancy throughout a processing network, whilst periodically injecting energy into the parallel resources to restitute the signal back into its original phase space [3]. This principle of *parallel restitution* is fundamentally superior to techniques based on modular redundancy, as then there is no intermittent collapse of parallel lines to unreliable single lines. Thinking about noise tolerance this way naturally leads to a thermodynamical interpretation through consideration of the *computation entropy* of parallel redundant logic gates, and, hence, to limits to noise tolerance based in physics.

Parallel restitution is not restricted to classical computation, as it also forms the basis of quantum error-correction schemes and reliable neural communication. Reconfiguration is also used in the brain, this being known as *plasticity* in neuroscience. The brain must be subject to the same physical limits as nanocomputers and so it may be fruitful to investigate neural systems with a view to back-engineering some of their underlying computational principles into nanoprocessors. Perhaps nanoelectronics will ultimately allow us to truly interface with the brain and investigate the limits of both technologies.

12.2 Nature of computational faults at the nano-scale

Trying to build 3D computer processors more complex than anything conceived of before, and with typical feature sizes on the scale of tens of nanometres, will inevitably present special problems. Foremost of these will be defects in manufacturing the nanoscale devices themselves as well as imperfections in connecting them up within a network properly. Furthermore, nanodevices are expected to be extremely fragile and vulnerable to damage by environmental factors. However, the problem of defects is essentially an engineering challenge and techniques may be developed in the future to manufacture nanoprocessors flawlessly. More fundamental is the effect of noise on device operation, this being set by the laws of physics itself. We will never be able to 'engineer' thermal noise out of a system completely, because this would violate the second law of thermodynamics; the only hope is that the laws of physics in turn allow us to compute in a noisy environment through ingeniously designed trade offs in space and time but even then within certain constraints.

12.2.1 Defects

Permanent defects in micro and nanoscale processors may be classified into those that occur during the manufacturing process and those that occur during the working life of the machine. These kinds of defects already affect present-day CMOS devices. At a coarse level, dust contamination during semiconductor manufacture may damage many transistors in a localized area of a chip, rendering it useless. At the materials level, defects may occur in the SiO_2 gate dielectric or in the Si/SiO_2 interface and if there are too many of these, then a MOSFET will malfunction and become a 'dead' device [40]. If p_f is the probability of manufacturing failure of a device and there are N devices comprising a processor, the probability of failure of the entire processor at manufacture, i.e. the probability at manufacture that the processor does not give all intended outputs for every possible input, is then approximately:

$$P_f = 1 - (1 - p_f)^N. \tag{12.1}$$

Simple screening tests are used to weed out processors with these kinds of manufacturing defects.

Also very serious are permanent defects that develop after manufacture, during processor operation. In MOSFETs, this is primarily due to damage by the electrons flowing through a device itself, this being known as *hot carrier interface* (HCI) state generation [40]. Damage again primarily occurs at interfaces in the device, i.e. through dielectric breakdown and defects at the Si/SiO_2 interface, and arises even when the operating voltage is very low. The mechanism by which relatively low-energy electrons can transfer enough energy to break the lattice bonds in these cases is still unknown [41]. For a MOSFET to become permanently dead, many defects have to accumulate and reach an average critical value. In some cases, the device is manufactured with very few defects and only after extended operation do enough defects accumulate to disable it. Probabilistically though, some devices are manufactured with a large number of defects but not enough to impair the device's operation and, hence, not enough to make a processor fail a screening test. After a short period of processor operation however, HCI state generation makes up the remaining number of defects needed to reach the critical value and then the device and, in turn, the processor fail. Burn-in testing is used to screen out processors with such 'lemon' devices. As CMOS technology is scaled down to the nanometre level, since the device dimensions are reduced, the critical number of defects that need to accumulate to cause device failure will drop. Thus, a major problem anticipated with nanoscale CMOS is that the incidence of manufacturing failures and lemons will increase exponentially with scaling down [40, 41].

The imparting of electronic current momentum to individual atoms in interconnects is another cause of permanent failures during operation. This phenomenon of *electromigration* will become particularly acute at the nanoscale, where the motion of individual atoms will have to be taken into account.

Finally, radiation will be a significant cause of permanent defects in future nanodevices. Heavy ions, emitted through radioactive decay of the device and chip packaging materials of CMOS processors themselves, can cause single-event latchups (SELs) [42]. This is where a transistor becomes stuck in a '0' or '1' position, recovery only occurring on shutting down and re-powering the processor. More serious physical damage can be caused by high-energy cosmic rays, which induce catastrophic failure of devices through gate rupture. The detailed mechanisms of permanent failure in novel nanodevices as a result of radiation will, of course, be different but even more significant as the energy imparted by the particles becomes proportionally larger.

It is important to understand the enormous gap between the reliability in the manufacture of present CMOS devices and the nanoscale devices that have been produced so far. The probability of manufacturing failure for CMOS devices is approximately 10^{-7}–10^{-8}, giving chip manufacturing yields of 5–90% for $N = 10^7$. In contrast, the use of molecular combing for connecting single-walled carbon nanotubes in between electrodes has a p_f of 4×10^{-1} [43] whilst for the manufacture of hybrid molecular devices $p_f \approx 5 \times 10^{-1}$. Even if there is a modest improvement in p_f for nanodevice fabrication, given that N will be much higher at around 10^{12}, some form of massively redundant reconfigurable architecture will be needed to manufacture nanoprocessors with even reasonable reliability.

12.2.2 Noise

Transient faults are momentary errors in the proper functioning of gates or transistors in computer processors. They are technically known as single-event upsets (SEUs) and, in practice, can persist in an affected device for one clock cycle or several. Such random and transient errors in device operation can be considered to be *noise* that a computer has to contend with [44]. The sources of this in present CMOS devices are various and include electromagnetic interference, device parameter variances, quantum tunnelling, radiation, shot noise and thermal fluctuations. These sources of noise are anticipated to affect nanoscale devices even more acutely.

Electromagnetic interference may arise from mutual inductance and parasitic capacitance between devices and interconnects, whilst parameter fluctuations between different devices on a chip are caused by variances in device dimensions and dopant levels during manufacture [45]. With current CMOS dielectric materials, quantum tunnelling currents start becoming significant when the gate oxide thickness drops below 3–5 nm [42], although future nanoscale MOSFETs are anticipated to use alternative high-dielectrics or sandwiched structures that will allow gate oxide thicknesses of 1 nm [45]. Atmospheric neutrons and alpha-particles from nuclear decay processes in the chip materials themselves induce SEUs through charge production in devices, thus transiently disrupting switching function [42]. They are a particular problem for processors intended for use in space where radiation levels are higher and usually special radiation hardened chips or processors based on silicon-on-insulator technology [46] must be employed. Past modelling projections suggested the intuitive result that as CMOS technology is scaled down to the nanometre level, the critical charge for an SEU developed by high-energy particles would decrease. Surprisingly though, the threshold linear energy transfer has stabilized for recent generations of processors as probably CMOS devices have now reached the limits of sensitivity to radiation; as the device volume is reduced, the probability of interaction with high energy particles diminishes. This present balance between decreased threshold energy and reduced interaction probability is an effect specific for both the device type (CMOS) and the length scale they are manufactured at. It is unknown which way the equilibrium will tip for future nanodevices.

However, a much more fundamental barrier to nanoscale information systems exists as a result of the *equipartition theorem* of thermodynamics [47]. For a system in thermal equilibrium, every independent degree of freedom that can store energy must have average energy $kT/2$ due to thermal fluctuations. This gives rise to 'white' thermal noise, i.e. noise with a flat power spectrum. For white noise, the time correlations between fluctuations are less than the Smoluchowski time, 10^{-10} s [48]. In electronic circuits and devices, such thermal fluctuations give rise to *Johnson–Nyquist noise*. Here, voltage fluctuations arise from the thermal motion of electrons, which then relax through the resistance of the device. The noise power is simply given by

$$\langle V_{\text{noise}}^2 \rangle = 4kTR\Delta f \qquad (12.2)$$

where k is Boltzmann's constant, T is temperature, R is the resistance and Δf the bandwidth of the measuring system.

Another source of noise in electronic systems that is customarily treated separately to thermal noise is *shot noise*. Classical as well as quantum shot noise effects have been modelled in nanowires, and found to modulate their conduction properties in important ways [49, 50]. Shot noise is usually distinguished from Johnson noise by the fact that it arises only when there is a dc current across a conductor, whilst the latter is present even if there is no current. If one considers a conductive channel carrying such a dc current, an electric field exists across it, creating a potential barrier. Diffusing electrons jump down the barrier much more readily than up against it and, thus, the forward current is much greater than the reverse current. As the time during which individual electrons arrive at any point in the channel is small compared to the duration between electron arrivals, the current can be considered to be a sum of δ-functions, this being a Poisson process. Fluctuations in barrier crossings will mean there will be variation in the arrival of individual electrons: this is the origin of what is generally understood to be shot noise. The noise power is given by

$$\langle I_{\text{noise}}^2 \rangle = 2q \langle I \rangle \Delta f \tag{12.3}$$

where q is the electronic charge and $\langle I \rangle$ is the mean current. However, Sarpeshkar *et al* [51] have shown that, in fact, both shot noise and Johnson noise occur through the same mechanism, namely thermal fluctuations. They showed that, in the subthreshold regime of MOS transistors, experimental measurements of white noise power corresponded precisely with the shot noise theoretical derivation. No extra term had to be added for thermal noise and, in fact, they derived the Johnson–Nyquist formula, (equation 12.2), from their shot noise equation by setting the barrier potential to zero and calculating the forward and reverse currents using Einstein's diffusion relation. They thus showed that *thermal noise is two-sided shot noise*, i.e. shot noise in both forward and reverse diffusion currents. Conventionally, scattering events in a conductive channel are labelled thermal noise, whilst fluctuations in barrier crossing are described as shot noise. However, both mechanisms result in a Poisson process with a specific electron arrival rate and are ultimately due to thermal fluctuations and the equipartition theorem.

Now that we have defined and characterized noise, we must think carefully about how it can affect information systems. We have mentioned representation, communication and computation as useful paradigms for considering this. However, the problem is, in fact, an integrated one with no clearcut separation: representation is the communication of 1 bit, whilst communication is computation without manipulation. As we build-up a theory of noise-tolerant information and computation, we will see that the optimum solution addresses the problem automatically.

Birge *et al* [4] considered, for nanoscale information systems, how statistical fluctuations in the representation of a computational state would affect the reliability of computation and how increasing the ensemble of state carriers could make an information-processing system more reliable. They analysed this from the viewpoint of molecular computation and found that redundancies in the thousands were needed for reliable representation and, thence, computation.

For nanoscale semiconductor devices, a serious issue would be fluctuations due to shot/thermal noise in the number of electrons representing a digital bit and here a reduction in the number of electron carriers would make state representation exponentially more unreliable. Kish [2] has recently looked at the effect of CMOS nanoscaling and thermal noise on system reliability. The *power dissipation problem* facing extremely-large-scale integration is the increasing power dissipation density and, hence, increasing temperature, of conventional CMOS circuits as they are scaled down and made more dense [52]. One of the possible solutions is to reduce the operating voltage and, hence, the power dissipation per clock cycle but Kish has shown how this would bring the logical threshold voltage too close to the level of thermodynamic fluctuations at a certain length scale. Since lowering the voltage reduces the number of electrons in a bit representation, computational noise arises as a result of noisy state representation, which is the same result obtained by Birge *et al* .

We could take the effects of noise further and consider how noisy computation occurs not just because the information fed to a nanodevice or nanocircuit is noisy (either because of unreliable single-bit representation or an erroneous binary word in a register) but because the device/circuit itself is noisy in processing even absolutely reliable input. After von Neumann, such computational noise is modelled by considering the most fundamental devices in a nanocomputer to be Boolean logic gates. If a transient error occurs in a gate computation, there occurs at actuation a bit-flip in the correct output given a specific input. Such a noisy gate fails with probability ϵ per actuation, each gate in a circuit failing independently: we thus solely assume the affects of 'white' noise. Computational noise may also be correlated ('coloured') but such effects are usually technology dependent, such as with common mode failures [53], and so we ignore them here. We may also presume that $\epsilon < 1/2$, since any gate that fails with probability $\epsilon > 1/2$ is equivalent to a $(1 - \epsilon)$-noisy version of its complement [22].

As devices are scaled to the nanometre level, we assume that the primary culprit of SEUs will be thermal noise. We can hence describe how the thermal noise in logic gates dimensionally scales down using a Boltzmann distribution,

$$\epsilon = \frac{1}{\mathcal{Z}} \exp\left(-\frac{\Delta E(r)}{kT}\right) \tag{12.4}$$

where $\mathcal{Z}$ is the partition function and $\Delta E(r)$ is the switching energy of each gate per clock cycle, which is a function of r, the gate size. As r approaches the nanoscale, $\Delta E(r)$ gets smaller either because of purely device-specific characteristics or through design in order to minimize the power dissipation density.

It is also interesting to consider briefly the corresponding problems of noise in quantum computation. Stated very basically, quantum computers use many of the ideas behind their reversible classical counterparts, with the added quantum ingredients of superposition of states and entanglement. A quantum bit ('qubit') can be represented as a superposition of the digital bits '0' and '1', $a|0\rangle + b|1\rangle$ and, thus, has an analogue nature as the complex coefficients a and b are continuous quantities. There are, thus, two main noise effects in quantum computation: one is from *decoherence* of qubits through their entanglement with the environment, the other, from *random unitary transformations* [54]. Noisy quantum logic gates may introduce random unitary transformations and may not, therefore, implement their unitary transformations precisely. This is potentially very easy to do, since such transformations form part of a continuum. Such errors in quantum computation on qubits can be characterized as a linear superposition of normal states, bit-flip errors $a|1\rangle + b|0\rangle$, phase errors $a|0\rangle - b|1\rangle$ and both errors simultaneously $a|1\rangle - b|0\rangle$. Noise plays an important part in the transition between quantum and classical computing, which we will discuss later.

The noise levels in present CMOS devices are extremely low, around 100 FITs (failures-in-time). One FIT is equivalent to one device SEU in 10^9 hr of device operation. This translates into a noise level for present CMOS devices of $\epsilon \approx 10^{-20}$. It is extremely unlikely that future nanodevices will be able to achieve such low levels of noise, especially as ϵ will grow exponentially with scaling down, as described by equation (12.4). We will now explain how we can modify computer architectures to make them more resilient to noise. We will also explore whether there is an ultimate limit to ϵ set by nature and, thus, a limit to scaling computation down to ever smaller dimensions.

12.3   Noise-tolerant nanocomputation

If a computation can be solved in polynomial time, then it is classified as belonging to the complexity class **P**. This complexity class is thought to be a subset of a larger class of decisional problems **NP**, where the solution requires additional information in the form of a certificate (factoring is an example of such a problem). **P** and **NP** are classes of *deterministic* algorithms, which means they perform the same set of operations every time they are executed on the same input. Once noise becomes a factor (i.e. $\epsilon > 0$), such algorithms are no longer deterministic but *random*. Such random algorithms, implemented by design or as a result of noise, are part of the greater complexity class **BPP** (*bounded-probabilistic polynomial time*). The noise-tolerant architectures that we will explore for nanocomputation will necessarily belong to this complexity class. A corresponding class exists encompassing fault tolerant quantum computation called **BQP**.

If long chains of computations are performed with noisy versions of **P** algorithms, the reliability of the output quickly drops to zero. If such algorithms are modified to be noise tolerant, then the output reliability may, within certain constraints, be bounded above a specific threshold no matter how long the computation is. In our model of noise-tolerant computation, processors are modelled by *noisy circuits*. Every circuit is composed of Boolean logic gates that form the *basis* of the circuit. Two types of circuits are possible. One is a *network*, where there is no restriction on the connectivity between the basis gates comprising the circuit. The other more restricted type is a *formula*, where the basis gates form a directed acyclic graph such that the output of each gate is used as the input to at most one gate. Each gate fails with probability ϵ, as described earlier, and the noise-tolerant circuit has resultant reliability $(1 - \delta)$: as for all **BPP** algorithms, we cannot say that the output of the circuit is definitely correct, only that it is $(1 - \delta)$-reliable. By the *Chernoff bound*, if a circuit computes with reliability $(1 - \delta) > 1/2$, then by repeating the algorithm, the reliability increases *exponentially* in the number of repetitions [55]. Thus, although the condition $(1 - \delta) > 1/2$ is sufficient to show that noisy computation is possible, here we will be practical and require our noise-tolerant computation to occur in as short a time as feasible and, thus, require δ to be extremely small.

We will see that noise-tolerant circuits necessarily require two ingredients. One is an increase in the circuit *size*, i.e. the total number of basis gates in the circuit. The second is an increase in the *depth* of the circuit, which translates into a slowdown in the time to perform the computation. These two compromises essentially mean that noise-tolerant circuits must be *redundant* and they correspond to a space and time tradeoff respectively. Although both tradeoffs are always needed, we could, for example, decrease the space tradeoff and increase that in time. However, here we will consider a 'fault-masking' implementation, in that we will try and implement our noisy nanoprocessors in as near real time as possible, with a minimum depth increase. We will also describe how noise-tolerant circuits have a threshold noise level ϵ for the basis gates, below which the circuit can compute for an arbitrary length whilst bounded with $(1 - \delta)$ reliability.

For designing practical noise-tolerant nanoprocessors, an additional constraint arises though. If to build a reliable 10^{12} device nanocomputer a redundancy of say 10^4 is required, then the nanoprocessor effectively only has 10^8 devices and is not much better than current CMOS processors. Although redundancy is essential, the noise-tolerant architectures we engineer for nanocomputers must minimize their impact on circuit size and speed in order to keep the nanocomputer technologically useful.

The theory of error-correcting codes has deep connections with noise-tolerant circuits that are not fully understood as yet. For example, the basic scheme of R-modular redundancy that we consider here is a hardware implementation of a simple repetition code. We will see that the optimum solution to noisy information and computation incorporates error-correcting codes with space and time redundancy, as has been shown by Spielman [38].

We first review briefly the history of noise-tolerant computation and develop crude models of R-modular redundancy and cascaded R-modular redundancy in the context of nanocomputation. We then describe the more optimal technique of parallel restitution and demonstrate the elements of noise threshold, increased circuit size and increased circuit depth that will be characteristic of future nanoprocessors.

12.3.1   Historical overview

The theory of noise-tolerant computation has an illustrious history. Wiener was the earliest to consider errors in computation and he suggested *triple modular redundancy* to combat it [30]. The father of reliable computation though is unarguably von Neumann and, in a series of lectures at Caltech in 1952 (later published in 1956), he set out the theoretical basis for the field [1]. His foresight was remarkable and he established the existence of a noise threshold and the necessity of increased circuit size and depth for reliable computation. His motivation was partly the unreliability of the then current computing machines but more so the problem of reliable computation in the brain. Around this same period, Hamming was also spurred by the unreliability of computers to develop the theory of error-correcting codes [14]. The late 1940s and early 1950s were a period of heightened optimism in the computer science community that the brain could be eventually emulated and understood in terms of automata. Inspired by Turing, the neural network theory of McCulloch and Pitts [56] and what he perceived to be a mathematical framework for understanding neuroscience, von Neumann considered the milieu of a warm, wet brain with its neurons of perceived unreliability arranged into networks of great logical depth. His assertion was that for the brain to operate reliably and perform long trains of computations, some form of fault tolerance was essential and, to this end, he mathematically investigated triple modular redundancy, cascaded triple modular redundancy and a part-analogue/part-digital procedure he called multiplexing.

After an initial lull, further extensive theoretical work was done by computer scientists on the problem of reliable computation [57]. Subsequent to von Neumann's somewhat applied treatise, Dobrushin and Ortyukov showed formally that to make a noisy Boolean circuit compute with an output of bounded reliability, the redundancy (or complexity) of the circuit had to increase by a factor of $\mathcal{O}(\ln N)$, where N is the number of gates in the network [35, 58]. Pippenger *et al* re-proved this theorem but found the original proofs of Dobrushin and Ortyukov to contain some flaws [59]. Their essentially correct arguments were later re-proved by Gàcs and Gàl [60]. A lot of effort was expended in trying to prove that a reliable circuit had to be only of size $\mathcal{O}(N)$ instead of $\mathcal{O}(N \ln N)$ and Pippenger managed to prove this for certain classes of Boolean functions [61]. In this work, Pippenger also managed to move away from von Neumann's random construction for a 'restoring stage' and develop an explicit construction based on expanding graphs with specific spectral properties. A major breakthrough was the application of information theory to Boolean computation to prove that a Boolean formula with noisy gates had to increase in logical depth to become reliable and that a threshold gate error $0 < \epsilon < 1/2$ existed, above which reliable computation could not be performed [21]. Feder extended this work by probabilistic methods to prove this result for networks as well as formulas [63]. Evans used more sophisticated information theoretic arguments to fix tighter bounds for formulas on the threshold gate error and the increase in logic depth and circuit size [22, 26]. The most positive bounds proved so far are based on the application of error-correcting codes to parallel computation, as has been done by Spielman [38]. Here, if w processors perform a parallel computation, then it can be made resilient to noise-prone processors by encoding the inputs and outputs by a generalized Reed–Solomon code and using instead $w \log^{O(1)} w$ parallel processors. The failure probability of the parallel computation over time t would then be at most $t \exp(-w^{1/4})$ and the depth increase in the computation in time would be $t \cdot \log^{O(1)} w$.

A shortcoming with the previously mentioned work is the failure to take into account noisy interconnects between gates. If the network size is large and the length of wires substantially long, this must be allowed for. The solution to this problem came through theoretical analysis of noise-tolerant cellular automata that used purely local procedures to protect against errors, hence dispensing with the need for wires. Toom developed the first non-ergodic fault tolerant cellular automaton that could compute and store information [62]. Later he developed an extremely simple transition rule (*Toom's Rule*) for reliable cellular automata: take the majority vote of your state and that of your northern and eastern neighbour [64]. Gàcs further developed reliable cellular automata with properties of 'self-organization', where even with an initially homogenous configuration a hierarchical organization can emerge during noisy computation [39, 65]. Von Neumann's initial work on reliable computation, mathematics and biology seems to have come full circle to more 'organic' architectures akin to his self-reproducing automata [66].

Von Neumann was not completely satisfied by his pioneering work however, as he was also highly influenced by the information theory of Shannon [13] and Szilard [67]. To cite his 1956 paper:

> Our present treatment of error is unsatisfactory and *ad hoc*. It is the author's conviction, voiced over many years, that error should be treated by thermodynamical methods, and be the subject of a thermodynamical theory, as information has been, by the work of L Szilard and C E Shannon. [1]

This statement has generally been understood to mean von Neumann believed noisy computation should be treated by information theoretic methods. Winograd and Cowan were the first to attempt this [16] before the work of Pippenger and Evans. Lately, however, opinion has gravitated more towards a physical interpretation of von Neumann's thinking, beyond information theory, where noisy computation is, in fact, bounded by physics and thermodynamics itself. Lloyd noted the elemental requirement of energy to correct errors in computation [27], whilst Sadek *et al* linked the fundamental principles of noise-tolerant computation to thermodynamics, entropy production and heat dissipation

[3]. Such work is being spurred by efforts to develop nanocomputers with neuromorphic architectures, operating at the very limits of what is physically possible. Von Neumann's original vision for his pioneering ideas are finally beginning to be realized.

Although the previously described body of work has been extremely important in proving what principles must underlie reliable classical computation, most of the results are theoretical in that they do not describe *how* reliable circuits may actually be constructed and, furthermore, only results of approximate numerical order are derived. Some numerical results exist for bounds on the gate error threshold, for example $\epsilon \simeq 0.0886$ for NAND gate formulas [68] and $\epsilon = 1/6$ for three-input Boolean functions [69] but these sorts of results are generally computed for an output unreliability $\delta < 1/2$, whereas practically for a nanocomputer we would need to develop architectures and threshold results typically for $\delta < 10^{-19}$. A large body of engineering literature exists on practical fault tolerant architectures but as semiconductor fabrication improved, these designs moved away from architectures implemented at the component level to fault tolerance implemented at the system level [36, 37]. For example, the Orbiting Astronomical Observatory flown in the 1960s implemented noise-tolerant architectures at the transistor level, whereas the latest systems in fly-by-wire computers on aircraft employ redundant processors and channels combined with software recovery. In contrast, nanocomputer architectures will need to implement noise tolerance at the device scale and so a return to fundamental theoretical work combined with applied thinking is required. Initial work looking at cascaded triple modular redundancy in single-electron tunnelling (SET) devices and quantum cellular automata (QCAs) was done by Spagocci and Fountain [70, 71], with the first comprehensive investigations of fault tolerant nanocomputation being done by Nikolić, Sadek and Forshaw [3,5,72,73], and Han and Jonker [74]. We now review this body of work, which mainly considers the classical von Neumann computer architecture in the context of nanoelectronics. It should be noted, however, that there is a growing opinion that alternative architectures that abandon the clocked processor approach may be more suited to reliable computation at the nanoscale [75,76].

12.3.2   *R*-modular redundancy

In his book *Cybernetics*, Wiener described the technique of triple modular redundancy (TMR) for making a computer more reliable. In its simplest form, three copies of the computer are made to work in parallel and their outputs are compared and the majority voted amongst. The combined system then works with a greater reliability, but with a factor of three tradeoff in redundancy and an increase in processing time due to the depth increase from the voting circuitry. This simplistic implementation is not optimal however, as the processors with which the majority voter makes the comparison are very error prone in themselves as they contain so many gates. If we assume the voters to be completely reliable, the best way to implement a TMR architecture would be to implement it at the smallest hierarchy possible, namely at the level of the logic gates themselves. Consider the TMR implementation for a NAND gate as an example. Every gate in the processor circuit is triplicated in parallel, with all input wires being fanned out to the three gates. A single majority gate then compares between the single-bit outputs of the NAND gates. The output voted at least two times out of three for each triplicated gate is considered correct. Although this approach is optimal in terms of protection against noise, it leads to a $3n$ decrease in speed, where n is the logical depth of the processor. The factor of three arises from the logical depth of each of the majority gates, assuming they are constructed from the universal set of gates AND, OR and NOT.

Naturally though, the majority gates must be constructed from the same noise-prone gates as the rest of the processor and, once this is factored in, we find that implementing TMR at the gate level is no longer optimal. This is because, at this scale, the maximum number of majority gates are utilized and if too many of them are relied upon to correct errors, since they are themselves noisy, they contribute to the degradation of the output. Hence, practically, TMR cannot achieve its maximum theoretical efficiency, because the theoretical optimum does not take into account the reliability of the intermittent 'encoding' and 'decoding' process. This is analogous to noisy channels in information theory, where the effects of noise in the encoding and decoding of information are ignored. Thus, there exists an intermediate hierarchical circuit level for implementing TMR that exists somewhere between the scale of single gates and the level of the entire processor. This, for example, might be at the level of a full-adder or an ALU.

TMR can be improved even further by generalizing the scheme such that each subunit is replicated instead $R = 5, 7, 9 \ldots$ times. This technique is known as R-*fold modular redundancy* (RMR), and is illustrated in figure 12.1(*a*). An A-bit input is fanned out an odd number of times R to the replicated processing clusters. The B-bit output from each of these clusters is voted between by a B-bit majority gate (which comprises B single-bit majority gates in parallel) to give a single B-bit output. To implement RMR, every subunit in a non-redundant processor is simply swapped with the construction described.

We now present a rather crude analysis of the efficacy of RMR for extremely-large-scale integration in nanoprocessors. It is basic in that we do not use a Markovian analysis to look at the effects of noise on information flowing through the logical depth of the processor and each of its subunits (we save this more sophisticated approach for the technique of parallel restitution in section 12.3.4). Instead we assume that every gate comprising a processor or one of its subunits *must* work correctly for it to give the correct answer in one clock cycle for a random input. This is too severe a constraint and, hence, the results obtained are an upper bound. The numerical results do, however, approximately describe the efficacy of RMR, and how it and variants of its approach are inadequate for noise-tolerant nanocomputation.

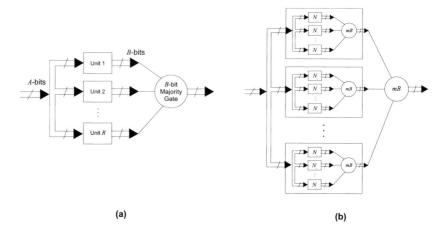

(a)   **(b)**

Figure 12.1. (*a*) In *R*-fold modular redundancy (RMR), processor subunits containing N_c gate resources are replicated an odd number of times *R* and voted amongst by a *B*-bit majority gate [5]. Each single-bit majority gate has *m* gate resources. The non-redundant processor contains a total of $N = C \cdot N_c$ gate resources, where *C* is the number of non-redundant subunits. In the RMR processor, this is increased to $N = C(RN_c + mB)$. (*b*) Cascaded *R*-fold modular redundancy (CRMR) is the concatenated version of RMR, where further RMR units are made from the RMR units themselves. From [5], *Nanotechnology* © IOP Publishing Ltd.

Consider an RMR processor containing *N* gate resources. The RMR architecture is implemented at the optimal circuit level, such that computational subunits containing N_c gate resources are replicated and voted amongst. Each single-bit majority gate consists of *m* gates or devices. It should be noted that if the majority gates are constructed from Boolean logic, *m* will scale quite sharply with *R*. We ignore this effect in our analysis and assume *m* to be fixed.

The probability P_{unit} that each non-redundant subunit gives the correct output per clock cycle is given by the following pessimistic equation:

$$P_{unit} = (1 - \epsilon)^{N_c} \approx 1 - N_c\epsilon \approx e^{-N_c\epsilon} \tag{12.5}$$

where the approximation assumes $\epsilon \ll 1$.

If we take into consideration the failure probability of the voter, then by combinatorial analysis the probability of correct output from the RMR version of each computational subunit $P_{unit-mr}$ is given by,

$$P_{unit-mr} = P_{voter}^B P_{group} + (1 - P_{voter}^B)[1 - P_{group}] \tag{12.6}$$

where P_{voter} is the probability of correct operation for each single-bit majority gate given by, $P_{\text{voter}} = (1 - \epsilon)^m \approx 1 - m\epsilon \approx e^{-m\epsilon}$ and

$$
P_{\text{group}} = P_{\text{unit}}^R + \binom{R}{R-1} P_{\text{unit}}^{(R-1)}(1 - P_{\text{unit}})
$$

$$
+ \binom{R}{R-2} P_{\text{unit}}^{(R-2)}(1 - P_{\text{unit}})^2
$$

$$
+ \cdots + \binom{R}{(R+1)/2} P_{\text{unit}}^{(R+1)/2}(1 - P_{\text{unit}})^{(R-1)/2}.
$$

The number of R-modular redundant subunits in the processor is given by $C = N/(N_c R + m B)$ and, thus, the output reliability of the entire processor per clock cycle is $(1 - \delta) = P_{\text{unit-mr}}^C$, since each subunit must operate flawlessly if the chip output is to be correct. Note that the latter expression effectively implies the *propagation* of errors when using the noise-tolerance technique of R-modular redundancy. We have ignored compensation errors between modularly redundant units which, for large C, is a good approximation.

By deriving $d\delta/dN_c = 0$ [5], it is possible to find the optimum size of N_c and, hence, the intermediate hierarchical circuit level for which implementation of R-modular redundancy optimizes the reliability of the processor. By making approximations and utilizing the fact that this technique is only particularly effective when the condition $N_c\epsilon \ll \ln 2$ is satisfied, the size of N_c which minimizes δ is found to be

$$
N_c \simeq \left[\frac{2mB}{(R-1)\binom{R}{(R+1)/2}\epsilon^{(R-1)/2}} \right]^{2/(R+1)}. \tag{12.7}
$$

The final expression for the processor reliability $(1 - \delta)$ is then

$$
(1 - \delta) \simeq 1 - \frac{N}{R}(mB)^{(R-1)/(R+1)} \left(\frac{R}{(R+1)/2} \right)^{2/(R+1)} \epsilon^{2R/(R+1)}
$$

$$
\times \left[\left(\frac{2}{R-1} \right)^{(R-1)/(R+1)} + \left(\frac{R-1}{2} \right)^{2/(R+1)} \right]. \tag{12.8}
$$

Table 12.1 shows the results of this analysis for $R = 1, 3, 5 \ldots, \infty$. As can be seen, there exists a *threshold noise level* ϵ_t below which the reliability of the processor can always be bounded by $(1 - \delta)$. In the limit of large R, this is $\epsilon_t \sim \sqrt{\delta/(mBN)}$. The minimum achievable processor failure probability is given by $\delta = NmB\epsilon^2$, which is suboptimal: theoretical results in computer science predict that, ideally, if $\epsilon < \epsilon_t$, then $\lim_{R\to\infty} \delta = 0$. We show in section 12.3.4 how this can be achieved with parallel restitution.

Table 12.1. Efficiency of R-modular redundancy at reducing computational failure in processors ($N_c\epsilon \ll \ln 2$). δ is the failure probability of the processor per calculation, whereas ϵ_t is the threshold gate noise below which the reliability of the computation is bounded by $(1 - \delta)$.

R	δ	ϵ_t	Optimum module size N_c
1	$N\epsilon$	δ/N	N
3	$N\epsilon \cdot 1.15(mB\epsilon)^{1/2}$	$\dfrac{0.9}{(mB)^{1/3}}(\delta/N)^{2/3}$	$\left(\dfrac{mB}{3}\right)^{1/2}(\epsilon)^{-1/2}$
5	$N\epsilon \cdot 0.8(mB\epsilon)^{2/3}$	$\dfrac{1.1}{(mB)^{2/5}}(\delta/N)^{3/5}$	$\left(\dfrac{mB}{20}\right)^{1/3}(\epsilon)^{-2/3}$
7	$N\epsilon \cdot 0.6(mB\epsilon)^{3/4}$	$\dfrac{1.3}{(mB)^{3/7}}(\delta/N)^{4/7}$	$\left(\dfrac{mB}{105}\right)^{1/4}(\epsilon)^{-3/4}$
9	$N\epsilon \cdot 0.5(mB\epsilon)^{4/5}$	$\dfrac{1.5}{(mB)^{4/9}}(\delta/N)^{5/9}$	$\left(\dfrac{mB}{500}\right)^{1/5}(\epsilon)^{-4/5}$
$\vdots$	$\vdots$	$\vdots$	$\vdots$
limit	$\sim N\epsilon(mB\epsilon)$	$\sim\dfrac{1}{(mB)^{1/2}}(\delta/N)^{1/2}$	$\sim 1/\epsilon$

12.3.3 Cascaded R-modular redundancy

Cascaded R-modular redundancy is an extension of the RMR technique such that there is concatenation of the modular redundant architecture to higher levels, by forming further RMR units from the RMR units themselves. This is illustrated in figure 12.1(*b*). The computational redundancy R_{total} of the system then depends on both the degree of primary replication R and the level of concatenation i and is, thus, given by $R_{\text{total}} = R^i$. The slowdown in computation is by a factor li, where l is the logical depth of the R-fold majority gates.

To simplify analysis, we will take R to be fixed at three and increase the redundancy purely by increasing i. The CRMR technique is then called *cascaded triple modular redundancy* (CTMR). For a TMR subunit, the reliability can be calculated from equation (12.6) to be

$$P_{\text{unit-mr}}^{(1)} \simeq (1 - \epsilon)^{mB}[P_{\text{unit}}^3 + 3P_{\text{unit}}^2(1 - P_{\text{unit}})]. \qquad (12.9)$$

If we concatenate the architecture, then the redundant subunit reliability to ith order $P_{\text{unit-mr}}^{(i)}$ is given by

$$P_{\text{unit-mr}}^{(i)} \simeq (1 - \epsilon)^{mB}[(P_{\text{unit-mr}}^{(i-1)})^3 + 3(P_{\text{unit-mr}}^{(i-1)})^2(1 - P_{\text{unit-mr}}^{(i-1)})]. \qquad (12.10)$$

The reliability of the entire processor per clock cycle is then

$$(1 - \delta) = \{P_{\text{unit-mr}}^{(i)}\}^C \qquad (12.11)$$

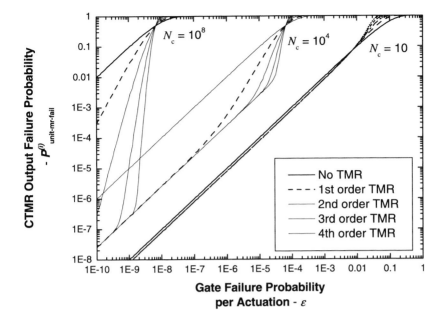

Figure 12.2. The output failure probability of a TMR/CTMR subunit ($P^{(i)}_{\text{unit-mr-fail}} = 1 - P^{(i)}_{\text{unit-mr}}$), comprising $3^i N_c$ gate resources and with B-bit outputs, as a function of the individual gate failure probability ϵ, using noisy majority gates. The groups of curves are for $N_c = 10$, $B = 2$, $m = 4$ (right), $N_c = 10^4$, $B = 64$, $m = 4$ (centre) and $N_c = 10^8$, $B = 64$, $m = 4$ (left)—see [5] and [72].

where C is the number of CTMR subunits,

$$C = \frac{2N}{2N_c R^i + mB(R^i - 1)}.$$

Figure 12.2 demonstrates the effectiveness of the CTMR technique. It shows that there is no advantage in using CTMR for subunits containing a small number of gate resources if the majority gates are made from the same gate resources as the units that they are monitoring. However, at least in principle, improvement is possible for subunits with large N_c.

There are three regions in each set of curves [5]:

(a) $N_c \epsilon > \ln 2$, where redundancy affords no advantage,
(b) $10^{-3} < N_c \epsilon < \ln 2$, where redundancy is most effective, and
(c) $N_c \epsilon < 10^{-3}$, where only first-order TMR offers an advantage.

In case (b), the effectiveness of redundancy scales as a power law with the order of CTMR. The subunit failure probability is given by

$$P^{(i)}_{\text{unit-mr-fail}} \propto (N_c \epsilon)^{2i}. \tag{12.12}$$

For case (c), the effectiveness of redundancy depends on the ratio mB/N_c. Starting from equation (12.10) it can be shown that in region (c) the failure probabilities are

$$P_{\text{unit-mr-fail}}^{(0)} \equiv 1 - P_{\text{unit}} \approx N_c\epsilon \qquad (12.13)$$

$$P_{\text{unit-mr-fail}}^{(i)} \approx \frac{mB}{N_c} N_c\epsilon = \frac{mB}{N_c} P_{\text{unit-mr-fail}}^{(0)}$$

where $i = 1, 2, \ldots$.

12.3.4 Parallel restitution

A fundamental limitation with modular redundancy techniques (noted by von Neumann) was that the reliability of any output bit from a TMR/CTMR processor could never be less than ϵ, since the output line had to originate from a gate somewhere. To overcome this, von Neumann developed a new type of noise-tolerant architecture called *multiplexing*. This increased the amount of parallelism in the subunits but now, instead of performing both the 'restoration' and 'decoding' with majority gates, he separated these functions and integrated a dedicated decoder/discriminator within every subunit. The problem was that von Neumann assumed these decoders to be noiseless just as in error-correcting codes, which was unrealistic. Nikolić *et al* investigated the efficacy of multiplexing for nanocomputation and found that redundancies in the thousands were needed for reliable operation [5, 72, 73]. Han and Jonker combined the technique of multiplexing with reconfiguration to combat purely permanent defects in processors [74]. Another fundamental problem with both modular redundancy and multiplexing is that these techniques intermittently fan single interconnect lines out to the redundant sub-units and fan the redundant lines back to single ones through voting or decoding, before fanning them out to the next subunit again. Thus, the parallelism in any operations is *not* maintained and this allows for the propagation of errors. Sadek *et al* developed a new noise-tolerant architecture called *parallel restitution*, based on von Neumann's multiplexing, which overcame the problems of error propagation and noiseless decoding [3]. Its essential feature was the maintenance of parallelism throughout a processing network and its resulting high efficacy makes it a viable candidate for nanoscale integration in the future. Their analysis also made primary connections between noise tolerance, entropy and thermodynamics, which set bounds on just how small one can make *any* classical computer because of the heat production that necessarily follows error correction.

Consider an information-processing network comprised of processing elements linked by interconnects. Parallel restitution is a method of modifying the network architecture to be noise tolerant by (1) replacing each interconnect with a parallel bundle of interconnects, (2) replacing each processing element with a parallel array of processing elements and (3) counteracting the decay of

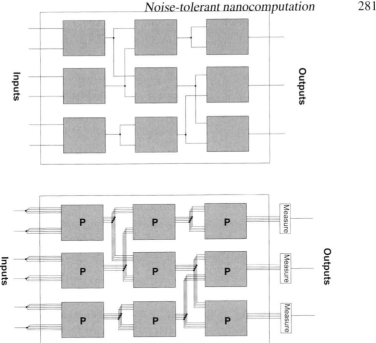

Figure 12.3. (*a*) An arbitrary network of information processing units and (*b*) the pararestituted equivalent, with parallelism $R_p = 4$. From [3], © 2004 IOP Publishing Ltd.

signals carried by the interconnect bundles through some form of signal restitution performed in parallel, this requiring energy investment.

The inputs to the processor are fanned out to bundles from single lines and at the final outputs, special circuits 'measure' the output bundles according to some threshold to determine what information they represent. This is then transmitted back down to single lines.

Figure 12.3(*a*) depicts an arbitrary network of processing units that together comprise a processor. Figure 12.3(*b*) then shows the equivalent *pararestituted* network, where parallel restitution has been implemented. In the example illustrated, the parallelism is $R_p = 4$. Figure 12.4 shows the principle behind the design of the pararestituted processing units, P in figure 12.3(*b*).

Of particular importance is the *maintenance of parallelism throughout the network*, this being the key to parallel restitution's incredible resistance to noise, because it prevents the propagation of errors.

Each communication bundle then represents the information carried by a single line. The information carried by the bundle can be thought of as a probability wave that interacts with the parallel processing elements, undergoing dissipation due to the increase in computational entropy brought about by the

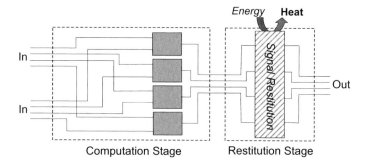

Figure 12.4. The principle behind the structure of the pararestituted processing elements P in figure 12.3 [3]. Any pararestituted component consists of a *computation stage* to perform the desired operation in parallel and a *restitution stage* to counteract dissipation of the information carried by the line bundles due to noise. The external energy is needed to pump the computation entropy to a low value. From [3], © 2004 IOP Publishing Ltd.

noisy processing elements. The action of intermittent signal restitution is then to counteract the dissipation of the probability wave to preserve the information carried by the bundles. This is done through the use of external energy to 'pump' the computation entropy to a lower value at the expense of the entropy of the entire processor. Signal restitution is then, in effect, a kind of signal 'refrigeration', using energy to keep the signal carried by a bundle in a preserved low entropy state. Signal restitution can be implemented through 'hardware' (amplification circuits) or 'software' (error-correcting codes). The restitution technique employed for the digital logic scheme uses amplification circuits, whilst Spielman [38] employed error-correcting codes in his noise-tolerant parallel architecture.

Although parallel restitution can be implemented at the level of processing elements of any complexity, the reliability of the entire processor is maximized if parallelism is implemented at the level of the smallest information-processing elements in the network. This is because the smallest processing elements will be more reliable in themselves than more complicated units built from those elements. Hence, in digital computers, implementation would be most effective at the level of Boolean logic gates, in quantum computers at the level of quantum gates and, in the nervous system, at the cellular level.

Although parallel restitution is presented in the context of digital computing, in fact the underlying principles are more general and apply to different kinds of information processing in addition. Thus, the basic principles of parallel restitution can be summarized generally as

(i)   parallel processing of information,
(ii) parallel transmission of information and
(iii) intermittent signal restitution performed in parallel.

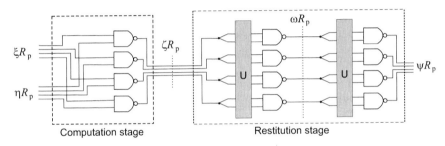

Figure 12.5. The design of a pararestituted NAND gate with $R_p = 4$ [3]. The computation stage is as in figure 12.4, the restitution stage being hardware-based and made from NAND gates, simplifying construction. Energy is pumped into the restitution circuit through FANOUT: part of this energy restitutes the input signal against noise and the rest is dissipated through data collapse at the NAND gates. The NAND restitution circuit must be iterated to compensate for its inverting effect. The number of activated lines in the input bundles are given by ξR_p and ηR_p respectively. The number of activated lines in the output bundle from the pararestituted NAND gate is ψR_p. The quantities ζR_p and ωR_p are the number of activated lines from the computation stage output bundle and the intermediate stage of the restitution circuit respectively. The boxes marked U are fixed permutations of the lines, that are randomly different for each pararestituted NAND gate and every restitution circuit within a pararestituted NAND gate. From [3], © 2004 IOP Publishing Ltd.

12.3.4.1 Implementation in logic networks

The scheme for implementing parallel restitution in nanoscale digital logic networks is now described. Although current microelectronic design entails the implementation of NAND, NOR and NOT gates in microprocessors, nanoelectronic-based processors, particularly those at the molecular scale, would probably benefit from construction entirely from NAND, the universal logic gate. This is because bottom-up techniques of assembly would probably be used and so confining the repertoire of components to interconnects and one type of logic element would offer advantage in this system of assembly. The scheme presented implements parallel restitution entirely through the use of NAND.

If one considers the arbitrary circuit in figure 12.3 to be a logic network and the constituent processing elements to be NAND gates, then the detailed design of the pararestituted NAND gate is as in figure 12.5. The parallel line bundles then represent the digital bits in single lines. Hence ideally, in the case of noiseless gates and input bits, each single line comprising the line bundle should carry the same value of '0' or '1' at each instantiation. The introduction of noise would then cause the lines carrying the bundle to exhibit non-uniform bit values and, therefore, the fraction of bundle lines carrying a particular bit (by default say '1' as this represents 'activation') would conform to a probability distribution.

It is assumed in this scheme that the measuring/decoding devices used to determine the bit characterization of the processor output bundles are composed of perfect devices. Presumably, they would be constructed by conventional means and be integrated into the macroscopic peripheral hardware. Hence, it is assumed there is perfect communication between the nanoprocessor outputs and the noiseless discriminators in the macro-world used to measure the output bundles. The discriminators work by measuring the fraction of activated lines in the output bundles, such that if greater than $R_p(1 - \Delta)$ lines carry bit '1', the bundle of lines replacing a single line are considered to carry '1' and the discriminator activates the final output line as such. If less than $R_p \Delta$ lines carry bit '1', the discriminator carries a value of '0' to the final output line. Δ is a *threshold level*, to be derived later for the specific case of a pararestituted NAND gate formula (generally, $0 < \Delta < 0.5$). The discriminator would call a malfunction if the fraction of activated lines were between Δ and $1 - \Delta$.

The computation stage of the pararestituted NAND gate is as in figure 12.4. The restitution stage though is hardware-based and, following von Neumann [1], is constructed from NAND gates. As it is irreversible, a NAND gate can be rigged to act as a nonlinear, although asymmetric, 'amplifier'. The NAND restitution circuit in figure 12.5 has the action of skewing the proportion of activated lines in a bundle even further to the ideal of all '0' or all '1'. Hence, its amplification properties perform the desired function of signal restitution. However, as the NAND gates also *invert* the input if used in this way, the NAND restitution circuit must be iterated to restore the original bit characterization of the line bundle. Correlation between the line bundles, which would destroy the amplification characteristics, is prevented through fixed permutations within the line bundles, that should be random between different pararestituted NAND gates and the NAND restitution circuits within each pararestituted gate. Energy is pumped into the restitution circuit through the FANOUT operations prior to the random permutations. A portion of the energy goes into restituting the input signal against noise but the rest is dissipated subsequent to data collapse at the second and third NAND logic layers. Note that none of the information carried by the input bundle is dissipated: as energy is pumped in whilst maintaining parallelism, only a part of the energy pumped in during FANOUT contributes to dissipation.

The number of restitution stages can be increased to boost the level of signal restitution. Markov chain analysis shows the computational state probability distribution to stabilize and reach a constant vector very quickly as the restitution depth increases. At the same time, increasing the level of restitution decreases speed and increases redundancy, energy consumption and heat generation. As a compromise between these factors, it is found that having only one level of restitution is sufficient, the rapid computational state vector stabilization allowing use of this minimal restitution depth. Adding the logic layer of the gate computation, the total logic depth of the processor in our scheme is then increased by a factor of three. Note that the resulting redundancy R, of the processor is then $R = 3R_p$ and the system speed is reduced threefold.

12.3.4.2 *Mathematical analysis*

The objective of mathematical investigation here is to determine the ultimate reliability of logic networks protected against noise by parallel restitution. Before this, as described in section 12.3.4.3, a set of equations are presented. These describe the set distribution of a pararestituted gate output bundle, given specific set distributions for the inputs and the gate error probability distribution. In addition, the equations must describe the evolution of a line bundle's activation set distribution under chaining of the pararestituted gates so as to be able to model networks. Hence, our final equations will give the system reliability as a function of the gate error probability per actuation ϵ $(0 < \epsilon < 0.5)$, the redundancy R and the system size, complexity and speed.

Several elements of von Neumann's method of multiplexing [1] are common to the scheme for parallel restitution in logic networks and, hence, some of his mathematical analysis is, in principle, applicable here. The most relevant of this was his explicit derivation through combinatorial analysis of the probability distribution for the output bundle activation fraction of a pararestituted NAND gate computation stage. This discrete distribution is given by

$$P_0(\zeta|\xi,\eta) = \frac{\left\{ \begin{array}{c} (\xi R_{\mathrm{p}})![(1-\xi)R_{\mathrm{p}}]!(\eta R_{\mathrm{p}})! \\ \times[(1-\eta)R_{\mathrm{p}}]! \end{array} \right\}}{\left\{ \begin{array}{c} [(1-\zeta)R_{\mathrm{p}}]![(\xi+\zeta-1)R_{\mathrm{p}}]! \\ \times[(\eta+\zeta-1)R_{\mathrm{p}}]! \\ \times[(2-\xi-\eta-\zeta)R_{\mathrm{p}}]!R_{\mathrm{p}}! \end{array} \right\}} \qquad (12.14)$$

where ξR_{p} and ηR_{p} are respectively the number of activated lines in the input bundles and ζR_{p} is the number of activated lines in the output bundle (see [3] for a derivation). Hence, ξ, η and ζ are the respective line activation fractions.

In deriving a stochastic distribution equation for the output of a pararestituted NAND gate, we may start with equation (12.14), this equation being exact for fixed fractions of activated lines at the inputs (i.e. non-stochastic inputs) where only the identity of the activated input lines are random. It thus describes the probability distribution of ζ conditional on the values of ξ and η.

Using conditional probability, it may be shown that

$$\begin{aligned} P(Z) &= P(X=0)P(Y=0)P(Z|X=0, Y=0) \\ &\quad + P(X=1)P(Y=0)P(Z|X=1, Y=0) \\ &\quad + \cdots + P(X=R_{\mathrm{p}})P(Y=R_{\mathrm{p}})P(Z|X=R_{\mathrm{p}}, Y=R_{\mathrm{p}}) \\ &= \sum_{X=0}^{R_{\mathrm{p}}} \sum_{Y=0}^{R_{\mathrm{p}}} P(X)P(Y)P(Z|X, Y) \end{aligned}$$

and, thus, this allows us to re-express equation (12.14) in terms of *stochastic* inputs as opposed to ones dependent on fixed values. Hence,

$$P_0(\zeta) = \sum_{\xi R_p=0}^{R_p} \sum_{\eta R_p=0}^{R_p} P(\xi)P(\eta)P_0(\zeta|\xi, \eta). \qquad (12.15)$$

where $P(\xi)$ and $P(\eta)$ are the discrete input probability distributions.

It now remains to introduce the effect of gate error, ϵ. We do this by systematically considering the effect of single, double, triple etc errors on the output activation fraction and adding the ensuing probabilities together. The resultant equation is

$$P(\zeta R_p = i|\epsilon) = \sum_{z=0}^{R_p} P_0(\zeta R_p = z)(1 - \epsilon)^{R_p-|z-i|}\epsilon^{|z-i|}$$
$$\times \left(\left| \frac{R_p}{2}\left(1 - \frac{z-i}{|z-i|}\right) - z \right| \frac{}{|z-i|} \right). \qquad (12.16)$$

Finally, we allow for a distribution of gate error probabilities by again using conditional probability and obtain the final expression for the output of a pararestituted NAND gate computation stage:

$$P(\zeta) = \int_0^1 P(\zeta|\epsilon)P(\epsilon)\,d\epsilon. \qquad (12.17)$$

$P(\epsilon)$ is assumed to be normally distributed with mean gate error probability $\bar{\epsilon}$.

Although they are exact, equations (12.14)–(12.17) are extremely computationally expensive to solve even for moderate values of R_p. Hence, our model replaces equations (12.14) and (12.16) with more approximate ones that are much simpler to compute and analyse. Essentially, we recognize the activation ('1' or '0') per actuation of the lines comprising a bundle to be a Bernoulli trial described by the binomial distribution. Thus,

$$P(\zeta|\xi, \eta, \epsilon) = \binom{R_p}{\zeta R_p} \bar{\zeta}^{\zeta R_p}(1-\bar{\zeta})^{(1-\zeta)R_p} \qquad (12.18)$$

where, for a NAND gate, $\bar{\zeta} = (1-\bar{\xi}\bar{\eta})(1-\bar{\epsilon})+(\bar{\xi}\bar{\eta})\bar{\epsilon} = 1+2\bar{\xi}\bar{\eta}\bar{\epsilon}-\bar{\xi}\bar{\eta}-\bar{\epsilon}$. $\bar{\zeta}$ here thus defines the average value of the computational function that is implemented by the pararestituted module. Table 12.2 gives a list of the corresponding $\bar{\zeta}$ for various other Boolean functions. However, treating the distributions this way inherently implies that the input activation fractions are binomially distributed and *not fixed* prior to conditioning upon $P(\xi)$ and $P(\eta)$. Although this does not significantly affect the results, it, strictly speaking, makes the analysis approximate.

Table 12.2. Binomial mean computational definitions for various Boolean operations. ϵ is the gate error probability per actuation and γ is the probability of communication error per unit length per bit transmission (overbar denotes mean value), [3], © 2004 IOP Publishing Ltd.

Function	Mean Input(s)	Mean Output(s)
NAND	$\bar{\xi}, \bar{\eta}$	$\bar{\zeta} = (1 - \bar{\xi}\bar{\eta})(1 - \bar{\epsilon}) + \bar{\xi}\bar{\eta}\bar{\epsilon}$
NOR	$\bar{\xi}, \bar{\eta}$	$\bar{\zeta} = (1 - \bar{\xi} - \bar{\eta} + \bar{\xi}\bar{\eta})(1 - \bar{\epsilon})$
		$+ (\bar{\xi} + \bar{\eta} - \bar{\xi}\bar{\eta})\bar{\epsilon}$
NOT	$\bar{\xi}$	$\bar{\zeta} = (1 - \bar{\xi})\bar{\epsilon} + \bar{\xi}(1 - \bar{\epsilon})$
CNOT	$\bar{\xi}, \bar{\eta}$	$\bar{\zeta}_1 = \bar{\xi}(1 - \bar{\epsilon}) + (1 - \bar{\xi})\bar{\epsilon},$
		$\bar{\zeta}_2 = (\bar{\xi} + \bar{\eta} - 2\bar{\xi}\bar{\eta})(1 - \bar{\epsilon})$
		$+ (1 - \bar{\xi} - \bar{\eta} + 2\bar{\xi}\bar{\eta})\bar{\epsilon}$
Wire	$\bar{\xi}$	$\bar{\zeta} = \bar{\xi}(1 - \bar{\gamma}) + (1 - \bar{\xi})\bar{\gamma}$

Equations (12.15), (12.17) and (12.18), therefore, completely describe the output probability distribution of a pararestituted NAND gate computation stage:

$$P(\zeta) = \binom{R_{\mathrm{p}}}{\zeta R_{\mathrm{p}}} \sum_{\xi R_{\mathrm{p}}=0}^{R_{\mathrm{p}}} \sum_{\eta R_{\mathrm{p}}=0}^{R_{\mathrm{p}}} \int_{\epsilon=0}^{\epsilon=1} \mathrm{d}\epsilon \, P(\xi)P(\eta) \frac{\exp\left\{ -\frac{(\epsilon - \bar{\epsilon})^2}{2\bar{\epsilon}(1 - \bar{\epsilon})} \right\}}{\sqrt{2\pi\bar{\epsilon}(1 - \bar{\epsilon})}}$$
$$\times (1 + 2\xi\eta\epsilon - \xi\eta - \epsilon)^{\zeta R_{\mathrm{p}}} (\xi\eta + \epsilon - 2\xi\eta\epsilon)^{(1-\zeta)R_{\mathrm{p}}}. \qquad (12.19)$$

Considering figure 12.5, it can be appreciated that each restitution circuit is merely a special case of the computation stage but where $\xi = \eta$. Hence, the restitution stage can be modelled by two more equations based on equation (12.19), but where $\xi = \eta$ is set to ζ and ζ set to ω and, correspondingly, to obtain ψ. These equations thus fully describe the output distribution of a pararestituted NAND gate for stochastic inputs and stochastic gate error probability, and allow one, in principle, to determine the output line-bundle distributions of any pararestituted NAND network.

Although the full set of equations presented, describe in principle, a pararestituted NAND gate, in practice they can be computationally expensive to solve for large values of R_{p}. In addition, when we come to consider modelling pararestituted networks to make a reliability analysis, it will be found that having a more straightforward method of computing these equations will be invaluable, since they also describe a simple linear network of pararestituted gates.

The evolution of $P(\zeta)$ to $P(\omega)$, and $P(\omega)$ to $P(\psi)$ is a stationary discrete-time Markov chain since the process is discrete time, has a countable state space, satisfies the Markov property $P(\zeta_n = i_n | \zeta_{n-1} = i_{n-1}, \zeta_{n-2} = i_{n-2}, \ldots, \zeta_0 =$

$i_0) = P(\zeta_n = i_n|\zeta_{n-1} = i_{n-1})$ and is stationary in that $P(\zeta_n = i_n|\zeta_{n-1} = i_{n-1}) = P(\zeta_{n+k} = i_{n+k}|\zeta_{n+k-1} = i_{n+k-1})$.

One can thus describe the restitution stage by a Markov chain with transition probabilities

$$p_{ij} = P(\zeta_j|\zeta_i)$$
$$= \left(\begin{array}{c} R_p \\ \zeta_j R_p \end{array} \right) \int_{\epsilon=0}^{\epsilon=1} d\epsilon \, P(\epsilon)(1 + 2\zeta_i^2 \epsilon$$
$$- \zeta_i^2 - \epsilon)^{\zeta_j R_p}(\zeta_i^2 + \epsilon - 2\zeta_i^2 \epsilon)^{(1-\zeta_j) R_p} \tag{12.20}$$

and an $(R_p + 1) \times (R_p + 1)$ transition probability matrix

$$\Phi = \begin{pmatrix} p_{00} & p_{01} & \cdots & p_{0R_p} \\ p_{10} & p_{11} & & \\ \vdots & & \ddots & \\ p_{R_p 0} & & & p_{R_p R_p} \end{pmatrix} \tag{12.21}$$

such that the final distribution vector $\mathbf{a_n}$ is given by

$$\mathbf{a_n} = \mathbf{a_0} \Phi^n \tag{12.22}$$

where $\mathbf{a_0} = \{P(\zeta_0 R_p = 0), P(\zeta_0 R_p = 1), \ldots, P(\zeta_0 R_p = R_p)\}$ is the initial distribution vector.

The distribution vector $\mathbf{a_n}$ is then periodic due to the inverting effect of each substage of a pararestituted NAND gate.

Thus given an initial distribution $\mathbf{a_0}$ defined by equation (12.19), one can use equations (12.20)–(12.22) with $n = 2$ to compute the final output distribution of a pararestituted NAND gate $P(\psi)$.

The long-run transition probability matrix, $\Pi = \lim_{n \to \infty} \Phi^n$, although periodic, is stable due to the absence of absorbing states in Φ. This thus mathematically describes how parallel restitution is inherently resistant to the effects of error propagation in noisy networks.

Although Sadek *et al* did not carry out the requisite numerical calculations, the aforementioned analysis can be easily adapted to investigate the effect of noisy wires on system reliability. The expressions are essentially the same except for ζ, which is given by the relation in table 12.2. Using the Markov chain equations, pararestituted networks of noisy gates interspersed with noisy wires can then be modelled.

12.3.4.3 Reliability analysis

Determination of the reliability of a pararestituted gate network requires knowledge of the following:

(i) the greatest logic depth of the network,

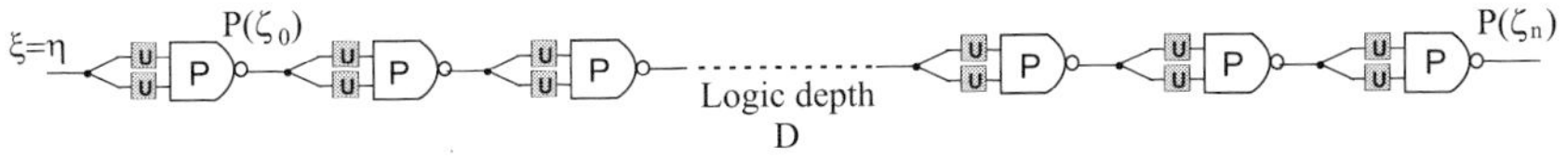

Figure 12.6. A simple linear network of pararestituted NAND gates, used to model the output bit bundle distribution of a pararestituted gate network with logic depth D, [3], © 2004 IOP Publishing Ltd.

(ii) the number of output bits,
(iii) the typical degree of noise on the inputs,
(iv) the evolution of a line bundle's activation distribution through logic,
(v)   the threshold activation level Δ and
(vi) the system speed.

Items (i) and (ii) are typically functions of the complexity and gate size N of a network. If one considers a relatively simple network to be a 'convolution' of logic, then the following simple formulae can be used.

The reliability of a single-bit output bundle per calculation P_{bit} is given by

$$P_{\text{bit}} = \sum_{\zeta_n R_{\text{p}}=0}^{\Delta R_{\text{p}}} P(\zeta_n) \tag{12.23}$$

where $n = 3D - 1$ and D is the greatest logic depth. This equation can, for small R_{p} ($\lesssim 100$), be computed for extremely high reliabilities by an error integral, through approximating $P(\zeta_n)$ by a normal distribution.

This equation assumes the simple linear NAND gate circuit depicted in figure 12.6, so as to consider the degradation of the input bundle activation distribution at the output in a *predictable* manner. The circuit depicted is, in fact, an abbreviated description of a branched chain network, where for a logic depth D, 2^D parallelized inputs with identical activation distributions branch down after input to 2^{D-1} pararestituted gates to a single pararestituted NAND gate. Hence, each gate receives its inputs from the outputs of two separate gates in a previous logic depth layer. Such a network tree is a graph and is, hence, describable by formulas. With random line permutation units U placed at the inputs of every pararestituted gate in the abbreviated circuit in figure 12.6, correlation between the branched parallelized input lines is removed and then the circuit is equivalent to the branched network described, both being subject to the same Markov chain equations presented in section 12.3.4.2.

$P(\zeta_n)$ in equation (12.23) is thus given by equations (12.20)–(12.22), where appropriate noise on the inputs $\bar{\xi} = \bar{\eta}$ for the system is assumed. It also assumes the inputs to the chain are prepared so as to make the intended output-bit bundle characterization '0', this being the regime that is most sensitive to error.

If there are B output bits and the clock frequency is v, the reliability of the pararestituted network is then given by

$$(1 - \delta) = P_{\text{bit}}^{Bv At} \tag{12.24}$$

where t is the time in seconds for which the system operates and A is the probability that the processor will be accessed per clock cycle (normally, $A \sim 0.1$). Generally, B and v will depend on the specific processor design, particularly the complexity.

It can be appreciated that the reliability of the pararestituted network is only positive when the condition $R_p \geq 1/\Delta$ is satisfied. If $R_p < 1/\Delta$, then $P = 0$. This is because the discriminator device at the final output bits would never be able to satisfy the bundle activation conditions, $\geq (1 - \Delta)R_p$, and, $\leq \Delta R_p$, for bit characterization '1' and '0', respectively.

Although nanoscale processors have yet to be constructed, it is possible to make reasonable assumptions about their probable overall architecture. They are unlikely to be vastly more sophisticated networks than current microprocessors, in that their logic depth would probably be constrained to ~ 20 by speed considerations as is the case currently. More likely, given that they would consist of $N_p \sim 10^{12}$ gate resources, they would perhaps be arranged as hundreds to thousands of parallel processors, each with about the capability and resource size of current microprocessors ($N_c \sim 10^7$).

Thus, a reasonable model of a nanoprocessor would be to modify equation (12.24) to

$$(1 - \delta) = P_{\text{bit}}^{cBv At} \tag{12.25}$$

where $c = N_p/(N_c R)$ is the number of parallel processors comprising the nanoprocessor. The range of redundancy $R = 3R_p$ that could be tolerated by the nanoprocessor is then $0 < R < c$, although for $N_p = 10^{12}$, R should be less (preferably much less) than 1000 to keep the nanoprocessor technologically useful.

To calculate the nanoprocessor's reliability, it is assumed that $B = 64$ and $D = 11$. It is also assumed that there is no noise on the inputs and, as $D = 11$, the δ-function start vector is defined by $\bar{\xi} = \bar{\eta} = 1$. In general, for odd D, one chooses the δ-function start vector to be $\bar{\xi} = \bar{\eta} = 1$ and, for even D, one chooses $\bar{\xi} = \bar{\eta} = 0$. This is to ensure consideration of the case of greatest error sensitivity where the output-bit characterization of the pararestituted NAND chain is '0'. As is shown in the next section, $\Delta = 0.07$.

12.3.4.4 *Threshold gate noise ϵ_t and threshold activation level Δ*

Although parallel restitution is highly tolerant to error propagation, there fundamentally must exist a threshold error ϵ, below which the bit characterization of the network line bundles remains coherent. Although von Neumann did not implicitly describe the method by which he determined it, he stated this threshold

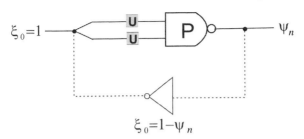

Figure 12.7. Iterative NAND gate circuit used to consider the worst possible case for error propagation in a pararestituted NAND gate network [3]. It is a graph described by formulas but, although the evolution of $P(\zeta_n)$ is Markovian, it is non-stationary. © 2004 IOP Publishing Ltd.

to be $\epsilon_t \simeq 0.0107$. We describe here how the threshold may be derived for a pararestituted NAND gate network and how, in turn, we may estimate the value for Δ, the threshold activation level by which the bit characterization of a line bundle is decided.

Consider the following set of graphical formulas describing the propagation of the mean output stimulation fraction in a pararestituted NAND gate, where the two inputs are fanned in.

$$\bar{\zeta}_n = 1 - \bar{\xi}_{n-1}^2 + 2\bar{\xi}_{n-1}^2\bar{\epsilon} - \bar{\epsilon} \tag{12.26a}$$

$$\bar{\omega}_n = 1 - \bar{\zeta}_n^2 + 2\bar{\zeta}_n^2\bar{\epsilon} - \bar{\epsilon} \tag{12.26b}$$

$$\bar{\psi}_n = 1 - \bar{\omega}_n^2 + 2\bar{\omega}_n^2\bar{\epsilon} - \bar{\epsilon} \tag{12.26c}$$

Due to the nature of the NAND function, the most sensitive error regime for a pararestituted NAND gate is $\bar{\xi} = \bar{\eta} \sim 1$: for two inputs to a NAND gate equal to 1, a bit flip error on either one of the inputs alters the output value from 0 to 1. As this regime has the greatest sensitivity to input noise, *it is the most important one that needs to be considered when computing a pararestituted NAND network's reliability.* Thus, to take account of the worst possible error propagation case in a pararestituted network, it is reasonable to consider the circuit in figure 12.7, where the Markov chain in figure 12.6 is modified such that the output vector is reflected about the basis set. This then is equivalent to considering equations (12.26a)–(12.26c) and iterating them such that

$$\bar{\xi}_n = 1 - \bar{\psi}_n. \tag{12.26d}$$

Starting with $\bar{\xi}_0 = 1$, it is then possible to consider the effect of error magnitude on the propagation of error in a pararestituted NAND gate network. Note, however, that this Markov process is then non-stationary.

When iterating formulas (12.26a)–(12.26d) for a large number of iterations and various values of gate error ϵ, it can be seen that if ϵ is less than a threshold value 0.010 77, then the bit characterization of the output is maintained. If the error is larger than this, then the bit characterization is lost and the value of $\bar{\psi}_{n \to \infty}$ diverges. For gate errors below the threshold, the probability of error in the bit characterization can be made arbitrarily small by making R_p appropriately large.

Considering the value of $\bar{\psi}_n$ at $\epsilon_t \simeq 0.01077$, one finds its value at the limit is $\simeq 0.07$. This limiting value has an important interpretation in that, effectively, if the output probability distribution $P(\psi)$ of a pararestituted NAND gate is fully within 7% of R_p, then, in view of the previous argument, the probability $P_{\text{bit-0}}$ of the line bundle representing bit '0' is 1. Likewise, if the distribution lies fully beyond 93% of R_p then the probability $P_{\text{bit-1}}$ of the bit characterization being '1' is 1. Thus,

$$P_{\text{bit-0}} = \sum_{\psi R_p = 0}^{\Delta R_p} P(\psi) \qquad (12.27a)$$

and

$$P_{\text{bit-1}} = \sum_{\psi R_p = (1-\Delta) R_p}^{R_p} P(\psi) \qquad (12.27b)$$

where Δ is the threshold line-bundle activation fraction.

Figure 12.8 depicts the phase portrait of equations (12.26a)–(12.26d) over the phase space of Δ and ϵ. The shaded region represents the phase space of the line-bundle activation fraction Δ and gate noise ϵ where the line-bundle bit characterizations are maintained after an infinite number of iterations. Thus, for each value of ϵ up to the threshold $\epsilon_t \simeq 0.010\,77$, there is a maximum value for Δ defined by the full curve. This curve is, hence, a repeller in phase space: values of ϵ and Δ outside this curve cause the bit characterization to explode to '1' regardless of what the correct value should be. Within the repeller curve, as $n \to \infty$, the value for the line bundle activation fraction moves towards those defined by the attractor in phase space, depicted by the broken curve. As can be seen, the maximum tolerable noise is $\epsilon_t \simeq 0.010\,77$ and the value of Δ at this threshold is $\simeq 0.073$.

12.3.5 Comparative results

Figure 12.9 shows the results of reliability calculations performed for a hypothetical nanoprocessor, as discussed in section 12.3.4.4, using the noise-tolerance techniques of RMR and parallel restitution with NAND gates. The plot shows the tolerable gate failure probability per actuation ϵ *versus* system redundancy R, such that the allowed reliability of the whole nanoprocessor is set to $P = 0.9$ over a continuous operation time of 10 years. In all calculations, the processor was assumed to have a gate size $N_p = 10^{12}$, $B = 64$ output bits, clock speed $v = 1$ GHz and access probability $A = 1$. A parallel subprocessor resource size of $N_c = 10^7$ with logic depth $D = 11$ was assumed.

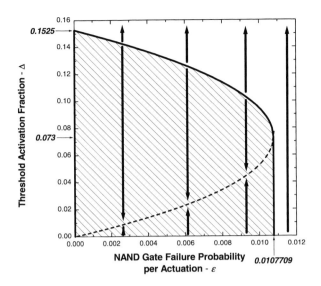

Figure 12.8. Phase portrait of equations (12.26a)–(12.26d) over the phase space of the line bundle activation fraction Δ and NAND gate noise ϵ [3]. The shaded area represents the phase space of Δ and ϵ over which correct bit characterization is maintained. The full curve is a repeller in phase space, whilst the broken curve is an attractor. © 2004 IOP Publishing Ltd.

The results show parallel restitution to improve the noise-tolerance of a gate network massively, by approximately 27 orders of magnitude. Even compared to RMR, the improvement is 13 orders of magnitude. With the minimum implementable value for R_p of 16 and, hence, $R \sim 50$, parallel restitution can relax the noise-tolerance bound of nanodevices in a nanoprocessor to $\epsilon \sim 10^{-4}$. This is compared to $\epsilon \sim 5 \times 10^{-17}$ with RMR and $\epsilon \sim 10^{-31}$ with no noise tolerance. With such relatively modest redundancy, the nanoprocessor in the model described would still be equivalent to about 2000 parallel desktop processors (each containing $\sim 10^8$ devices).

Parallel restitution is then very effective in protecting against computational noise and seems to be a viable candidate for reliable nanocomputation [3]. Architectures based on modular redundancy, in contrast, are simply too ineffective. Existing microprocessors typically have about 10^8 transistors divided roughly equally between logic and memory, whilst memory chips currently contain up to 10^9 devices. It is expected that improvements in CMOS technology will lead to an order of magnitude increase in these numbers before the technology reaches its limits. Any new nanoscale computing technology will then have to provide at least an additional ten-fold improvement in performance over future CMOS technology for it to be useful. With parallel restitution, it

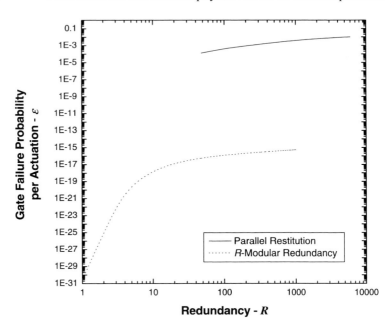

Figure 12.9. Results of reliability analysis for R-modular redundancy and parallel restitution, showing a plot of gate failure probability per actuation ϵ as a function of system redundancy R, such that the system reliability over 10 years is fixed at 90% [3]. The processing network modelled was assumed to have a gate size of $N_p = 10^{12}$, $B = 64$ output bits and speed $v = 1$ GHz. For the modular redundancy calculations, the majority gates were assumed to consist of $N_m = 4$ gate resources. A parallel subprocessor size of $N_c = 10^7$ with logic depth $D = 11$ was assumed for the parallel restitution calculations. The two curves together suggest the form of a universal limiting curve defined by a fundamental theory of noise in information processing. © 2004 IOP Publishing Ltd.

has been shown that nanoscale computers, comprising of order 10^{12} device resources, can be reliably made two orders of magnitude more powerful than future CMOS if the noise in constituent logic gates can be engineered to not exceed $\epsilon \sim 10^{-4}$. This noise level is a goal for nanotechnologists to achieve. The resulting overhead has been calculated to be a modest $R = 50$ and the increase in logic depth to be $n = 3$. This is a big improvement on majority vote schemes and von Neumann's multiplexing, where redundancies in the thousands are required [5, 72]. These optimum numerical results are comparable to the most optimistic theoretical bounds obtained so far by Spielman. This is perhaps expected, as Spielman's technique is a specific implementation of the parallel restitution paradigm. Parallel restitution massively outperforms modular redundancy approaches, as parallelism is maintained throughout the system, and, hence, information in signals that can be used to help restitute the computational outcome is not needlessly lost on data collapse during intermittent majority voting. It is also energetically more efficient, since when energy is pumped into the signal whilst it is parallel, more of the energy can go into restitution instead of being dissipated in the needless data collapse. Logic depth and circuit size increases are necessary, in agreement with the theoretical computer science results described earlier. A threshold gate error of $\epsilon = 0.0107$ was derived, below which the processor reliability can be made arbitrarily large through increasing circuit size. Further analysis is needed to determine whether this technique could also handle permanent errors or whether reconfiguration would need to be integrated.

12.4 Physical limits of nanocomputation

As the limits of CMOS technology are approached, and as Moore's law starts to be challenged, attempts have been made to set technology-independent bounds on computation [23, 27, 28] and nanocomputation in particular [77]. Physics itself is the ultimate computer and so the limits to computation must be decisively derived from its underlying laws. Many of the limits derived may seem irrelevant to our current engineering capabilities but the underlying principles of what physics tells us about the bounds to information processing are just as relevant under more practical conditions. Frank has shown through the consideration of physical limits that the optimum nanocomputational architecture for classical computing is a reversible, highly parallel 3D mesh of processors that removes entropy from the system ballistically [77]. More generally, limits on computational speed, communication and memory have been derived from relativity, thermodynamics, statistical mechanics and quantum field theory. With a few exceptions, one notable theme that has been overlooked by researchers is the effect of errors on computation and its fundamental basis in the physical limits of computing. Sadek *et al* have shown how noise effects and fault tolerance are essential considerations and delineated their basis in thermodynamics [3]. We now review this body of work and consider the suggestions it makes as to the direction we should take for the future development of 3D nanoscale information systems.

12.4.1 Information and entropy

Before we consider speed, communication and memory in detail, let us review the concepts of information and entropy. Shannon, building on Szilard's work, showed that the measure of information was a thermodynamic quantity almost identical to thermodynamic entropy, which we call *Shannon information* [13]. This is given by $H(X) \equiv - \sum_x p(x) \log_2 p(x)$, where $p(x)$ is the probability distribution for the occurrence of symbols x in a message. The unit of information is the *bit*. Most now agree that these two measures are, in fact, one and the same thing [28, 77, 78]. One way to look at information is as a measure of the 'surprise' at learning something new. If an event has a very low probability of occurrence and we learn about it, then the information content of this is very high. Information is just one part of the total, fixed *physical information* of a closed system. The remainder consists of *Boltzmann* or *thermodynamic entropy*. Boltzmann's breakthrough was in describing the canonical entropy of thermodynamics via a microcanonical description, namely as the logarithm of the total number of discrete particle states of a system, $S = k \ln \Omega$. The unit of Boltzmann entropy is the *nat* (J K^{-1}). What then is the relationship between information and entropy? The laws of physics are ultimately reversible and so one cannot destroy information, otherwise reversibility would disappear. Information is a measure of the states that we *do* know about in that, in principle, we can track the trajectories of the constituent particles backwards because of the reversibility of physics. With information, we can always track its evolution from the past or into the future. Now, if information were to interact with states of which we have no knowledge, then the information would become lost to the observer but still exist physically. The observer would lose track of the information and have no way to follow its trajectory reversibly. The information has now become entropy. When heat flows from a system into a heat bath, information interacts with a very complex system that we know nothing about and so transforms from a state of being knowable through observation to being unknowable. This is the origin of the second law of thermodynamics.

The quantum generalization of the probability distributions in information theory, $p(x)$, are *density matrices* ρ [79]. In turn, the quantum generalization of the Shannon entropy, *von Neumann entropy*, can be constructed from these: $H = -\mathrm{Tr}\rho \ln \rho$. Just as classical information becomes entropy after interacting with a heat bath, quantum information decoheres after becoming entangled with the environment, maximizing the von Neumann entropy.

12.4.2   Speed, communication and memory

The first limit to computation we must consider is the very structure of spacetime itself [77]. If n elements are arranged in a computational mesh for sorting, then if they are arranged only two-dimensionally, the time taken to sort them is of order $\mathcal{O}(n^{1/2})$. In contrast, migration to three dimensions can reduce the computational time to $\mathcal{O}(n^{1/3})$. The optimum computer is, thus, a 3D one. The maximum speed of a 3D computer, in turn, depends on several factors. First, the communication of signals cannot propagate faster than the speed of light, c. In practice, if a signal is propagated along a dissipative channel, then the drop-off in speed with distance y scales as y^2, unless there is periodic regeneration by, for example, logic gates. The other bound on speed is the maximum rate of computation itself. The most basic definition of a state (like '0' and '1' in a digital system) is given by quantum mechanics, which states that different quantum states are only completely distinguishable if they are orthogonal, as in a Hilbert space. We can thus fundamentally define a gate operation in computer science as a transition that takes one orthogonal quantum state to another. This process is ultimately governed by the following equation for the time Δt it takes to make this state change:

$$\Delta t = \frac{h}{4E} \tag{12.28}$$

where h is Planck's constant and E is the average energy of the quantum system [80]. Quantum logic gates that have been constructed in optical, atomic and NMR systems, in fact, already attain this limit [27].

Having found the limits to speed, we can thus appreciate the assertion by Frank for the optimum architecture of a 3D computer [77]. If each mesh takes time t to perform some useful information processing and the interconnect signals between mesh elements have velocity v, then the optimum distance to place each of the mesh elements apart is $d = vt = v/f$. The very high limit to the frequency f of dynamical evolution means that the ultimate 3D computer is, in fact, a nanocomputer. Below a certain length scale, the speed of computation is increased even further by migrating to reversible computing.

We have seen how information can be quantified by computing the entropy of a system. If we then compute the entropy per unit volume of an area of space, $\sigma = S/V$, we then have a measure of memory storage. Bekenstein has derived the ultimate limit to memory storage based on general relativity and quantum field theory [81], with a similar bound being computed by Bremermann solely based on Shannon's theory and the energy–time uncertainty relationship [82]. It is a stupendous quantity ($\sim 10^{39}$ bits Å^{-3}) and lower, but still enormous, bounds have been derived independently by Smith and Lloyd through consideration of quantum field theory alone [23, 27]. Storing information via field theoretic states is, however, completely unstable and the best hope for stable, optimally dense information storage is to use single atoms to store bits, for example via orbital electron states [44]. Even if one conservatively assumes a single atom can store 1 bit, the theoretical storage capacity of 1 cm^3 of solid material would be about 10^{24} bits of information [28].

Now that we know the limits to speed and memory density, we can consider the limits to communication. If a volume with ultimate memory density σ is bounded by surface area A and we assume the maximum speed of transmission c, then the maximum rate of communication allowed by physics is $C = \sigma A c$ bits s^{-1}. We have discussed in previous sections how the most generalized concept of computation is a *probabilistic* one, in that one can only ever say a computation is $(1 - \delta)$-reliable. In the same vein, communication can also be viewed as part of the greater complexity class **BPP** and then the most generalized description is to say a message is $(1 - \delta)$-reliable. The described upper bound to C assumes $\delta = 0$. If we allow for a small increase in δ, then *Shannon's first coding theorem* states that C can be increased even further by encoding information in long strings [13]. Now, if we consider communication in the presence of noise, the bound to the capacity C is made tighter and is given by *Shannon's second coding theorem*,

$$C = \max_{p(x)} I(X, Y) \tag{12.29}$$

where $I(X, Y) = H(X) + H(Y) - H(X, Y)$ is the mutual information [83]. If there is noise ϵ in a communications channel, as long as the bit rate C' across the channel is kept below the capacity given by equation (12.29), by coding with strings of arbitrarily long length (i.e. of arbitrary redundancy R), the reliability of the communication can be made to approach unity: $\lim_{C'<C; R\to\infty} \delta = 0$. The optimum manner for implementing this redundancy is to match the symbol distribution $p(x)$ to the noise in the channel and the specifics of how this is done is the subject of coding theory [15].

We have seen how the limits to communication can be treated by a theory based in thermodynamics and physics. However, the constructions we have described in section 12.3 are more akin to specific examples of error-correcting codes and, ideally, we would like to have a theory for computation analogous to Shannon's theory. We now consider how the theory of parallel restitution may allow us to do just that.

12.4.3 Thermodynamics and noisy computation

Research into the thermodynamics of information processing was inspired by the paradox of *Maxwell's Demon* and the apparent possibility for the violation of the second law of thermodynamics by intelligent beings [29]. The work by Turing on 'mechanical' computation led to a popularization of the notion of intelligence as an emergent property of automata. Thus, investigation of entropy, energy and thermodynamics began in the context of information processing, where the Demon instead became a computer with finite memory. The resolution of the Demon paradox came through the realization that there is an energetic cost of $kT \ln 2$ not in measuring a bit of information but only in erasing it [12, 18]. The question then began to be asked of just how much energy computation theoretically needed, especially as heat dissipation has always been a huge problem in digital computing. Bennett showed that any computation could be rendered in a logically reversible manner by carrying forward the computation using reversible steps to the stage of obtaining the answer, copying the result to a blank tape and then reversing the computing operations to obtain the input data again [17]. He showed that such a logically reversible computation, if performed quasistatically, would be able to consume arbitrarily little energy. Even at normal speeds, the technique requires far less power than the traditional irreversible computing format and experimental processors have already been produced based on this [28]. Independently, Toffoli proved that reversible cellular automata were computationally universal, whilst Fredkin developed the CCNOT 'conservative' logic function for logically reversible networks [84].

An important development was the realization of the intimate connection between noise and reversible computing. Fredkin developed the *billiard ball model* for universal computation, where bits are represented by an infinite 2D gas of perfectly elastic spheres that ballistically traverse through a periodic potential. Such *ballistic computation* can operate at finite speed and with no energy consumption/dissipation but only if the system is completely isolated from all thermal degrees of freedom. Indeed, Feynman calculated that even ignoring thermal noise, the uncertainty principle would affect such a system after merely 17 sphere deflections [44]. In contrast, in Bennett's scheme for *Brownian computation*, the thermal noise of the system is strongly coupled to the information-bearing degrees of freedom and exert a viscous drag. Thus, with such reversible computation, there is a dissipation per step proportional to the speed of the computation. Hence, not only can computation be performed in the presence of noise but also be driven by it in the reversible case. However, a tradeoff then appears between speed and energy.

A plot of the results in figure 12.9 suggests the existence of a thermodynamical limiting curve setting bounds on the limits of performing reliable computation in the presence of noise, much as for communication in information theory. Can we formulate a Shannon-like computation theory based on entropy quantifying the tradeoffs in computing resources, speed and energy? What will this tell us about the limits of reversible computing and the transition to the quantum regime? We now discuss some of the ideas arising from parallel restitution that might take us in this direction.

12.4.3.1 Computation entropy

In considering thermodynamics and computing, we have to define just what we mean by computation. In their monograph on information theory and noisy computation, Winograd and Cowan considered that a module, having as input the random variable X and output Y, performs computation if the conditional entropy was greater than zero (i.e. $H(X|Y) > 0$) [16]. If the module merely allowed communication, then $H(X|Y) = 0$. Thus, their interpretation of computation was a process causing information loss. They then modelled noisy computation by a module that performs perfect computation followed by a module that acted as a noisy channel. Information loss occurs as a result of both modules and so they constructed a *computation capacity*, C^*, that quantified the mutual information across both the computation and noisy channel modules. They then investigated maximizing the computation capacity for Boolean functions to derive information theoretic bounds on noisy computation in logic networks.

Pippenger and Evans' approach to applying information theory to noisy computation was to decompose Boolean functions into disjoint paths, make those paths noisy and then compute lower bounds on the loss of information caused by all the noisy paths [21, 22, 26].

The approach adopted by Sadek *et al* abandoned consideration of information entropy in looking at noisy computation [3]. One can view computation, in the very general sense, to be the thermodynamical flow of energy through a system with time, under the control of physical laws. If one then considers the different kinds of energy (chemical, potential etc), one can think of *information energy*. Information processing as it is conventionally understood is then the flow of this kind of energy through a system as time progresses, the laws of physics acting as the engine of computation. In essence, computation can then be viewed as the 'manipulation' of information. This manipulation may be reversible or irreversible in that it destroys information and dissipates heat. Winograd and Cowan's model for computation of $H(X|Y) > 0$, of course, implicitly assumes the irreversible case and, thus, is not a good general model.

Hence, a computer is a thermodynamical engine with energy input from an external source and the information input itself, generating work in the form of information manipulation (which may be mapped by a mathematical function) and possibly generating waste heat that flows into a thermal bath [18, 25]. If one considers a single logic gate in contact with a heat bath, that is so small in scale that it is part of a thermalized distribution, this is rather like considering a single gas molecule in a container in contact with a heat bath. Now, just as one can analyse the thermodynamics of an *ensemble* of gas molecules, one can also analyse the thermodynamics of an ensemble of logic gates trying to perform the same computation. This ensemble is the parallel ensemble of logic gates in parallel restitution. The computational ensembles in this scheme are interconnected by Markov chains. Can one apply the concepts of state space and entropy to a computing ensemble in an analogous way that it is applied to a gas?

If we consider $P(\zeta_n)$ given by equations (12.19)–(12.22), we can construct an entropy from the probability distributions:

$$H_c(\zeta_n) = - \sum_{\zeta_n R_p = 0}^{R_p} P(\zeta_n) \log_2 P(\zeta_n). \tag{12.30}$$

Figure 12.10 shows plots of equation (12.30) for $n = 1$ as a function of the input phase space and for values of gate noise ϵ from 0 to 0.5. The function can be seen to take on a minimum when there is no noise in the computing elements ($\epsilon = 0$) and the input noise does not necessarily affect the computational result. If both of the inputs are definitely 0 or 1, then the function is zero. If one of the inputs is definitely 0, then the function is still zero, even if there is noise on the other input, since for a NAND gate the output is always 1 if one input is fixed at 0. Conversely, if one input is fixed at 1, then the entropy function rises to a maximum if the other input has a 0.5 probability of being either a 0 or 1, since the output is sensitive to the other input in this regime. The function is at its maximum when there are perfectly noisy inputs. Also note the effect of increasing the computational noise from $\epsilon = 0$ to 0.5. The phase space that defines the computational function begins to collapse, until at $\epsilon = 0.5$ the entropy is a maximum regardless of the noise on the inputs. These results show equation (12.30) to be an entropic function directly analogous to thermodynamic entropy, with an interpretation of 'perfect order' for $H_c(\zeta_n) = 0$ and 'maximum disorder' for the maximum value of $H_c(\zeta_n)$. The function in this case then seems to describe the entropy of the computation, independent of the information

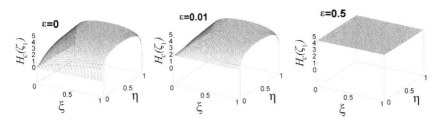

Figure 12.10. The computation entropy $H_c(\zeta_1)$ of a pararestituted NAND gate computation stage plotted over the phase space of its inputs, ξ and η [3]. It has a minimal value where the gates are noiseless and where noise on the inputs has no effect on the output of the computational function. Noisy inputs and noisy gates increase the computation entropy to a maximum.

content of the message being processed or indeed of the mode of computation (reversible or irreversible). The entropy here is, therefore, a minimum when our 'knowledge' of the possible computational output states that the information-processing function is supposed to give are narrowed down to the fewest options, and a maximum in the converse case [44]. Equation (12.30) hence describes what we call the *computation entropy*. It is a measure of the computational degrees of freedom, as described by Gershenfeld [25].

However, it should be noted that if the output state space is reduced, although this narrows down one's knowledge about what the possible output states are given the same input and although this decreases the computation entropy, the computational system will become more prone to noise. Thus, decreasing R_p (and, hence, R) in pararestituted systems increases susceptibility to noise.

It is interesting to consider whether the formulation of parallel restitution in terms of computation entropy can lead to a Shannon-like theory. With further work, it may be possible to construct a computation capacity, which would somehow be related to both the redundancy cutoff (derived from the threshold activation level Δ) and threshold gate noise ϵ_t. The theory might be extended further to quantify the tradeoffs in space, time and energy and the limits to computation in the presence of noise could then be fundamentally derived. In the limit of $\epsilon = 0$, ballistic computation would be optimal, whilst, in the limit of $\epsilon = 0.5$, Brownian computation would be.

Equation (12.30) has been formulated in terms of logical entropy and we have discussed how it can equate to thermodynamic entropy as long as the probabilities used in the descriptions refer to the same states and as long as the system described is part of a thermalized distribution [25,28,29,77,78,85]. Hence, to make the link with our thermal computing ensemble and thermodynamics, a *thermodynamic computation entropy* can be postulated:

$$S_c(\zeta_n) = -k \sum_{\zeta_n R_p = 0}^{R_p} P(\zeta_n) \ln P(\zeta_n). \qquad (12.31)$$

12.4.3.2 Noise, heat and nanoprocessor scaling

One can consider the effect of signal restitution on the computation entropy by analysing equation (12.19) for the computation stage and equations (12.20)–(12.22) for the restitution stage. The greater the degree of restitution is, the greater the logical depth and value of n will be. The conditional computation entropy $H_c(\zeta_n|\zeta_1)$ then always increases with n, because equations (12.20)–(12.22) describe a stationary Markov process [86], which is consistent with the second law of thermodynamics:

$$H_c(\zeta_n|\zeta_1) \geq H_c(\zeta_{n-1}|\zeta_1). \qquad (12.32)$$

This describes the inevitable entropy increase of the entire processor. In contrast, the computation entropy $H_c(\zeta_n)$ becomes non-increasing with n and maintains a non-uniform low value provided the gate noise ϵ remains below threshold. Hence, the interpretation is that parallel restitution keeps the computation entropy at a steady minimum by using external energy to pump the entropy to a low value. The cost is an increase in entropy of the entire processor. This inevitably leads to the production of heat.

Considering equation (12.32), the thermodynamic computation entropy given by equation (12.31) and the equation $dQ = T dS$, a relation may then be formulated in order to quantify the heat current associated with signal restitution for reliable noisy computation in a single pararestituted logic gate:

$$\Delta Q = T \left[S_c(\zeta_n|\zeta_1) - S_c(\zeta_{n-1}|\zeta_1) \right] \qquad (12.33)$$

where T is the temperature of the heat bath with which the computing ensemble is in contact.

If Δt is then the time between computational steps, equivalent to $1/f$ where f is the clock frequency, then the power of this heat current can be described:

$$\frac{\Delta Q}{\Delta t} = \frac{T[S_c(\zeta_n|\zeta_1) - S_c(\zeta_{n-1}|\zeta_1)]}{\Delta t}$$
$$\Rightarrow \Delta P = f T[S_c(\zeta_n|\zeta_1) - S_c(\zeta_{n-1}|\zeta_1)]. \qquad (12.34)$$

This is a *fluctuation–dissipation* relationship [25, 47]: as the noise ϵ increases, the entropy difference $S_c(\zeta_n|\zeta_1) - S_c(\zeta_{n-1}|\zeta_1)$ increases, as can be appreciated by studying the Markov chain relations (12.20)–(12.22). Increasing noise fluctuations thus increases the power dissipation. Using information theory and thermodynamics, Touchette and Lloyd have shown that there is a limit to the minimum amount of energy dissipation necessary with open-loop control [87]. This result is similar to the one described above, since restitution is a kind of open-loop gain mechanism for controlling errors.

Of course, the computation process itself does not necessarily lead to an entropy increase as it can be reversible. Significantly though, this is the case only if the computational elements and the inputs are 100% reliable. Once noise is introduced, to keep the reversible computation at a high reliability, irreversible signal restitution that consumes energy and increases the overall entropy is required [3,18,27,52]. *Energy consumption, entropy increase and heat production are, hence, fundamental consequences of reliable but noisy classical computation.*

These points are highly significant for nanocomputation. If processors are scaled down to the nanoscale and the physical limits of computation, the noise in the interconnects and computational devices will increase exponentially, in accordance with equation (12.4). At the same time, the ability of the processor to dissipate heat will fall due to the decrease in the surface area (quantized heat conduction would be the ultimate limitation). It has been suggested that nanocomputation should be based on reversible computation to get round the problems of heat dissipation [88] and several schemes for reversible nanoelectronic computation have been proposed [89–91]. However, we have shown that as long as there is noise, then signal restitution, with its energy consumption and consequent entropy increase and heat production, will be required. The resultant temperature increase would feed forward to increase the noise further. If then with scaling computation down there are fundamental increases in noise, increases in heat production and decreases in the ability to dissipate heat, the power dissipation density increases and a problem arises. At some length scale, a tradeoff between these factors must occur so that there is reliable operation but no melting of the processor. Indeed, below some length scale, it might be that it is impossible to construct useful *classical* computers at all. The solution to this problem may lie in the migration to quantum computing.

12.4.4   Quantum fault tolerance and thermodynamics

Classical computing, with its basis in digital logic, uses thermodynamics as its physical engine. Interactions between pieces of information can only occur locally and, in the limits of minimal speed and energy, these interactions can be reversible. Another type of computation is possible where, instead, a quantum superposition of orthogonal states is allowed to evolve which then interfere with one other. The interference sets up a distribution of states that differs from that formed if a single path of evolution had been followed and this can be used to perform powerful computations. Interactions can be non-local, through the use of entanglement. Gate manipulations are necessarily reversible, these being *unitary* operations. Such *quantum computation* can theoretically solve in polynomial time some problems that classical Turing computers would take exponential time to perform, such as the simulation of quantum systems and factorization.

We have already discussed the potentially serious effects of noise in quantum computation. Until the work of Shor demonstrating the possibility of *quantum error correction* [92], practical implementation of quantum computing seemed impossible. Quantum error-correcting codes are, in fact, very closely related to their classical counterparts. For example, the seven-qubit Steane code is the quantum implementation of a Hamming code [93]. The Hamming code uses seven bits of information to encode four bits. The initially independent four bits are correlated with one another through the addition of redundant bits. In the case of the Hamming code, this is done by using seven-bit codewords that annihilate in modular two arithmetic when operated on by a parity check matrix. If an unknown bit-flip occurs anywhere in the Hamming code due to noise, it is possible to measure what error occurred and correct it *without ever knowing what the encoded data are*. This is the key to why the Hamming code can be extended to quantum information, since if decoding was ever necessary to correct errors, in the quantum case this would lead to state collapse of the encoded qubit. In the Steane code, a single qubit is encoded and expanded to a seven-qubit block. The quantum state of the original qubit is stored in the code block through entanglement between the seven qubits. As no information about the original state is stored in any one qubit, decoherence of one of the qubits does not destroy the original information. By using *ancillary bits*, the exact location and type of error that occurred can be determined, this being known as the *syndrome*. Measurement of the syndrome is made by the ancilla and a 'decision' is made as to what type of quantum operation should be made to the code block to correct the error. The entire encoding, correction and decoding procedure is unitary (reversible) and, in principle, requires no energy, with the exception of the 'decision' which is a classical entropy-producing majority vote.

Quantum error correction does not by itself, however, ensure the possibility of noise-tolerant computation. This requires the encoding, computation, error correction and decoding all to be implemented in a fault tolerant manner, such that noise affecting any of these operations can still permit information processing with bounded reliability. Several such schemes for fault tolerant universal quantum gates and quantum circuits have been discovered, through the careful construction of parallelized circuits incorporating periodic error correction [54]. Such *transversal gate construction* [55] is, in fact, the quantum analogue of Spielman's technique and parallel restitution [3, 38]. Of course, direct implementation of parallel restitution, as has been described for NAND gates, cannot be carried over to quantum fault tolerance. This is due to the use of irreversible gates for restitution, the utilization of FANOUT for encoding and the direct measurement that occurs in decoding. Respectively, these violate the condition for unitary operations and the no-cloning theorem, and collapse the qubit state. These problems are tackled in quantum fault tolerance by using quantum coding to encode qubits and introduce redundancy, storing the original qubit state in the 'topology' of the parallel qubits. The gate computation is performed by a parallel array of suitably modified quantum gates that can compute on encoded states. Periodic measurement of the syndrome by the ancilla with any necessary error correction performs the function of restitution. Decoding of the final code block recovers the computed state. It has also been shown that a threshold gate error exists, as in the classical case, and that such a threshold would only exist if there were constant parallelism at every time-step (i.e. between every consecutive operation on the qubits) [94]. Computations of arbitrary length can be performed in the presence of subthreshold noise by employing *concatenated* codes, in much the same manner as cascaded R-modular redundancy.

The whole idea behind quantum fault tolerance, parallel restitution and noise-tolerant computing, in general, is to use redundancy to encode computation into a specific phase space. The effect of noise as a result of interaction with a heat bath is to cause the computational state to drift out of its encoded phase space. The action of restitution using work is then to compress the computing state back into the allotted phase space and reject entropy to the environment. The computation itself strictly does not require energy but an important question is whether restitution necessarily does. We have discussed how this has important implications for nanocomputation because of the positive feedback effect of exponentially increasing noise in scaling to the nanometre level. The resultant exponential increase in energy consumption to pump away entropy further escalates noise because of rising temperature. If the laws of physics somehow permit 'intrinsic' fault tolerance that also requires arbitrarily little energy, then this would break the classical noise barrier and allow molecular level device scaling.

Although noise-free quantum computation is certainly reversible, in practice it has been shown by Gea-Banacloche that once failure rates are factored in and classical fields are used to control the logic gates, the energy requirements become substantial [96] and even higher than in classical computation [97]. Steane has also noted that fault tolerant quantum computation acts as a thermodynamical engine, just as in the classical case [95]. As shown in figure 12.11, entanglement between the environment and quantum register increases entropy and causes errors. Measuring the syndrome using the ancilla absorbs the unwanted entropy, and the majority vote decision during verification then dissipates the entropy back into the environment. The majority gate needs an energy source to operate and pump the entropy against its natural direction of flow.

What is unclear is whether any of these observations are truly fundamental for noise-tolerant quantum computing. Quantum error-correcting codes are very much based on their classical progenitors and so it may be that the requirement for energy is a vestige of implementing classical ideas. It should also be kept in mind that noise in quantum computing is of a different and deeper nature than in classical systems. For instance, Minsky has noted how, contrary to popular view, it is actually classical physics that is uncertain and indeterminate and quantum physics that is certain and stable: any error in the inputs to a classical dynamical system amplifies with time to cause unstable behaviour (unless there is energy input and damping), whereas quantum dynamical systems,

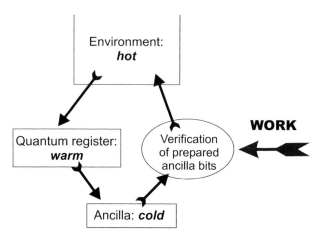

Figure 12.11. The thermodynamics of current schemes for quantum fault tolerance is identical to parallel restitution (see figure 12.4). The quantum computation is reversible and so strictly it does not require energy. However, the entropy generated through the effects of noise necessitates an energy source to pump it out of the system via a thermodynamical engine and maintain the computation entropy at a low value. The surrounding heat bath generates noise and, thus, errors in the quantum register. The ancillary bits are used to measure the syndrome, which in the process absorbs the entropy. Verification of the ancilla using majority voting requires energy input and data collapse causes the extracted entropy to be dissipated to the heat bath. After Steane [95].

for example atoms and molecular bonds, can remain stable for billions of years [98]. These questions are intimately related to the quantum-classical transition and the quantum interpretation problem. Gea-Banacloche and Kish have argued for the characterization of a quantum and classical regime for digital computation based on energy consumption [97] but their arguments are based on the use of classical fields to control quantum gates, where reliability of control requires a greater ensemble $\bar{n}$ of photons 'representing' the field. A possibly more fundamental and quantitative observation has been made by Aharonov that the length scale of the classical–quantum transition is determined by noise [99]. Aharonov argues that below a critical noise threshold the entanglement length of quantum systems remains infinite, whereas above the threshold the entanglement length becomes finite. Noise then induces a phase transition from the quantum to classical scale, its exact form being unknown. Further work shedding light in this area is that of Kitaev and Preskill [54]. We noted that the function of fault tolerance is to keep the computational state within a specific phase space and that this is done by forming a global topology and attempting to keep that topology constant. In this vein, Kitaev has suggested the use of Aharonov–Bohm interactions for *topological quantum fault tolerance*. The Aharonov–Bohm effect can be illustrated by considering the transport of an electron around a perfectly shielded magnetic solenoid. In this case, the electron wavefunction will acquire a phase change of exactly $\exp(ie\Phi)$ with every circumnavigation, where e is the electronic charge and Φ is the magnetic flux within the solenoid. This phase change is not dependent on the exact path taken by the electron, only on the number of circumnavigations. If the circular orbit is deformed, the phase change is completely unaffected. Thus the phase change is intrinsically fault tolerant. Such topological properties can be used as the basis for fault tolerant quantum computing and Preskill has, in turn, asked the question of whether nature itself is intrinsically fault tolerant. If so, then restitution at the quantum level would not require work, as long as the noise was within a specific threshold.

Thus, in addition to its exponential speed up of certain classical computing problems, quantum computation may have an equally important implication for nanocomputation. Thermodynamical noise at finite temperature forms an insurmountable length-scale barrier for classical computing. To develop even denser nanocomputer architectures below this length scale will possibly necessitate abandonment of classical computation and migration to quantum computational architectures, for which the energy requirements in the subcritical noise regime will be much lower. Such architectures might entail computation within single molecules, where entanglement effects and thermal noise are, in some way, shielded [100]. There thus may be three noise regimes: the classical one due to thermodynamical noise, the quantum one due to entanglement with the environment and, most fundamentally, the regime of zero-point field fluctuations. Topological fault tolerance shields nature from the fray of quantum field fluctuations to give conventional quantum mechanics. Decoherence, in turn, causes the transition to the classical regime. More work needs to be done on the topics described, especially on the thermodynamics of quantum computation and entanglement [101].

12.5   Nanocomputation and the brain

Von Neumann's investigations into computation, information (classical and quantum), entropy and the reliability problem were not solely directed to the development of our present-day information-processing machines. The wider problem of interest to him was information processing in biological and neural systems and how, perhaps, investigations here could lead to technological developments in computation [1, 66, 102–104]. The drive towards nanocomputers with extremely large scales of integration as in the brain makes these concepts even more poignant today.

Investigations in neuroscience are beginning to dispel the conventional view of neural computation solely arising through neural-network-type computation, where individual neurons operate according to relatively simple rules. The computational architecture of the brain is now understood to be multi-hierarchical, extending from the gross anatomical level right down to the nanoscale and the level of individual proteins [105]. As our computing technology approaches the nanoscale and our understanding of the brain approaches this level as well, we are beginning to find that some of the same fundamental physical limits and necessary architectural solutions are also present in neural systems. This suggests that back engineering some aspects of neural technology into the nanoprocessors of tomorrow or even synthesizing the two might bring computational advantages.

12.5.1 Energy, entropy and information

The physics of energy-efficient information and computation in the presence of noise has been found to be the basis of neural architectures in nature and not just of artificial systems. Animal life has been exquisitely designed through a process of natural selection to optimize their biological design so as to maximize functions that aid survival whilst minimizing the associated costs of energy and materials [106]. The adult human brain consumes 20% of the body's metabolic energy, this figure being a massive 60% in neonates. These figures are similar in the nervous systems of other animals and, thus, the metabolic demands of the brain have likely shaped the course of evolution to give rise to neural architectures that minimize energy consumption. In some species, such as insects, the metabolic demands of the retina are a very significant proportion of the nervous system. For example, the Blowfly retina taxes 8% of the insects' energy demand and, hence, systems such as these have been under intense experimental and theoretical scrutiny [107, 108].

Laughlin was the first to show that neural coding in the visual system is optimized to maximize the information capacity with respect to natural scenes of the organism [109]. For the graded analogue signals generated by photoreceptors, the information capacity can be quantified by Shannon's equation for a Gaussian channel:

$$C = \Delta f \log_2 \left[1 + \frac{S(f)}{N(f)} \right] \qquad (12.35)$$

where the signal $S(f)$ and noise $N(f)$ are both Gaussian. Δf is the signal bandwidth.

Subsequently, Levy and Baxter analysed the representational capacity of neural networks and found that real nervous systems do not fire at their maximum of 200 Hz but around 20 Hz [110]. They explained this by positing that neural systems employ coding schemes that maximize the ratio of representational/informational capacity to energy cost. Laughlin investigated neural coding further and noted the importance in the tradeoff between redundancy, noise and energy, just as with digital computers. If there is excessive noise, for example as in retinal photoreceptors because of photon shot noise and thermodynamical channel noise, information is preferentially transmitted along many parallel pathways of weak informational capacity instead of through a brute increase in the number of synapses converging onto a neuron [107]. This can be appreciated through consideration of equation (12.35). The capacity only increases as the logarithm of the signal $S(f)$ and so increasing the number of converging synapses onto a post synaptic neuron is not very efficient, especially as the energetic cost is high in doing so. However, the capacity increases linearly with bandwidth Δf, corresponding to the degree of parallelism. This biological principle is identical to the digital case of R-modular redundancy and parallel restitution. The data collapse and convergence in modular redundant fault tolerance strategies inefficiently increases energy consumption and is not as powerful for noise tolerance as parallel restitution where many parallel pathways are used in combination with topological signal restoration. Parallel restitution is, thus, not only the basis of optimal classical and quantum fault tolerance but also of energy-efficient and noise-tolerant neural computation [3].

12.5.2   Parallel restitution in neural systems

Von Neumann pondered whether his fault tolerant construction of multiplexing was implemented in some form in the brain and, indeed, his 1956 paper was aimed just as much at neural information processing as electronic computation. It has been reported that through collaboration with physiologists, he found the innervation of auditory hairs in the cochlea to follow an information mechanism similar to multiplexing [111]. The question of fault tolerance in the brain is a very important one in view of the high levels of latent noise.

   One of the most fundamental sources of neural noise arises from the thermodynamical switching of ion channels [112]. If single-ion channels were relied upon for signal communication or processing, the reliability would of course be zero. For the purposes of signalling, this is countered by using redundant ion channels arranged in a cylindrical axon. Sadek *et al* have noted how this strategy for reliable neural signalling is again a form of parallel restitution [3]. The digital state space of the ion channel conformations and, hence, of the membrane potential is maintained through an energy-consuming positive feedback mechanism called the *action potential*. The parallel ion channels comprise the computation stage, whilst the action potential mechanism at the nodes of Ranvier in myelinated axons form the restitution stage. This periodically counteracts the effects of noise whilst maintaining parallelism and, hence, restores the signal. There exists a voltage threshold, similar to the discriminator threshold described in section 12.3.4.4, which determines whether the analogue membrane potential signal is to be decoded into a '1' (spike) or '0' (no spike). The informational coding is different to digital computing however, in that encoding is via the arrival times between individual spikes [113]. Such frequency-modulated coding also helps protect against the amplitude noise that ultimately arises from thermal fluctuations.

   Stochastic simulations of patches of axonal membrane with noisy ion channels have shown that axons begin to generate spontaneous action potentials when the membrane area is reduced below a critical value [114]. This interesting result is phenomenologically analogous to the digital implementation of parallel restitution; reducing 'R_p' reduces the output state space and makes the system more prone to noise. In the digital model, the critical value is $R_p = 16$, corresponding to a 'computation capacity', below which the computational state vector of the line bundles lose their coherence. At this capacity, the noise threshold ϵ_t equals the ion channel noise. More accurate stochastic simulations of axons with proper cylindrical geometry have expanded on this result and shown that spontaneous action potentials begin to generate if nerve axon diameters are reduced below 100 nm [115]. This length-scale limit arises from limits to the packing density of ion channels. Here, the number of circularly arranged ion channels then equals the critical limit set by the computational capacity.

Signalling costs make up the bulk of the brain's energy consumption, about 50% overall, and this rises to 80% in cortical grey matter (because of the greater complexity of interconnection here and lack of myelin) [116, 117]. Such energy requirements ultimately arise from the need to pump energy into the action-potential-based restitution stage, through recharging of the membrane potential via Na^+–K^+ ATPase pumps. The power dissipation of the brain is only 10 W, this partially being a limit constrained through evolution because of the inability of biological tissue to withstand higher temperatures. We can see thus far that the problems, constraints and solutions in neural systems are almost identical to their digital counterparts. Power dissipation (and thus energy consumption) is also a constraint on terascale integration in nanoprocessors, in addition to noise and redundancy. The problem in electronic processors though seems to be somewhat reversed. Signalling is not the major culprit in heat dissipation but computation itself. By contrast, in the brain, signalling over large distances dissipates the most energy, whilst in studies thus far no energy budget has been allocated to computation, probably as it is ill defined and quite small. What is the thermodynamics of *computation* in the brain as opposed to communication?

Sarpeshkar considered von Neumann's hypothesis of the brain using both analogue and digital representations for computation and, using a MOS technology model, he showed that digital computation was optimal for noise-tolerant computation, whilst analogue computation had the most processing power and used the least energy [118]. Hence, by taking single analogue lines and partially distributing them over multiple digital lines, he showed that computation optimized between noise tolerance, energy consumption and redundancy was possible. Signal restitution was performed by intermittent analogue–digital–analogue conversions implemented in parallel. This 'hybrid' form of computation is, of course, parallel restitution and Sarpeshkar's conclusions are identical to Laughlin's observations [107]. In the digital scheme in section 12.3.4, we looked at the optimum way to implement noise tolerance by moving to a more analogue format, whilst considering the worst case redundancy and heat dissipation that results. Sarpeshkar, in contrast, has come from the other direction and digitized an analogue computation whilst trading off the various factors. This work was inspired by efforts to develop an energy-efficient silicon cochlea based on neuromorphic architectures [119].

An example of this kind of computational approach is the architecture of synaptic integration in the Blowfly retina. Here, each retinal pixel consists of six electrically coupled photoreceptor cells that impinge on a single large monopolar cell (LMC) [107]. The LMC is a spiking synapse and converts the graded potential of the photoreceptors into a series of spike trains. Each parallel photoreceptor samples light from the same point in space and makes 220 identical synaptic connections to the LMC, thus there are 1320 parallel connections in all. At the LMC, there is mixing of the signals from each of the synapses through integration of their post-synaptic potentials (PSPs). The integrated graded PSP is converted into a series of action potentials by the threshold properties of the LMC. Thus, multiple computational units transduce input information into a parallelized output signal that impinges onto a restitution unit. The restitution stage correlates the signals by mixing, as with the digital scheme, and restitutes the signal against noise by generating action potentials. Parallel pathways of weaker informational capacity are more energy efficient and noise tolerant than a single convergent pathway.

Despite these noise-tolerant architectures, neurons have a surprisingly high probability of failure. In culture, synaptic failure can approach $\epsilon = 0.9$ [114]. This phenomenon is most likely a computational feature and not a compromise in design. What seems to contribute to coding is the probability of the failure of the synapses themselves. Modulation of this failure probability then affects downstream events such as synaptic plasticity. Levy and Baxter, through application of information theory to synaptic computation, have also found that synaptic failures make computation more energy efficient [120].

The brain performs different types of information processing at multiple hierarchies, at every level the mode of computation being optimized for the task, given the constraints of noise, energy and resources [105]. Axons transmitting pararestituted action potentials interconnect distant regions of the brain, whilst in regional processing areas, organized neural processing layers appear. At smaller scales, complex neural networks then move down to the level of synaptic integration and computation in single synapses. Each neuron itself, with its complex dendritic inputs, performs sophisticated information processing. It even seems likely that computation may involve complex biochemical pathways within each neuron [114] or, more speculatively, occur at the level of small ensembles of molecules [121] or even single protein molecules. Every level matches its device physics and information-processing strategies to maximize information transmission and processing as needed, whilst minimizing energy consumption and not unduly overdamping noise. For example, information transmission over long ranges down to the level of synaptic integration seems to use parallel restitution but, below this scale, different strategies are utilized. This is as thermal noise is so high, that use of parallel restitution would utilize excessive energy for entropy pumping, increasing power dissipation and energy consumption greatly.

In some cases, mechanical transmission is used, as in the triads of muscle, where ryanodine and dihydropyridine receptors are physically linked; a voltage-sensitive channel at the membrane opens a calcium channel at the sarcoplasmic reticulum through a mechanical link to release the second messenger calcium. Such nanoscale mechanical links have been put forward as the basis for reversible nanologic devices [90]. Another strategy is to use reversible computation based on *diffusion*. We saw in section 12.4.3 that for $\epsilon \rightarrow 0.5$, Brownian computation is optimal for energy-efficient computing. Collaborative work carried out by one of us (AS) is suggesting that this is, indeed, the case for neural computation at the molecular level [122]. The precise mechanism and purpose of short-term synaptic depression in neurons is still unknown but work by us has strongly suggested that such depression arises from the depletion of calcium ions in the synaptic cleft and replenishment via diffusion from the extracellular space. Such diffusion at low concentrations and in small volumes dissipates very little energy. This possible explanation suggests the purpose of short-term synaptic depression as a reversible computation/transduction primitive in the brain. Intracellular synaptic proteins such as synaptotagmin are highly sensitive to intracellular calcium concentrations and, thus, variations in calcium concentration that are intimately dependent on the previous firing history of a neuron may play an important role in molecular-scale neural computation. The brain might use such modest energy for computation itself because of its basis in localized molecular reversible primitives. Egelman and Montague have also noted that such depletion and diffusion effects of calcium could also play a computational role at a higher hierarchical scale in view of the tightly compacted nature of neural tissue [123]. Depletion of calcium would occur not only in the synaptic cleft but also in the extracellular space between neurons, leading to a cellular-automata-like mechanism.

12.5.3   Neural architecture and brain plasticity

The most striking aspect of neural architecture is its 3D arrangement. *Rent's rule* is an empirical equation found to apply to single chips as well as to the complex circuits comprising them, which gives the number of required output terminals T according to the number of terminals t on the constituent components [79]. In the best case, $T = tN^{2/3}$, where N is the number of components. The exponent holds approximately true for logic chips but is relaxed to 0.1 for memory chips. The problem with purely 2D chips architectures is that, as the scale of integration of a processor is increased, the perimeter of the chip where the output terminals are located only increases as the square root of the area, whilst the number of output terminals increases as the two-third power, a much bigger scaling factor. Hence, eventually, a limit is reached where the output terminals cannot be jammed together further. In contrast, with migration to 3D architectures, the surface area available for terminal placement increases as the two-third power of the volume, precisely matching the scaling factor in Rent's rule. Thus, with 3D architectures, the phenomenon of terminal jamming does not occur, hence making it the ideal format for extremely large scales of component integration. The brain takes full advantage of this effect. The cortex is comprised of a thin layer of *grey matter* a few millimetres thick that comprises the dendrites, synapses, unmyelinated axons and cell bodies of neurons (i.e. the computational components and local wiring). The cortical grey matter in the 'new brain' is comprised of six layers and, thus, has a more complex organization than the grey matter in older areas such as the hippocampus which comprises three layers. Beneath the layer of grey matter are the 'system buses' of the brain, the *white matter*, which comprises high-speed myelinated axons that interconnect different regions of the grey matter. The amount of white matter scales approximately as the four-third power of grey matter volume [117], an observation which is perhaps based in Rent's rule. The proportion of wiring by volume in neural architectures ranges from 40% to 60% and is much lower than the 90% typical of VLSI systems. It has been suggested that this is due to the efficient use of 3D design that minimizes communication delay but maximizes component density [117]. This has been the inspiration for the development of neuromorphic 3D nanoelectronic systems as with the CORTEX project that can take full advantage of the spatio-temporal limits to computation set by nature.

Another fascinating aspect of neural architecture is its latent ability to adjust and reconfigure its connectivity, this being known as *plasticity* in neuroscience. The molecular basis of memory is thought to arise from this, due to the Hebbian adjustment of synaptic strengths via the mechanisms of long-term potentiation (LTP) and long-term depression (LTD), but recent work has suggested that it may instead arise through long-term changes in dendritic excitability [124]. Current VLSI architectures are purely static and we have already seen in [5] how the ability to reconfigure computer architectures will be an essential feature in future nanoelectronics. Does the brain use plasticity to reconfigure around defective or damaged areas just as with our designs for future nanoprocessors? This is certainly the case and, indeed, it seems to perform this feat even more efficiently than our projected technologies. For many years it was thought that brain damage that occurred in adulthood was, for the greater part, unrecoverable and that the brain's architecture became fixed after a period of developmental modelling up until adolescence. Several lines of study have fundamentally altered this view. In patients who suffer a stroke, part of the brain becomes permanently damaged and scarred subsequent to occlusion of the blood supply from a clot. In serious cases, there is a resultant immediate loss in neural function corresponding to the area of the brain affected (for example speech, limb movement, sight etc). However, in the weeks and months that follow an episode, the brain implements active mechanisms of recovery that include adaptive plasticity [125]. If the primary neural pathways in a brain area have become permanently damaged but there are surviving synapses that usually perform only a supporting role, these will become strengthened to reform the neural network. Thus, if say a part of the speech centre was affected subsequent to a stroke and some weak pathways here survive, speech function will recover in time through a moderate degree of relearning by the patient. If there is more serious damage and no neural tissue in this area survives, recovery of function is still possible through an even more remarkable process. Brain areas that usually perform completely different functions can adapt to take on the lost function in addition to their original ones. Naturally though, this takes more time and the lost ability has to be more or less completely relearned by the patient.

The most astonishing instance of reconfiguration and plasticity in the brain is that following *hemispherectomy*, a neurosurgical treatment for intractable epilepsy [126]. For reasons that are not understood, plasticity in childhood is much greater than in the adult years, which explains our greater ability to learn new languages and skills when we are younger. Thus, in children under eight years with severe intractable epilepsy in one half of the brain due to Rasmussen's syndrome, congenital vascular injuries or other cause, a radical procedure can be considered. To prevent the epilepsy spreading to the other half of the brain, complete removal of the affected half of the cortex can be performed. These patients literally then have only have half a brain. In the immediate aftermath of surgery, all functions performed by the removed half are lost. However, in the weeks and months following surgery, there is massive reorganization of the remaining half of the cortex so as to take on almost all the resultant lost functions. The outcome of surgery is usually very good, with normal intellectual development and function, the only disability being slight hemiparesis on the affected side. It should be noted that even abilities such as language which reside in the left-hand side of the brain are recovered if the left-hand half is removed. Such a wonderful ability, to be able to 'kill' 50% of the devices in a computer and still retain complete functionality, will eventually need to be a feature of nanoelectronic processors.

It is still unknown what the precise mechanisms of plasticity in these cases is. Experiments in primates have shown that if a digit is surgically removed, the topographical brain area that represents it is invaded by neighbouring regions [127]. It is likely that the formation of completely new synaptic connections is involved. In addition, neural stem cells can introduce small numbers of new neurons to form essential connections when there is severe damage [128]. Brain plasticity, hence, seems to be a combination of physical reconfiguration, 'self-repair' and software re-coding.

12.5.4 Nano-scale neuroelectronic interfacing

We are quickly learning that the 'devices' of the brain, the myriad types of neurons, are not simple, passive devices as was previously thought. Each is a highly complex and powerful processor in itself [114], having up to 10^4 highly branched dendritic inputs that perform very complex information processing, with myriad sites for synaptic contact which include dendritic spines. The pre- and post-synaptic boutons themselves are highly engineered computing structures, performing many computing functions through biochemical pathways, diffusion phenomena and single protein effects. The many different types of ion channel all over the neuron perform sophisticated signal shaping, switching and

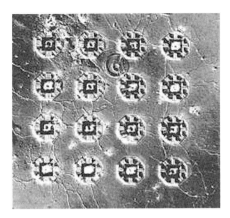

Figure 12.12. The neurochip, under development at the Pine Lab and Micromachining Laboratory, California Institute of Technology, allows microelectronic interfacing with individual neurons in a biological neural network [130] (image by courtesy of Pine Lab, Caltech). Reproduced from Maher *et al* [130], © Elsevier 1999.

communicative duties. Theoretical attention has even turned to the microtubules within dendrites, and the possible role of proteins here in localized quantum superposition effects over biologically relevant timescales [129]. The 'biophysics of computation' is turning out to be richer and more complex than ever thought, with the limitations of experimental techniques hampering its elucidation.

Current techniques for electrical recording from single neurons entail the use of 20 μm diameter ionic microelectrodes, that are almost as large as the neurons under investigation themselves. Using these, it is almost impossible to record from more than two or three neurons in a neural network simultaneously, making it difficult to study plasticity and network function. Likewise, in the field of neuroprosthetics, the use of macroscopic electrodes for interfacing with the brain limits the functionality of such devices, as each electrode is merely coarsely connected to tens of thousands of neurons. These kinds of problems as well as the vision for constructing hybrid organoelectronic computers have spurred efforts to develop a *neuroelectronic interface*.

Much work has been done in this field ranging from the capacitive stimulation and recording of neurons using planar microelectrode arrays [131], to the development of micromachined silicon multielectrodes [132]. The brain communicates not only electrically but also chemically and micromachined interfaces have even been constructed that act as chemical synapses, releasing as little as 1 pl of neurotransmitter per stimulation [133]. Most interesting is the work at the Pine Lab, Caltech, to develop neuroelectronic arrays that can interface with individual neurons in a complex neural network, recording and stimulating single nerve cells that are addressed and do not move about as with simpler planar arrays. These employ micromachined wells and cages that allow migration of immature neurons into the wells but only allow the outgrowth of neural processes such as dendrites and axons at maturation.

Work is even progressing on the development of neuromorphic and neurocomputational chips that can be implanted in the human brain to replace the function of a damaged region [134]. Berger at the University of Southern California is currently performing experiments to replace the hippocampus of rats with a chip replacement. No one understands the function of the hippocampus but its input–output state space is well understood through experimental recording and this has been programmed into the chip. The chip can also be optimized subsequent to implantation to adapt more closely to the user.

The miniaturization and increasing scales of integration of computing technology will eventually allow us to interface with the brain in the same manner that the brain interfaces with itself. This ultimate achievement will require the development of a *nanoscale neuroelectronic interface*, where stimulation and recording can be achieved with the individual structures in a single neuron, including the thousands of dendrites, dendritic spines and synaptic terminals. We have discussed in this chapter the limits to computation arising from noise and entropy production and the limits to information processing imposed by physics and the tradeoffs between space, time and energy. These same limits constrain the operation of neural systems, which also compute at the nanoscale. It is then interesting to ask, if information processing in the brain occurs at the nanoscale or even molecular scale, what are the strategies employed for reliable computation, and what is the device physics? Interfacing with the brain at this level will allow us to determine this experimentally. If computation occurs at such small scales in neurons and if we find that our classical strategies fail above those scales, then we might learn a vast amount about how to do nanocomputation from studying neural systems.

12.6 Summary

The future usefulness of various nanoelectronic devices may be seriously limited if they cannot be made in large quantities with a high degree of reliability. It is theoretically possible to make very large functional circuits, even with one dead device in 10, but only if the dead devices can be located and the circuit reconfigured to avoid them. However, this technique would require a large redundancy factor. If it is not possible to locate the dead devices or circuits have to be noise tolerant, then one of the other two techniques would have to be used: modular redundancy/majority vote scheme or parallel restitution architecture. The latter was found to be very effective against computational noise. For the cost of a modest overhead of $R = 50$ and an increase in logical depth of $n = 3$, a nanoscale computer consisting of 10^{12} devices would run reliably if the error probability of constituent logic gates does not exceed 10^{-4}.

Parallel restitution implements redundant communication and computation, and signal restitution in a highly parallel fashion. Comparing this system to ensembles in statistical mechanics enabled the introduction of a new quantity—the computational entropy—that quantifies the entropy of the computational input–output state space. The rectification of noise-induced errors creates power dissipation, which has to be added to the 'normal' power dissipation created by the operation of the circuit devices. Since noise will increase with the scaling down of the computing devices, dissipation density will quickly increase, creating an additional problem for the further miniaturization of information-processing devices.

The issues of noise, entropy, energy consumption and heat dissipation are also topical in theoretical neuroscience. The ultimate neural system—the brain—performs different types of information processing at multiple hierarchies, under the constraints of noise, energy and resources. Some of the information processing might even occur at the nanoscale and, if this is true, the question is what strategy is employed for reliable computation. We might learn a vast amount about how to do nanocomputation from studying neural systems, this being von Neumann's original inspiration [102]. The creation of a nanoscale 3D, parallel, reconfigurable, noise-tolerant, energy-efficient and computationally powerful processor might ultimately emulate some aspects of the animal brain.

The nanocomputers of the future will begin to approach the limits to information processing set by physics and much attention will have to be paid to their architecture and mode of operation. Noise, defects and entropy production will become extremely serious problems if current design methodologies are employed. The nanoprocessor of the future will borrow concepts from highly efficient neural systems and integrate 3D architectures based on classical reversible and quantum computation. These architectures will incorporate ingenious and ultra-efficient tradeoffs in space, time and energy consumption to resist the effects of noise whilst incorporating ballistic removal of unwanted entropy. They will also be reconfigurable, having the ability to rewire around small or even large regions of defective circuitry and, furthermore, have the ability to adapt their circuitry to optimize to the computational task at hand. In addition to the massive requisite applied physics and engineering effort, further fundamental investigation of the physical limits to efficient computation in the presence of noise and defects will be required to make this vision reality.

References

[1] von Neumann J 1956 Probabilistic logics and the synthesis of reliable organisms from unreliable components *Automata Studies* ed C E Shannon and J McCarthy (Princeton, NJ: Princeton University Press) pp 329–78

[2] Kish L B 2002 End of Moore's law: thermal (noise) death of integration in micro and nano electronics *Phys. Lett.* A **305** 144–9

[3] Sadek A S, Nikolić K and Forshaw M 2004 Parallel information and computation with restitution for noise-tolerant nanoscale logic networks *Nanotechnology* **15** 192–210

[4] Birge R R, Lawrence A F and Tallent J R 1991 Quantum effects, thermal statistics and reliability of nanoscale molecular and semiconductor devices *Nanotechnology* **2** 73–87

[5] Nikolić K, Sadek A and Forshaw M 2002 Fault tolerant techniques for nanocomputers *Nanotechnology* **13** 357–62

[6] Singh R, Poole J O, Poole K F and Vaidya S D 2002 Fundamental device design considerations in the development of disruptive nanoelectronics *J. Nanosci. Nanotechnology* **2** 363–8

[7] Heath J R, Kuekes P J, Snider G S and Stanley Williams R 1998 A defect-tolerant computer architecture: Opportunities for nanotechnology *Science* **280** 1716–21

[8] Lach L, Mangione-Smith W H and Potkonjak M 1998 Low overhead fault tolerant FPGA systems *IEEE Trans. VLSI Syst.* **6** 212–21

[9] Mange D, Sipper M, Stauffer A and Tempesti G 2000 Towards robust integrated circuits: The embryonics approach *Proc. IEEE* **88** 516–41

[10] Goldstein S C 2001 Nanofabrics: Spatial computing using molecular electronics (Goteborg, Sweden) *Proc. 28th Annual Int. Symp. on Computer Architecture (ISCA)* (Los Alamitos, CA: IEEE Computer Society) pp 178–91

[11] Challet D and Johnson N F 2002 Optimal combinations of imperfect objects *Phys. Rev. Lett.* **89** 028701

[12] Landauer R 1961 Irreversibility and heat generation in the computing process *IBM J. Res. Develop.* **5** 183–91

[13] Shannon C E 1948 A mathematical theory of communication *Bell Syst. Tech. J.* **27** 379–423, 623–56

[14] Hamming R W 1950 Error detecting and error-correcting codes *Bell Syst. Tech. J.* **29** 147–60

[15] Hamming R W 1986 *Coding and Information Theory* 2nd edn (Englewood Cliffs, NJ: Prentice-Hall)

[16] Winograd S and Cowan J D 1963 *Reliable Computation in the Presence of Noise* (Cambridge, MA: MIT)

[17] Bennett C H 1973 Logical reversibility of computation *IBM J. Res. Develop.* **17** 525–32

[18] Bennett C H 1982 The thermodynamics of computation—a review *Int. J. Theor. Phys.* **21** 905–40

[19] Bennett C H 1988 Notes on the history of reversible computation *IBM J. Res. Develop.* **32** 16–23

[20] Landauer R 1987 Computation: A fundamental physical view *Phys. Scr.* **35** 88–95

[21] Pippenger N 1988 Reliable computation by formulas in the presence of noise *IEEE Trans. Inf. Theory* **34** 194–7

[22] Evans W S 1994 Information theory and noisy computation *PhD Thesis* University of California, Berkeley

[23] Smith W D 1995 Fundamental physical limits on computation *Technical Report* NEC Research Institute, Princeton, NJ http://external.nj.nec.com/homepages/wds/fundphys.ps

[24] Landauer R 1996 The physical nature of information *Phys. Lett.* A **217** 188–93

[25] Gershenfeld N 1996 Signal entropy and the thermodynamics of computation *IBM Syst. J.* **35** 577–86

[26] Evans W S and Schulman L J 1999 Signal propagation and noisy circuits *IEEE Trans. Inf. Theory* **45** 2367–73

[27] Lloyd S 2000 Ultimate physical limits to computation *Nature* **406** 1047–54

[28] Frank M P 2002 The physical limits of computing *IEEE Comp. Sci. Eng.* **4** 16–26

[29] Leff H S and Rex A F (ed) 2003 *Maxwell's Demon 2: Entropy, Classical and Quantum Information, Computing* (Bristol: IOP)

[30] Wiener N 1948 *Cybernetics* (New York: The Technology Press and Wiley)

[31] Moore E F and Shannon C E 1956 Reliable circuits using less reliable relays *J. Franklin Inst.* **262** 191–208, 281–97

[32] Dennis N G 1974 Reliability analysis of combined voting and standby redundancies *IEEE Trans. Rel.* **R-23** 66–75

[33] Mathur F P and de Sousa P T 1975 Reliability models of NMR systems *IEEE Trans. Rel.* **R-24** 108–12

[34] Mathur F P and de Sousa P T 1975 Reliability modelling and analysis of general modular redundant systems *IEEE Trans. Rel.* **R-24** 296–9

[35] Dobrushin R L and Ortyukov S I 1977 Upper bound on the redundancy of self-correcting arrangements of unreliable functional elements *Problemy Peredachi Informatsii* **13** 56–76

[36] Depledge P G 1981 Fault tolerant computer systems *IEE Proc.* **128** 257–72

[37] Siewiorek D P 1991 Architecure of fault tolerant computers: an historical perspective *Proc. IEEE* **79** 1708–34

[38] Spielman D A 1996 Highly fault tolerant parallel computation *Proc. 37th Annual IEEE Conf. on Foundations of Computer Science (Burlington, VT)* (Piscataway, NJ: IEEE) pp 154–63

[39] Gàcs P 1997 Reliable cellular automata with self-organization *Technical Report* Department of Computer Science, Boston University, Boston, MA 02215 ftp.cs.bu.edu/fac/gacs/papers/long-ca-ms.ps

[40] Hess K, Haggag A, McMahon W, Cheng K, Lee J and Lyding J 2001 The physics of determining chip reliability *IEEE Circuits Devices* **17** 33–8

[41] McMahon W, Haggag A and Hess K 2003 Reliability scaling issues for nanoscale devices *IEEE Trans. Nanotechnology* **2** (1) 33–8

[42] Johnston A H 1998 Radiation effects in advanced microelectronics technologies *IEEE Trans. Nucl. Sci.* **45** 1339–54

[43] Joachim C, Gimzewski J K and Aviram A 2000 Electronics using hybrid-molecular and mono-molecular devices *Nature* **408** 541–8

[44] Feynman R P 1999 *Feynman Lectures on Computation* (London: Penguin Books)

[45] Meindl J D, Chen Q and Davis J A 2001 Limits on silicon nanoelectronics for terascale integration *Science* **293** 2044–9

[46] Irom F, Farmanesh F H, Swift G M, Johnston A H and Yoder G L 2003 Single-event upset in evolving commercial silicon-on-insulator microprocessor technologies *IEEE Trans. Nucl. Sci.* **50** 2107–12

[47] Gershenfeld N 2000 *The Physics of Information Technology* (Cambridge: Cambridge University Press)

[48] Magnasco M O 1993 Forced thermal ratchets *Phys. Rev. Lett.* **71** 1477–81

[49] Scherbakov A G, Bogachek E N and Landman U 1997 Noise in three-dimensional nanowires *Phys. Rev.* B **57** 6654–61

[50] Chen Y C and Di Ventra M 2003 Shot noise in nanoscale conductors from first principles *Phys. Rev.* B **67** 153304

[51] Sarpeshkar R, Delbrück T and Meade C A 1993 White noise in MOS transistors and resistors *IEEE Circuits Devices* **9** 23–9

[52] Waser R (ed) 2003 *Nanoelectronics and Information Technology: Advanced Electronic Materials and Novel Devices* (Weinham: Wiley-VCH)

[53] Mitra S, Saxena N R and McCluskey E J 2000 Common-mode failures in redundant VLSI systems: a survey *IEEE Trans. Rel.* **49** 285–95

[54] Preskill J 1998 Fault tolerant quantum computation *Introduction to Quantum Computation and Information* ed H K Lo, S Popescu and T P Spiller (River Edge, NJ: World Scientific) p 213

[55] Nielsen M A and Chuang I L 2000 *Quantum Computation and Quantum Information* (Cambridge: Cambridge University Press)

[56] McCulloch W S and Pitts W 1943 A logical calculus of the ideas imminent in nervous activity *Bull. Math. Biophys.* **5** 115–33

[57] Pippenger N 1990 *Developments in the Synthesis of Reliable Organisms from Unreliable Components (Proc. Symp. Pure Math. 50)* (Providence, RI: American Mathematical Society) pp 311–23

[58] Dobrushin R L and Ortyukov S I 1977 Lower bound for the redundancy of self-correcting arrangements of unreliable functional elements *Problemy Peredachi Informatsii* **13** 82–9

[59] Pippenger N, Stamoulis G D and Tsitsiklis J N 1991 On a lower bound for the redundancy of reliable networks with noisy gates *IEEE Trans. Inf. Theory* **37** 639–43

[60] Gàcs P and Gàl A 1994 Lower bounds for the complexity of reliable boolean circuits with noisy gates *IEEE Trans. Inf. Theory* **40** 579–83

[61] Pippenger N 1985 *On Networks of Noisy Gates (IEEE Symposium on Foundations of Computer Science 26)* (New York: IEEE) pp 30–8

[62] Toom A L 1974 Nonergodic multidimensional systems of automata *Problems Inform. Transmission* **10** 239–46

[63] Feder T 1989 Reliable computation by networks in the presence of noise *IEEE Trans. Inf. Theory* **35** 569–71

[64] Toom A L 1980 Stable and attractive trajectories in multicomponent systems *Multicomponent Random Systems* ed R L Dobrushin and Ya G Sinai (New York: Marcel Dekker)

[65] Gàcs P 1986 Reliable computation with cellular automata *J. Comput. Sys. Sci.* **32** 15–78

[66] von Neumann J 1966 *Theory of Self-Reproducing Automata* (Urbana, IL: University of Illinois Press)

[67] Szilard L 1929 Über die entropieverminderung in einem thermodynamischen system bei eingriffen intelligenter wesen *Z. Physi.* **53** 840–56

[68] Evans W and Pippenger N 1998 On the maximum tolerable noise for reliable computation by formulas *IEEE Trans. Inf. Theory* **44** 1299–305

[69] Hajek B 1991 On the maximum tolerable noise for reliable computation by formulas *IEEE Trans. Inf. Theory* **37** 388–91

[70] Spagocci S and Fountain T 1999 Fault rates in nanochip devices *Electrochem. Soc. Proc.* **99** 354–68

[71] Spagocci S and Fountain T 1999 Fault rates in nanochips *Electrochem. Soc. Proc.* **98** 582–96

[72] Nikolić K, Sadek A and Forshaw M 2001 Architectures for reliable computing with unreliable nanodevices *Proc. 2001 1st IEEE Conf. on Nanotechnology (Maui, Hawaii, IEEE-NANO)* pp 254–9

[73] Forshaw M, Nikolić K and Sadek A 2001 Fault tolerant techniques for nanocomputers *Third Year Report MEL-ARI 28667, EU ANSWERS (Autonomous Nanoelectronic Systems With Extended Replication and Signalling) Project* University College, London http://ipga.phys.ucl.ac.uk/research/answers/reports/3rd_year_UCL.pdf

[74] Han J and Jonker P 2003 A defect- and fault tolerant architecture for nanocomputers *Nanotechnology* **14** 224–30

[75] Beckett P and Jennings J 2002 Towards nanocomputer architecture *Proc. 7th Asia–Pacific Computer Systems Architecture Conf., ACSAC'2002, Melbourne, Australia (Conf. on Research and Practice in Information Technology)* vol 6, ed F Lai and J Morris (Australian Computer Society)

[76] Peper F, Lee J, Abo F, Isokawa T, Adachi S, Matsui N and Mashiko S 2004 Fault tolerance in nanocomputers: a cellular array approach *IEEE Trans. Nanotechnol.* at press

[77] Frank M P and Knight T F 1998 Ultimate theoretical models of nanocomputers *Nanotechnology* **9** 162–76

[78] Jaynes E T 1957 Information theory and statistical mechanics *Phys. Rev.* **106** 620–30

[79] Feynman R P 1998 *Statistical Mechanics: A Set of Lectures* (Reading, MA: Addison-Wesley)

[80] Margolus N and Levitin L B 1998 The maximum speed of dynamical evolution *Physica* D **120** 162–7

[81] Bekenstein J D 1981 Universal upper bound on the entropy-to-energy ratio for bounded systems *Phys. Rev.* D **23** 287–98

[82] Bremermann H J 1982 Minimum energy requirements of information transfer and computing *Int. J. Theor. Phys.* **21** 203–7

[83] Shannon C E 1948 Communication in the presence of noise *Proc. IRE* **37** 10–21

[84] Toffoli T 1982 Conservative logic *Int. J. Theor. Phys.* **21** 219–53

[85] Schneider T D 1994 Sequence logos, machine/channel capacity, maxwell's demon, and moleculer computers: a review of the theory of molecular machines *Nanotechnology* **5** 1–18

[86] Cover T M and Thomas J A 1991 *Elements of Information Theory* (New York: Wiley)

[87] Touchette H and Lloyd S 2000 Information-theoretic limits of control *Phys. Rev. Lett.* **84** 1156–9

[88] Hall J S 1994 Nanocomputers and reversible logic *Nanotechnology* **5** 157–67

[89] Merkle R C 1993 Reversible electronic logic using switches *Nanotechnology* **4** 21–40

[90] Merkle R C 1993 Two types of mechanical reversible logic *Nanotechnology* **4** 114–31

[91] Merkle R C and Drexler K E 1996 Helical logic *Nanotechnology* **7** 325–39

[92] Shor P W 1995 Scheme for reducing decoherence in quantum memory *Phys. Rev.* A **52** 2493–6

[93] Steane A M 1998 Quantum error correction *Introduction to Quantum Computation and Information* ed H K Lo, S Popescu and T P Spiller (River Edge, NJ: World Scientific) p 184

[94] Aharonov D and Ben-Or M 1997 Fault tolerant quantum computation with constant error *Proc. 29th Annual ACM Symp. on the Theory of Computing (El Paso, TX)* (New York: ACM Press) pp 176–88

[95] Steane A M 2003 Extracting entropy from quantum computers *Ann. Inst. Henri Poincaré* **4** S799–809

[96] Gea-Banacloche J 2002 Minimum energy requirements for quantum computation *Phys. Rev. Lett.* **89** 217901

[97] Gea-Banacloche J and Kish L B 2003 Comparison of energy requirements for classical and quantum information processing *Fluct. Noise Lett.* **3** C3–7

[98] Minsky M 2002 *Feynman and Computation: Exploring the Limits of Computers* (Boulder, CO: Westview) ch 10, pp 117–30

[99] Aharonov D 2000 Quantum to classical phase transition in noisy quantum computers *Phys. Rev.* A **62** 062311

[100] Joachim C 2002 Bonding more atoms together for a single molecule computer *Nanotechnology* **13** R1–7

[101] Popescu S and Rohrlich D 1998 The joy of entanglement *Introduction to Quantum Computation and Information* ed H K Lo, S Popescu and T P Spiller (River Edge, NJ: World Scientific) p 29

[102] von Neumann J 1958 *The Computer and the Brain* (New Haven, CT: Yale University Press)

[103] Arbib M A 1964 *Brains, Machines and Mathematics* (New York: McGraw-Hill)

[104] Cowan J D 1990 *Von Neumann and Neural Networks (Proc. Symp. Pure Math. 50)* (Providence, RI: American Mathematical Society) pp 243–73

[105] Shepherd G M (ed) 2004 *The Synaptic Organization of the Brain* 5th edn (Oxford: Oxford University Press)

[106] Laughlin S B 2001 Energy as a constraint on the coding and processing of sensory information *Curr. Opinion Neurobiol.* **11** 475–80

[107] Laughlin S B, de Ruyter van Steveninck R R and Anderson J C 1998 The metabolic cost of neural information *Nature Neurosci.* **1** 36–41

[108] Abshire P and Andreou A G 2001 Capacity and energy cost of information in biological and silicon photoreceptors *Proc. IEEE* **89** 1052–64

[109] Laughlin S B 1981 A simple coding procedure enhances a neuron's information capacity *Z. Naturforsch.* **36c** 910–12

[110] Levy W B and Baxter R A 1996 Energy-efficient neural codes *Neural Comput.* **8** 531–43

[111] Aspray W 1990 *The Origins of John von Neumann's Theory of Automata (Proc. Symp. Pure Math. 50)* (Providence, RI: American Mathematical Society) pp 289–309

[112] White J A, Rubinstein J T and Kay A R 2000 Channel noise in neurons *Trends Neurosci.* **23** 131–7

[113] Rieke F, Warland D, de Ruyter van Steveninck R and Bialek W 1999 *Spikes: Exploring the Neural Code* (Cambridge, MA: MIT)

[114] Koch C 1999 *Biophysics of Computation: Information Processing in Single Neurons* (New York: Oxford University Press)

[115] Faisal A A, Laughlin S B and White J A 2004 *Does Channel Noise Limit Axon Diameter?* submitted to *Proc. Natl Acad. Sci., USA*

[116] Ames A 2000 CNS energy metabolism as related to function *Brain Res. Rev.* **34** 42–68

[117] Laughlin S B and Sejnowski T J 2003 Communication in neuronal networks *Science* **301** 1870–4

[118] Sarpeshkar R 1998 Analog versus digital: Extrapolating from electronics to neurobiology *Neural Comput.* **10** 1601–38

[119] Sarpeshkar R 1997 Efficient precise computation with noisy components: extrapolating from an electronic cochlea to the brain *PhD Thesis* California Institute of Technology

[120] Levy W B and Baxter R A 2002 Energy-efficient neuronal computation via quantal synaptic failures *J. Neurosci.* **22** 4746–55

[121] Bialek W 2000 Stability and noise in biochemical switches *Advances in Neural Information Processing Systems* vol 13, ed T K Leen, T G Dietterich and V Tresp (Cambridge, MA: MIT) pp 103–9

[122] Sadek A S and Laughlin S B 2004 Personal communication

[123] Egelman D M and Montague P R 1998 Calcium dynamics in the extracellular space of mammalian neural tissue *Biophys. J.* **76** 1856–67

[124] Frick A, Magee J and Johnston D 2004 Long-term potentiation is accompanied by enhancement of the local excitability of pyramidal neuron dendrites *Nature Neurosci.* **7** 126–35

[125] Azari N P and Seitz R J 2000 Brain plasticity and recovery from stroke *Am. Sci.* **88** 426–31

[126] Vining E P, Freeman J M, Pillas D J, Uematsu S, Carson B S, Brandt J, Boatman D, Pulsifer M B and Zuckerberg A 1997 Why would you remove half a brain? The outcome of 58 children after hemispherectomy—the Johns Hopkins experience: 1968 to 1996 *Pediatrics* **100** 163–71

[127] Purves D *et al* (ed) 2004 *Neuroscience* 3rd edn (Sunderland, MA: Sinauer Associates)

[128] Eriksson P S, Perfilieva E, Bjork-Eriksson T, Alborn A M, Nordborg C, Peterson D A and Gage F H 1998 Neurogenesis in the adult human hippocampus *Nature Medicine* **4** 1313–17

[129] Hagan S, Hameroff S R and Tuszyński J A 2002 Quantum computation in brain microtubules: decoherence and biological feasibility *Phys. Rev.* E **65** 061901

[130] Maher M P, Pine J, Wright J and Tai Y C 1999 The neurochip: a new multielecrode device for stimulating and recording from cultured neurons *J. Neurosci. Methods* **87** 45–56

[131] Fromherz P 2003 Semiconductor chips with ion channels, nerve cells and brain *Physica* E **16** 24–34

[132] Spence A J, Hoy R R and Isaacson M S 2003 A micromachined silicon multielecrode for multiunit recording *J. Neurosci. Methods* **126** 119–26

[133] Peterman M C, Mehenti N Z, Bilbao K V, Lee C J, Leng T, Noolandi J, Bent S F, Blumenkranz M S and Fishman H A 2003 The artificial synapse chip: a flexible retinal interface based on directed retinal cell growth and neurotransmitter stimulation *Artificial Organs* **27** 975–85

[134] Berger T W, Baudry M, Brinton R D, Liaw J S, Marmarelis V Z, Park A Y, Sheu B J and Tanguay A R 2001 Brain-implantable biomimetic electronics as the next era in neural prosthetics *Proc. IEEE* **89** 993–1012

Index

organic molecules, 61
field emission, 140
flash memory, 43
folded substrates, 44
forced cooling, 44, 45
Fresnel zone plates, 131
fullerene, C_{60}, 139, 141

gas, 44, 45
gold bump, 51
Green's function, 93

hearing aids, 43
heat, 1, 4, 5, 6, 7, 9
heaters, 50
heatsink, 43, 44, 52
heatsinking, 50
heat spreading layer, 50
high-performance, 1, 11
homeotropic, 232
horizontally, 3
hybrid molecular, 91
hybrid molecular/electronic, 2
hypercube, 2

image processing, 1
images, 52
inductive signalling, 48
inductors, 48
information, 295, 296, 308
information theory, 263, 272, 273,
    297, 308, 311
inhomogeneous, 52
interconnect, 2, 3, 125
interconnection technologies, 52
interfacial layer, 217
ionic species, creation of, 211
Irvine sensors, 49, 50
isotopically enriched, 51
isotropic, 47

kilowatts, 1, 4

Langmuir–Blodgett, 112
laser flash, 51

laser vaporization, 140, 141, 142
lasers, organic, 128
lateral geniculate nucleus (LGN),
    14, 15
layers, 2, 3
lift-off, 120
liquid, 44, 45, 47, 50
lithography, nanoimprint, 118
local autonomy, 22, 30
logic depth, 271, 284
lysine, 225

magnetoelectronics, 72
matrix Semiconductor, 46
median filter, 23, 24, 25, 27
median filtering, 52
megawatt, 14
memory, 13, 19, 20, 30, 31, 32, 37,
    38, 39, 43
memory, magnetic, 127
mesh, 49
metal deposition, 47
metallic connections, 47
microbridges, 49
microcontact imprinting, 118
mobile phones, 43
molecular alignment model, 209
molecular computing, 2
molecular conductance, 97
molecular crossbar, 110
molecular diode, 90
molecular electronics, 44
molecular electronics, 90
molecular interconnections,
    163–165, 177–180,
    187–191, 195–200
molecular logic, 99
molecular memory, 99
molecular-scale, 1, 5
molecular switch, 74, 110
molecular transistor, 99
molecular wires, 47, 203
molecular wiring, 1, 2, 3, 4, 5

CL

621 .
381
THR
5000208057